International Meeting on High-Dimensional Data-Driven Science (HD3-2017)

Journal of Physics: Conference Series
Volume 1036

Kyoto, Japan
10 - 13 September 2017

Editors:

Masayuki Ohzeki
Teppei Ikeya
Tatsu Kuwatani
Yoshinori Nakanishi-Ohno
Masamichi J. Miyama
Tomoyuki Obuchi

ISBN: 978-1-5108-6588-4
ISSN: 1742-6588

Printed from e-media with permission by:

Curran Associates, Inc.
57 Morehouse Lane
Red Hook, NY 12571

Printed by Curran Associates, Inc. (2018)

For permission requests, please contact the Institute of Physics at the address below.

Institute of Physics
Dirac House, Temple Back
Bristol BS1 6BE UK

Phone: 44 1 17 929 7481
Fax: 44 1 17 920 0979

techtracking@iop.org

Additional copies of this publication are available from:

Curran Associates, Inc.
57 Morehouse Lane
Red Hook, NY 12571 USA
Phone: 845-758-0400
Fax: 845-758-2633
Email: curran@proceedings.com
Web: www.proceedings.com

Table of contents

Volume 1036

International Meeting on "High-Dimensional Data-Driven Science" (HD3-2017)

10–13 September 2017, Mielparque Kyoto, Japan

Accepted papers received: 21 May 2018
Published online: 27 June 2018

Preface

Papers

IOP Conf. Series: Journal of Physics: Conf. Series **1036** (2018) 011001 doi:10.1088/1742-6596/1036/1/011001

International Meeting on "High-Dimensional Data-Driven Science" (HD^3-2017)

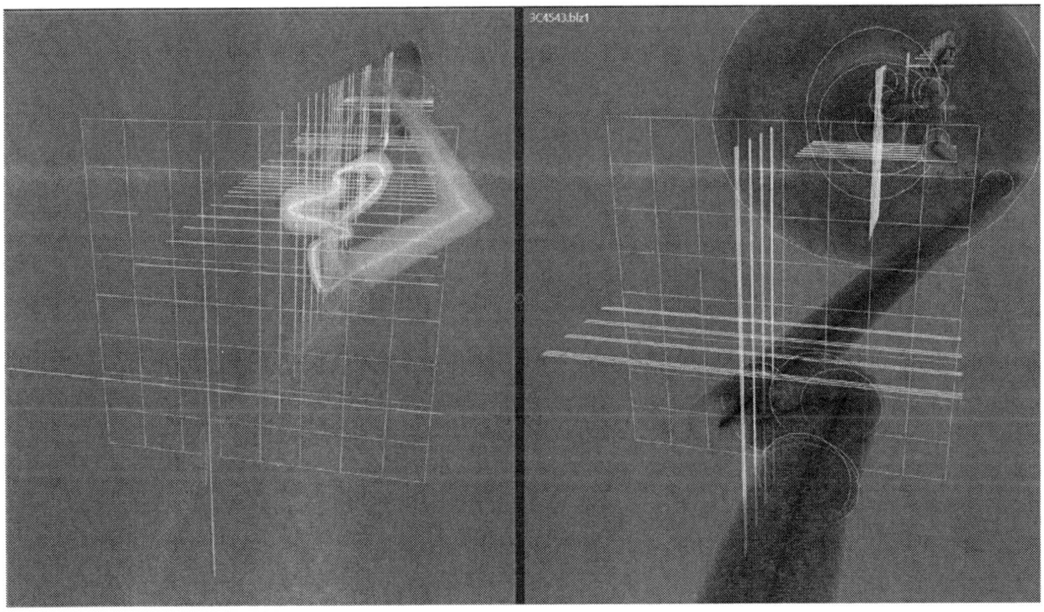

Kyoto, Japan

September 10-13, 2017

IOP Conf. Series: Journal of Physics: Conf. Series **1036** (2018) 011001 doi:10.1088/1742-6596/1036/1/011001

Preface:

In the Big Data era, high-dimensional data-driven science (HD^3 Science) has attracted considerable attention. In natural science, highly advanced experimental and measurement technologies yield high-dimensional data intractable for conventional analyses based on researchers' intuition. However, recent progress in information science provides a variety of methodologies through which the problems of high-dimensional data are addressed. The scientific principle of HD^3 Science is to establish a universal modeling platform where the motivation of natural science is smoothly conveyed with high-dimensional data to information science. The previous meeting for HD^3-2015 focused on "sparse modeling" as a principal component of the platform and its wide application to nuclear magnetic resonance spectroscopy in chemistry, magnetic resonance imaging in medicine, and astronomical interferometry in astronomy.

The scope of this HD^3-2017 meeting is the latest development and future perspective of HD^3 Science. Sparse modeling is effective for extracting explanatory variables and interactions from complex systems and has broadened its horizons to include planetary, geo, and life sciences. In addition to experimental science, computational science, such as molecular dynamics in life science, ab initio calculation in material science, and computational fluid dynamics, may provide another testing ground for HD^3 Science. Hierarchical modeling technologies will be essential for relating experimental and computational high-dimensional data to reveal multiscale phenomena. HD^3-2017 offers a forum where diverse viewpoints are brought together to strongly promote HD^3 Science.

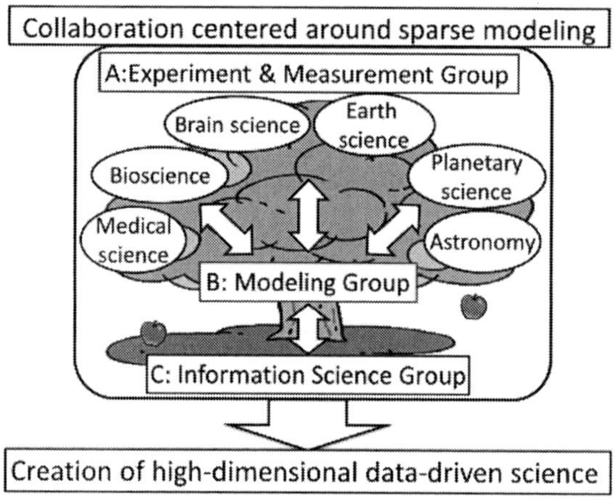

The meeting was held in Kyoto 10-13 Sep. 2017. We are pleased to publish 22

IOP Conf. Series: Journal of Physics: Conf. Series **1036** (2018) 011001 doi:10.1088/1742-6596/1036/1/011001

papers contributed by invited speakers in this volume of *Journal of Physics: Conference Series.* We hope that this volume will promote further development of High-Dimensional Data-Driven science.

Editors:

Masayuki Ohzeki, (Tohoku University, Editor in Chief)
Teppei Ikeya (Tokyo Metropolitan University)
Tatsu Kuwatani (Japan Agency for Marine-Earth Science and Technology)
Yoshinori Nakanishi-Ohno (The University of Tokyo)
Masamichi J. Miyama (Tohoku University)
Tomoyuki Obuchi (Tokyo Institute of Technology)

Organizing Committee:

Koji Hukushima (The University of Tokyo) (General Chair)
Yoshiyuki Kabashima (Tokyo Institute of Technology)
Kenji Nagata (National Institute of Advanced Industrial Science and Technology)
Yoshinori Nakanishi-Ohno (The University of Tokyo)
Masayuki Ohzeki (Tohoku University)
Masato Okada (The University of Tokyo) (Head Investigator of the Innovative Area)
Tomohisa Okada (Kyoto University)
Yonghao Yue (The University of Tokyo)

IOP Conf. Series: Journal of Physics: Conf. Series **1036** (2018) 011001 doi:10.1088/1742-6596/1036/1/011001

Peer review statement

All papers published in this volume of *Journal of Physics: Conference Series* have been peer reviewed through processes administered by the proceedings Editors. Reviews were conducted by expert referees to the professional and scientific standards expected of a proceedings journal published by IOP Publishing.

IOP Conf. Series: Journal of Physics: Conf. Series **1036** (2018) 012001 doi:10.1088/1742-6596/1036/1/012001

ES-DoS: Exhaustive search and density-of-states estimation as a general framework for sparse variable selection

Yasuhiko Igarashi[1,2,3], Hiroko Ichikawa[4], Yoshinori Nakanishi-Ohno[1,5], Hikaru Takenaka[2], Daiki Kawabata[2], Satoshi Eifuku[6], Ryoi Tamura[7], Kenji Nagata[1,8] and Masato Okada[2,3]

[1]PRESTO, JST, Honcho 4-1-8, Kawaguchi, Saitama, 332-0012, Japan
[2]Graduate School of Frontier Sciences, The University of Tokyo, Kashiwanoha 5-1-5, Kashiwa, Chiba, 277-8561, Japan
[3]Research and Service Division of Materials Data and Integrated System, National Institute for Materials Science, Sengen 1-2-1, Tsukuba, Ibaraki, 305-0047,Japan.
[4]Liberal Arts, Faculty of Science and Technology, Tokyo University of Science, Yamazaki 2641, Noda, Chiba, 278-8510, Japan
[5]Graduate School of Arts and Sciences, The University of Tokyo, Komaba 3-8-1, Meguro, Tokyo 153-8902, Japan
[6]Department of Systems Neuroscience, School of Science, Fukushima Medical University, 1Hikarigaoka, Fukushima City, Fukushima 960-1295, Japan
[7]Graduate School of Medicine and Pharmaceutical Sciences, University of Toyama, Sugitani 2630,Toyama, Toyama, 930-0194, Japan
[8]Articial Intelligence Research Center, AIST, Aomi 2-3-26, Koto, Tokyo, 135-0064, Japan

E-mail: okada@k.u-tokyo.ac.jp

Abstract. In this paper, we propose an exhaustive search with density-of-states estimation (ES-DoS) method for sparse variable selection in a wide range of learning tasks with various learning machines. We applied this ES-DoS method to synthetic and real data as an example of the regression and classification problems and discuss the results in this paper. The most important aspect of our ES-DoS method is to extract not only the optimal solution but also density of states (DoS) in terms of machine learning and data-driven science. Mapping the solutions of various approximate methods or scientists' hypotheses onto the DoS, we can comprehensively discuss and evaluate these methods and hypotheses. Our ES-DoS method opens the way for sparse variable selection in various fields, which promotes the high-dimensional data-driven science.

1. Introduction

It is important to find effective explanatory variables for every machine-learning task, and this process is called variable selection, also known as subset selection [2–4]. What is essentially needed in variable selection is, besides what criterion, such as cross validation error, should be used, exhaustive search (ES) in which all possible sets of explanatory variables are evaluated and compared with one another in terms of a given criterion [5]. The history of ES in variable selection can be traced back a half a century [5,6], though it was mentioned in an early study [7] that ES was not necessary or desirable due to the proposal of stepwise regression algorithms [8].

Since then, it has been well known that in the variable selection problem, the number of all possible variable subsets is 2^N, taking into account whether each variable is selected. The binomial theorem gives $2^N = {}_N\mathrm{C}_0 + {}_N\mathrm{C}_1 + \cdots {}_N\mathrm{C}_N$. This means that searching all the states naively in increasing or decreasing order of the number of explanatory variables used in each subset never reduces the computational complexity. Under such circumstances, in the linear-regression task, some reduction was accomplished by sharing part of the evaluation process between neighboring subsets and by applying search-tree techniques for efficient evaluation [9]. More recently, stochastic search methods, such as Markov chain Monte Carlo (MCMC, [10]), have been applied [11–14] in association with the development of Bayesian approaches [15–17]. From the viewpoint of sparse modeling, it is worthwhile to mention that an algorithm searching all possible subsets, the number of whose elements is less than or equal to a certain integer K, was proposed [18]. The motivation with this algorithm is closely related to that with a recently proposed method called K-sparse ES (ES-K), which searches all possible subsets that have exactly K variables [19].

In this study, although previous studies have adopted ES only to search for the best subset of variables, it is argued that the 'density of states (DoS)' should be keenly appreciated to make further progress in ES. To emphasize the difference, we call our method as an exhaustive search (ES) and density of states (DoS) estimation (ES-DoS) method. The DoS represents a frequency distribution of criterion values taken by all the subsets, where a 'state' corresponds to a subset based on the analogy with statistical mechanics. Obviously, the DoS can be obtained by carrying out ES, but it is often intractable in the case of a large number of explanatory variables. Therefore, we previously developed an efficient method called the approximate ES (AES) method for estimating the DoS while avoiding a combinatorial explosion [1,20]. The AES method uses the multiple-histogram method [21] in combination with the exchange Monte Carlo (EMC) method [22], also known as parallel tempering [23]. One often falls into utter confusion when approximate algorithms, such as relaxation and greedy, provide different results. This might be caused by insufficient data in the first place. We discuss case studies to maintain that such problems can be solved by ES-DoS method for sparse variable selection.

The rest of this paper is organized as follows: Section 2 formulates ES-DoS method to be applicable to sparse variable selection in machine learning. In Section 3, as an example of the classification problem, we discuss applying the ES-logistic regression (ES-LoR) and ES-support vector machine (ES-SVM) method to synthetic and real data, respectively. In Section 4, we also discuss applying the ES-linear regression (ES-LiR) method to synthetic data as an example of the regression problem. Finally, the topics covered in the preceding sections are summarized, and the prospects of data-driven science and ES method are discussed.

2. Formulation

2.1. Machine learning

This section formulates ES-DoS method to make it available to a wide range of learning tasks with various learning machines, as shown in Fig. 1. The formulation accounts for supervised learning, such as regression and classification, but a similar formulation is considered applicable to unsupervised learning such as clustering with Gaussian mixture models.

In general, machine learning, such as regression and classification, is to make a learning machine $f(\mathbf{x};\mathbf{w})$ learn the relation between input $\mathbf{x} = (x_1, x_2, \ldots, x_N)^{\mathrm{T}}$ and output y as follows:

$$y = f(\mathbf{x};\mathbf{w}). \tag{1}$$

Each component of input x_i and that of output y is called an explanatory variable and an objective variable, respectively. The variable $\mathbf{w}$ denotes a set of parameters of the learning machine. As shown in the middle of Fig. 1, there are various learning machines, and which learning machine is used should be determined appropriately in accordance with each objective

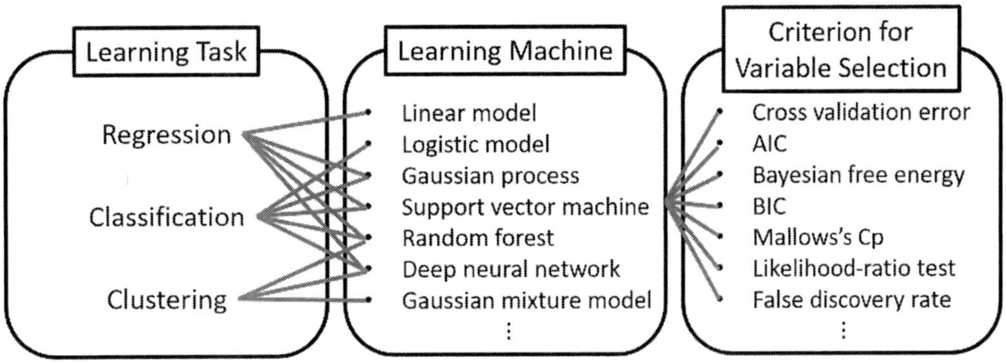

Figure 1. Constitution of exhaustive search. Exhaustive search is conducted on every triplet of learning task, learning machine and criterion for variable selection.

task and computational theory [1]. For example, a linear model and logistic model are used in regression and classification tasks, and the coefficients $\mathbf{w} = (w_1, w_2, \ldots, w_N)^{\mathrm{T}}$ are the parameters of the models as follows:

$$f(\mathbf{x}; \mathbf{w}) = \sum_i w_i x_i, \tag{2}$$

$$f(\mathbf{x}; \mathbf{w}) = 1/(1 + \exp(-\sum_i w_i x_i)). \tag{3}$$

When a Gaussian process is used as a learning machine, the parameters $\mathbf{w}$ characterize a kernel function [24]. In the case of a neural network, the parameters $\mathbf{w}$ represent weights.

The learning is, given p pairs of learning data,

$$\mathbf{X} \equiv \{\mathbf{x}_\mu\}_{\mu=1}^{p}, \tag{4}$$
$$\mathbf{Y} \equiv \{y_\mu\}_{\mu=1}^{p}, \tag{5}$$

defined as finding parameters $\mathbf{w}$ in machine learning to minimize a loss function. An example of a loss function is the following squared loss often used in regression tasks:

$$\epsilon(\mathbf{w}; \mathbf{X}, \mathbf{Y}) \equiv \frac{1}{p} \sum_{\mu=1}^{p} (y_\mu - f(\mathbf{x}_\mu; \mathbf{w}))^2. \tag{6}$$

In classification tasks, the logistic loss

$$\epsilon(\mathbf{w}; \mathbf{X}, \mathbf{Y}) \equiv \frac{1}{p} \sum_{\mu=1}^{p} \log(1 + \exp(-y_\mu f(\mathbf{x}_\mu; \mathbf{w})), \tag{7}$$

is often used, where $y_\mu \in \{-1, 1\}$.

2.2. State indicator

The problem of variable selection is discussed using an indicator $\mathbf{c} = (c_1, c_2, \ldots, c_N)^{\mathrm{T}} \in \{0, 1\}^N$ to represent a set of explanatory variables [5, 13], where $c_i = 1$ indicates that the i-the explanatory variable x_i belongs to the set and $c_i = 0$ indicates that it does not. In addition, each realization of the indicator $\mathbf{c}$ is referred to as a 'state' based on the analogy with statistical mechanics. In the cases of a linear model and logistic model, the input-output relation described by the learning machine in state $\mathbf{c}$ is written as

$$y = f(\mathbf{x} \circ \mathbf{c}; \mathbf{w} \circ \mathbf{c}), \tag{8}$$

where the symbol $\circ$ is the Hadamard product defined as $(\mathbf{x} \circ \mathbf{c}) = x_i c_i$. This formulation can be generalized to every learning machine as follows:

$$y = f_{\mathbf{c}}(\mathbf{x}(\mathbf{c}); \mathbf{w}(\mathbf{c})), \tag{9}$$

where $\mathbf{x}(\mathbf{c})$ and $f_{\mathbf{c}}(\mathbf{x}(\mathbf{c}); \mathbf{w}(\mathbf{c}))$ represent the selected variables and corresponding learning machine in state $\mathbf{c}$, respectively, and $\mathbf{w}(\mathbf{c})$ is a subset of $\mathbf{w}$ omitting all irrelevant components to $\mathbf{x}(\mathbf{c})$.

2.3. Criterion for variable selection

Although various criteria for variable selection have been proposed, as shown on the right of Fig. 1, we discuss the use of cross validation error (CVE) and Bayesian free energy (BFE) among the most popular criteria. Cross validation error is used to evaluate the predictability of selected explanatory variables and is asymptotically equivalent to the Akaike information criterion (AIC, [25]) in regular learning machines [26]. Bayesian free energy is defined as the negative logarithm of Bayesian posterior distribution, from which the Bayesian information criterion (BIC) is derived by approximation [27].

2.3.1. Cross validation error How to calculate CVE with respect to each state $\mathbf{c}$ is explained. The CV procedure is composed of three steps. First, the entire set of learning data $(\mathbf{X}, \mathbf{Y})$ is divided into training data $(\mathbf{X}_{\mathrm{tr}}, \mathbf{Y}_{\mathrm{tr}})$ and test data $(\mathbf{X}_{\mathrm{te}}, \mathbf{Y}_{\mathrm{te}})$. Second, the learning machine $f_{\mathbf{c}}(\mathbf{X}(\mathbf{c}); \mathbf{w}(\mathbf{c}))$ is optimized with respect to $\mathbf{w}(\mathbf{c})$ using only training data. The parameter set optimized at this training step is written as $\hat{\mathbf{w}}_{\mathrm{tr}}(\mathbf{c})$. Finally, how accurately the trained machine can describe the input-output relation of test data is evaluated using a loss function, such as Eqs. (6) and (7), to obtain CVE:

$$\mathrm{CVE}(\mathbf{c}) = \epsilon(\hat{\mathbf{w}}_{\mathrm{tr}}(\mathbf{c}); \mathbf{X}_{\mathrm{te}}(\mathbf{c}), \mathbf{Y}_{\mathrm{te}}). \tag{10}$$

The CVE value depends on the method of dividing data, and CV is often conducted repeatedly with different partitions of data to obtain a typical CVE value. For example, in M-fold CV, all data are randomly divided into M subsets of the same size, and the mean of M CVE values is calculated by carrying out CV M times using each subset as validation data only once.

2.3.2. Bayesian free energy A conditional probability of the indicator $\mathbf{c}$ given learning data $(\mathbf{X}, \mathbf{Y})$, which is called a posterior distribution, $P(\mathbf{c}|\mathbf{X}, \mathbf{Y})$, is derived from the joint probability of learning data, parameters of a learning machine, and state indicator,

$$P(\mathbf{X}, \mathbf{Y}, \mathbf{w}(\mathbf{c}), \mathbf{c}). \tag{11}$$

The parameters $\mathbf{w}(\mathbf{c})$ are marginalized out to obtain

$$P(\mathbf{X}, \mathbf{Y}, \mathbf{c}) = \int d\mathbf{w}(\mathbf{c}) P(\mathbf{X}, \mathbf{Y}, \mathbf{w}(\mathbf{c}), \mathbf{c}). \tag{12}$$

The product rule of probability is applied to the left-hand side to obtain

$$P(\mathbf{c}|\mathbf{X}, \mathbf{Y}) P(\mathbf{X}, \mathbf{Y}) = \int d\mathbf{w}(\mathbf{c}) P(\mathbf{X}, \mathbf{Y}, \mathbf{w}(\mathbf{c}), \mathbf{c}). \tag{13}$$

Again, the product rule is applied to the joint probability in the right-hand side to obtain

$$P(\mathbf{c}|\mathbf{X},\mathbf{Y})P(\mathbf{X},\mathbf{Y}) = \int d\mathbf{w}(\mathbf{c})P(\mathbf{Y}|\mathbf{X},\mathbf{w}(\mathbf{c}),\mathbf{c})P(\mathbf{X})P(\mathbf{w}(\mathbf{c})|\mathbf{c})P(\mathbf{c}), \tag{14}$$

where $P(\mathbf{w}(\mathbf{c})|\mathbf{c})$ represents in a probabilistic way the fact that $\mathbf{w}(\mathbf{c})$ is a subset of $\mathbf{w}$ omitting all irrelevant components to $\mathbf{x}(\mathbf{c})$, and $P(\mathbf{Y}|\mathbf{x},\mathbf{w}(\mathbf{c}),\mathbf{c})$ represents the input-output relation, Eq. (9), with an appropriate noise model. It should be noted that $\mathbf{X}$ is independent of $\mathbf{c}$ and $\mathbf{w}(\mathbf{c})$. By setting the prior distribution of $\mathbf{c}$, $P(\mathbf{c})$ to the uniform distribution for simplicity, we obtain

$$P(\mathbf{c}|\mathbf{X},\mathbf{Y}) \propto \int d\mathbf{w}(\mathbf{c})P(\mathbf{Y}|\mathbf{X},\mathbf{w}(\mathbf{c}),\mathbf{c})P(\mathbf{w}(\mathbf{c})|\mathbf{c}). \tag{15}$$

It should be noted that $P(\mathbf{X},\mathbf{Y})$ and $P(\mathbf{X})$ are constant with respect to $\mathbf{c}$.

Bayesian free energy is defined as the negative logarithm of a posterior distribution,

$$\mathrm{BFE}(\mathbf{c}) = -\log \int d\mathbf{w}(\mathbf{c})P(\mathbf{Y}|\mathbf{X},\mathbf{w}(\mathbf{c}),\mathbf{c})P(\mathbf{w}(\mathbf{c})|\mathbf{c}), \tag{16}$$

up to a constant. The formula of BFE is similar to that of free energy in statistical mechanics in that both have an integral with respect to many variables. This is because BFE is called a 'Bayesian free energy'.

2.4. Exhaustive search with density-of-states estimation

The discussion in Subsections 2.1–2.3 are summarized as follows: When a learning machine is made to learn an input-output relation, such as Eq. (1), using p pairs of input and output data $(\mathbf{X},\mathbf{Y})$, variable selection is carried out by evaluating each $\mathbf{c}$ in a criterion such as CVE and BFE. Therefore, ES compares all the 2^N states to find the optimal one without fail.

Previous studies on ES were aimed only at the best subset of explanatory variables, namely, a 'ground state'. However, it does not necessarily mean that they have brought out the full potential of ES. In this study, therefore, we argue that the DoS should be appreciated to make further progress in ES. When we emphasize the DoS, we especially call the method as an exhaustive search (ES) and density of states (DoS) estimation (ES-DoS) method. The DoS, a concept borrowed from statistical mechanics, represents a frequency distribution of criterion values within all the states. More formally, the DoS for each criterion H, such as CVE and BFE, is defined as follows:

$$D_H(E) = \sum_{\mathbf{c}} \mathbf{1}(E = H(\mathbf{c})), \tag{17}$$

where the function $\mathbf{1}$ takes the value 1 if the argument is satisfied or 0 if not. By definition, the DoS is successfully obtained with ES.

We previously proposed an approximate method for estimating the DoS with respect to CVE and BFE to circumvent a combinatorial explosion involved in ES due to a number of explanatory variables, and later named the method the approximate ES (AES) method [1, 20]. The AES method applies the multiple-histogram method [21] to histograms composed of state samples generated from probability distributions

$$p_\beta(\mathbf{c}) \propto e^{-\beta H(\mathbf{c})}. \tag{18}$$

Based on the analogy with statistical mechanics, each of the distributions is called a 'canonical distribution' with a corresponding 'temperature' $T = \beta^{-1}$, and H, which represents a criterion, is regarded as a 'Hamiltonian'. It is convenient to use the EMC method for sampling from canonical distributions with different temperatures [22]. Similarly, an approximate version of the aforementioned ES-K method, i.e., AES-K, was developed [19].

In practical situations, one often comes across the following problem. When approximate algorithms, such as convex relaxation and greedy methods, give different solution states to the same task, it is confusing as to which should be trusted, and to make matters worse, it might be caused by insufficient data in the first place. As will be introduced in what follows, we have tried to solve such problems with close observation of the DoS.

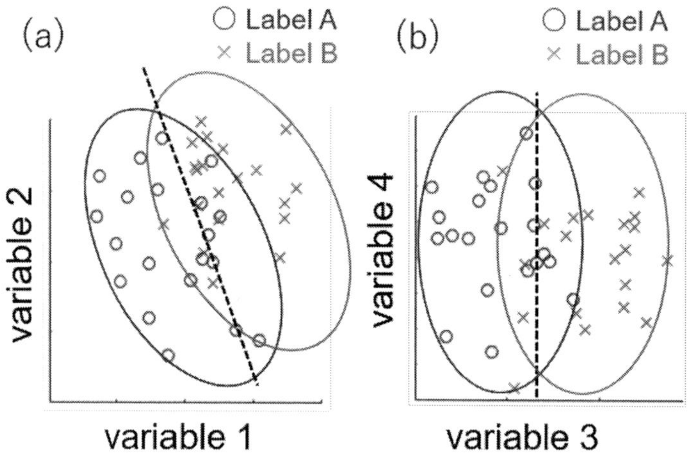

Figure 2. 4-dimensional synthetic data for classification

3. Classification

3.1. ES-SVM and ES-LoR

In this section, we discuss the ES method for classification shown in Fig. 1. Let us start the discussion about the importance of variable selection by applying the ES method to the synthetic data illustrated in Fig. 2. Figure 2 represents the task of binary classification in that the elements have either +1 or -1. We discuss the task from the aspect of the liner classification. The data represented in Fig. 2(a) are discriminated with both variables 1 and 2 clearly better than only with variable 1 or only with variable 2. On the other hand, in the case of Fig. 2(b), the data are discriminated with only variable 3 better than with both variables 3 and 4 because variable 4 is irrelevant for classification. In the case of Fig. 2(a), discriminative performance improves by using the two variables, while in the case of Fig. 2(b), discriminative performance improves by using only one variable. Although these can be easily found by the distribution of two variables, they cannot be found by the marginal distribution of each variable because the marginal distributions are not different. This means that the marginal distribution does not work for classification and that we have to search the multivariate distribution for classification, which causes an exponential increase in search space in the case of high dimensional input. This difficulty is solved using our ES method. In this section, we consider a support vector machine (SVM) as a machine-learning algorithm for classification and introduce ES for SVM (ES-SVM). The discriminative function and error function of an SVM are used based on a linear SVM as

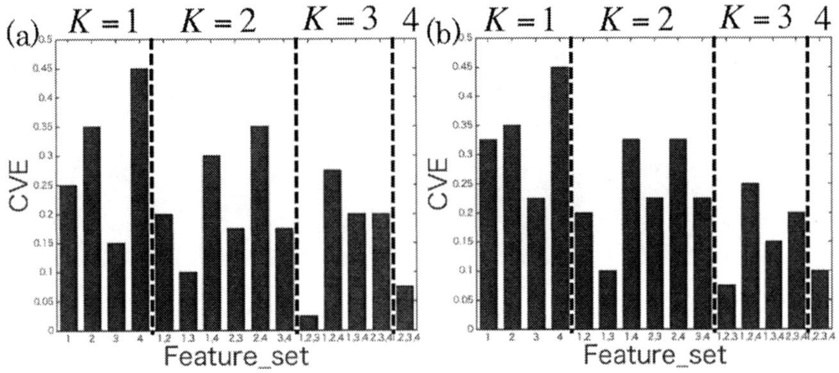

Figure 3. Classification accuracy obtained using (a) ES-SVM and (b) ES-LoR

follows,

$$y = f_{\mathbf{c}}(\mathbf{x}(\mathbf{c}); \mathbf{w}(\mathbf{c})) = f(\mathbf{x}; \mathbf{w} \circ \mathbf{c}) = \mathrm{sgn}(w_0 + \sum_i^4 c_i w_i x_i), \tag{19}$$

$$\epsilon(\mathbf{w}; \mathbf{Y}, \mathbf{X}) \equiv \frac{1}{p} \sum_{\mu=1}^{p} \max(0, 1 - y_\mu(w_0 + \sum_i^4 c_i w_i x_{i,\mu})) + \lambda \sum_{i=1}^{4} w_i^2. \tag{20}$$

For the criteria for variable selection, we used CVE($\mathbf{c}$), which are derived from 10-fold CV as follows:

$$\mathrm{CVE}(\mathbf{c}) = \frac{1}{p_{\mathrm{te}}} \sum_{\mu \in \mathrm{te}} \mathbf{1}(y_\mu(w_0 + \sum_i^4 c_i w_i x_{i,\mu}) < 0), \tag{21}$$

which represents the incorrect identification for test data, where p_{te} is the number of test data sets. We fixed $\lambda = 5.0$ in Eq. (20). Figure 3 shows the CVE for synthetic data obtained with the ES-1, -2, -3, and -4 methods. Among all the combination of variables ($2^4 - 1 = 15$ combinations), the combination consisting of $\{1, 2, 3\}$ had the lowest CVE, which is coincident with the above discussion. Since the combinations including variable 4 had a higher CVE than those excluding variable 4, this variable reduces the discriminative performance as noise. Cross validation error represents prediction error and is not appropriate for model selection. However, depending on the circumstance, we can appropriately select variables by CVE.

Similarly, we can consider logistic regression as a machine-learning algorithm for classification and introduce ES-LoR. We used Eqs. (3) and (7) as the discriminative function and error function of logistic regression, respectively. The results are shown in Fig. 3(b), where we used CVE($\mathbf{c}$) derived from Eq. (21) for the criteria for variable selection. Similar to the results of ES-SVM, the combination consisting of $\{1, 2, 3\}$ had the lowest CVE among all the variable combination.

3.2. Application of the ES-SVM to the real data

We applied ES-SVM for the real data obtained from the neurophysiological experiment with macaque monkeys conducted by Eifuku et al. [20, 28, 29]. In this experiment, Eifuku et al. recorded the neuronal activities of 23 neurons in the anterior inferior temporal cortex (AIT) of macaque monkeys when the monkeys were performing a pattern-recognition task requiring

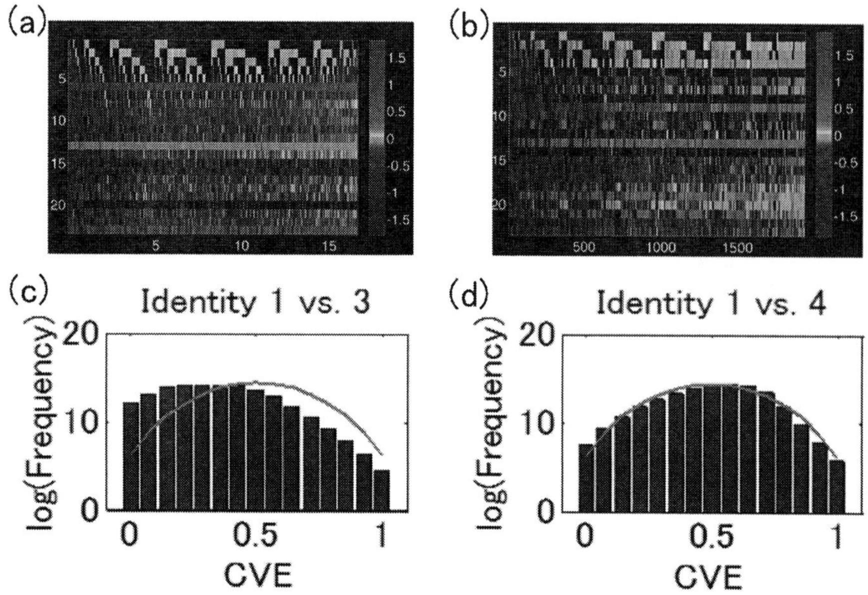

Figure 4. Weight diagram and density of states (DoS). (a) and (b): Weight diagram for discrimination of identities 1 vs. 3 and 1 vs. 4, respectively. Vertical axes indicate neuron index and horizontal axes indicate combination index. Each column represents subset of selected explanatory variables for calculating decision boundary. Colored cells in each column represent $\mathbf{c}_{opt}$ with CVE=0 and color gradient indicating $\mathbf{w}(\mathbf{c}_{opt})$. (c) and (d): DoS for discrimination of identities 1 vs. 3 and 1 vs. 4, respectively. Horizontal axes show CVE and vertical axes show logarithm of frequency. Curved line in each figure shows logarithm of frequency for random guessing.

the identification of facial images. The monkeys observed the 28 face images consisting of 4 different facial identities recorded from 7 different view and required to identify the same person irrespective of different facial views. We calculated the firing rate of 23 neuronal activities for each face image and investigated which neuron contributed to view-invariant face identification. It is a task of binary classification in that two out of four facial identities should be discriminated using the neuronal firing rate as the input data. We can pose the problems of linear discrimination among four identities; thus, the number of possible problems, that is the pairs of identities that should be discriminated, are ${}_4C_2 = 6$. Hereafter, when we consider the problem of discrimination between identities nos. 1 and 3, the problem will be noted as "the discrimination of identity 1 vs. 3."

In this study, we investigated the variable selection in the real data by using the weight diagram and DoS, focusing on the discrimination of identities 1 vs. 3 and 1 vs. 4. They were investigated by applying the ES-SVM by Nagata et al. [20]. For the ES-SVM applied to 23 explanatory variables, the SVM classifications and CVE calculations by leave-one-out cross validation (LOOCV) were conducted for the $2^{23} - 1 = 8,388,607$ possible variable subsets.

3.2.1. Analysis of the weight diagram and density of states (DoS) The weight diagram illustrates the weight coefficients of each explanatory variable $\mathbf{w}(\mathbf{c})$ for calculating the decision boundary. Using the weight diagram, we visualized the weight coefficients $\mathbf{w}(\mathbf{c}_{opt})$ of each explanatory variable determined by the optimal subsets $\mathbf{c}_{opt}$ with CVE $= 0$.

Figure 4 shows the weight diagram of optimal subsets for the discrimination of identities (a)

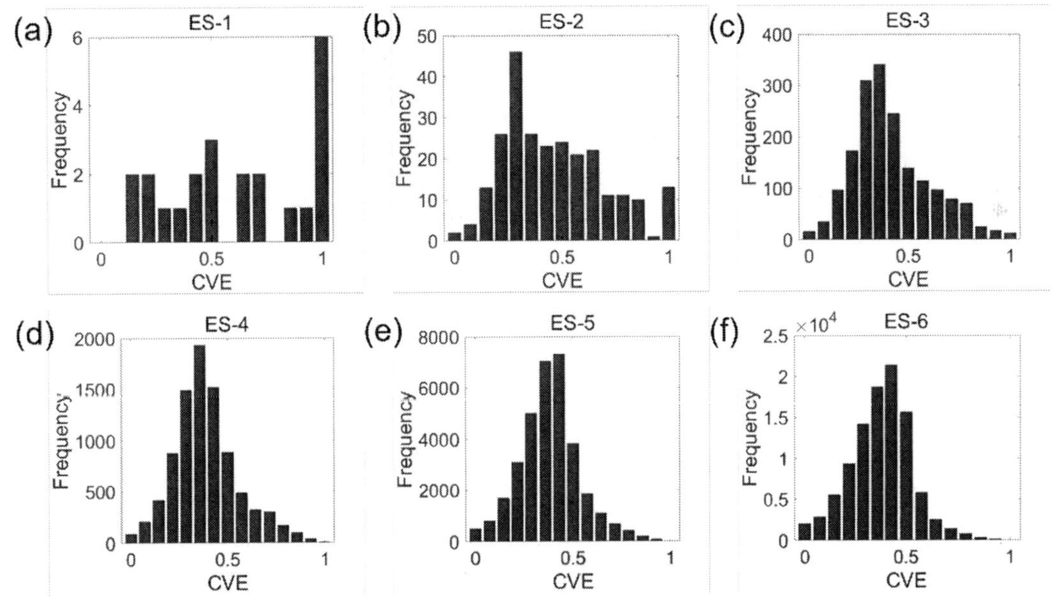

Figure 5. DoS for discrimination of identity 1 vs. 3. Each figure shows DoS for $K = 1, \ldots, 6$. Vertical axes indicate frequency and horizontal axes indicate CVE.

1 vs .3 and (b) 1 vs 4. In each weight diagram, the vertical axis is the vector indicating the index of the explanatory variable (e.g. neuron index) with length of $|\mathbf{x}|$. The horizontal axis is the vector indicating the index of the optimal subset. Since the ES-SVM for the discrimination of identity 1 vs. 3 found 166,412 optimal subsets, the length of the horizontal axis in Fig. 4 (a) is 166,412. Similarly, the ES-SVM for the discrimination of identity 1 vs. 4 found 1,938 optimal subsets, and the length of the horizontal axis in Fig. 4 (b) is 1,938. A column in the weight diagram indicates an optimal subset c_{opt}. In each column, a colored cell indicates that the explanatory variable consists of the optimal subset and its indicator $c_i = 1$. The color of the cell indicates a weight coefficient $\mathbf{w}(\mathbf{c}_{opt})$ for calculating the decision boundary. The red/yellow gradient indicates positive value and blue gradient indicates negative value. In each column, the non-colored cell, i.e., black cell, indicates that the explanatory variable was not selected as the optimal subset and $c_i = 0$. As shown in Fig. 4, the weight diagram enabled us to capture the whole picture of optimal subsets that cannot be provided by seeking one of the optimal subsets.

As shown in a previous study [20], we compared the DoS of the CVE obtained from the ES-SVM with DoS for random guessing. It is represented by binomial distribution DoS, where the given data y were randomly labeled regardless of input data $\mathbf{x}$. Figure 4 (c) shows the DoS obtained from the ES-SVM on the discrimination of identity 1 vs.3. Compared with the DoS for random guessing, the DoS is asymmetric and left-skewed distribution. This result suggests that ES-SVM successfully classified the neuronal activity data into identity 1 or 3 more frequently rather than the chance level, and our data include significant information for binary classification of identity 1 vs. 3. On the other hand, Fig. 4(d) shows the DoS obtained from the ES-SVM on the discrimination of identity 1 vs. 4. Similar to the DoS for random guessing, the DoS is a symmetric distribution, Nagata et al. [20] discussed that ES-SVM successfully classified the neuronal activity data into identity 1 or 4 at the chance level, and the data did not include significant information for binary classification of identity 1 vs. 4.

In the present study, we applied ES-K to the discrimination of identities 1 vs. 3 and 1 vs. 4.

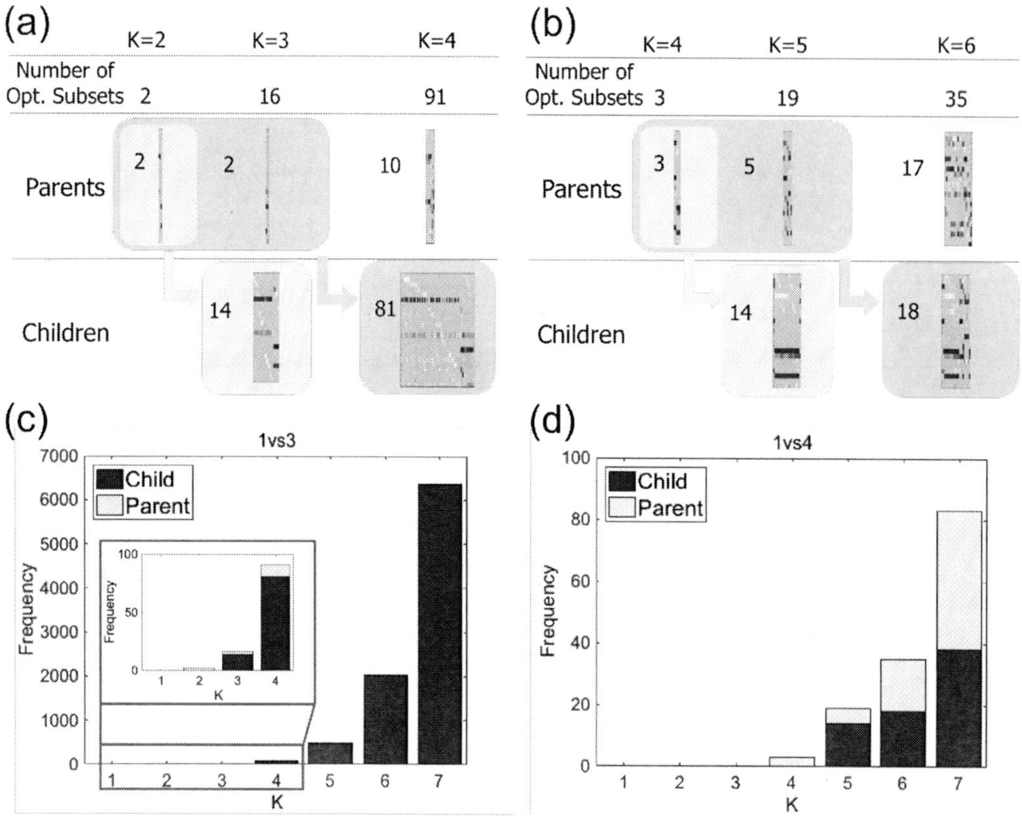

Figure 6. (a) and (b) Illustrations of "parents-children" relation observed in optimal subsets for discrimination of identities 1 vs 3 and 1 vs 4, respectively. (c) and (d) Proportion of "parents" and "children" observed in optimal subsets for discrimination of identities 1 vs 3 and 1 vs 4, respectively. Vertical axes indicate frequency and horizontal axes indicate K in each ES-K-SVM. In each bar, pale yellow indicates frequency of ES-K' s "parents" and dark blue indicates frequency of ES-$(K-1)$'s "children".

In order of increasing K, we investigated the structure of optimal subset $\mathbf{c}_{opt}$, which consists of K explanatory variables. First, we used ES-1 for the discrimination of identity 1 vs. 3. Figure 5(a) shows the DoS obtained from ES-1. We did not find the optimal subset with CVE $= 0$. Figure 5(b) shows the DoS obtained from ES-2. We found the two optimal subsets, neurons nos. 6 and 13 and neurons nos. 16 and 20. Hereafter, we note the variable indices (that is, neuron index) included in the subset as $\{6, 13\}$ and $\{16, 20\}$. As shown in Fig. 5(c), we found the 16 optimal subsets from ES-3. Out of these optimal subsets, 11 subsets contained $\{6, 13\}$ and 3 contained $\{16, 13\}$. In other words, these 14 subsets were supersets of the optimal subsets found with ES-2. When the optimal subsets found with ES-K are supersets of the optimal subsets found with ES-$(K-1)$, we called the relation the "parent-children" relation.

Figure 6(a) illustrates an overview of the "parent-children" relation. We call the two optimal subsets found with ES-2, $\{6, 13\}$ and $\{16, 20\}$, as the ES-2 "parents". We also call the 14 optimal subsets found with ES-3, $\{6, 13, \alpha\}$ and $\{16, 20, \beta\}$ as ES-2's "children", where α and β indicates the arbitrary neuron index. The case in which $K \geqq 3$, we call the optimal subsets found with ES-K, which are not ES-$(K-1)$'s children as ES-K s "parents". For example, for the above discrimination of identity 1 vs. 3, the two optimal subset found with ES-3, $\{13, 16, 21\}$ and $\{13, 16, 23\}$, are ES-3 ' s "parents". Therefore, the optimal subsets found with ES-K are composed of ES-($K-1$)'s "children" and ES-K ' s "parents". The proportion of ES-($K-1$)'s "children" and ES-K ' s "parents" obtained with ES-K at each K is shown in Fig. 6(c). Irrespective of the increment of K, most of the optimal subsets were ES-($K-1$)'s "children". This result indicates that the optimal subsets found by relatively small K (here, K=2) can be described as "parents" and most of the optimal subsets found by relatively large K (here, $K \geqq 3$) consist of the "parents". Since the optimal subsets commonly contain a smaller number of variables as "parents", it is important for feature selection to find the "parents". Next, we applied ES-K to the discrimination of identity 1 vs. 4. In this case, we could not find an optimal subset in the case of $K = 1, 2, 3$. As Fig. 6(b) illustrates, we found the three optimal subsets from ES-4. In contrast to the discrimination of identity 1 vs. 3, along with the increment of K, the number of optimal subsets found with ES-K increased. We found the 5 optimal subsets from ES-5 and 17 optimal subsets from ES-6. Therefore, the proportions of ES-($K-1$)'s "children" in the optimal subset found in each K were low at each K,

as shown in Figure 6(d). Finally, the proportion of ES-($K-1$)'s "children" among the 1,936 optimal subsets obtained with ES-SVM (that is, ES-K-SVM conducted every K from 1 to 23) was about 50%. This result indicates that the optimal subsets for the discrimination of identity 1 vs. 4 did not consist of the "parent-children" relation. Since the "parents" that constantly took part in the multiple optimal subsets are not stable, it is difficult to determine which exploratory variable should be selected.

In summary, in the discrimination of identity 1 vs.3, the optimal subsets are related to each other in the 'parent-children" relation, and the DoS distribution was a left-skewed distribution and differed from the DoS for random guessing. In the discrimination of identity 1 vs.4, the optimal subsets did not construct the 'parent-children" relation, and the DoS distribution overlapped with the DoS for random guessing. Therefore, the existence of the "parent-children" relation in the classification problem suggests that we can select the "parent" as an important explanatory variable for classification. This is closely related with the difference in the DoS distribution from the DoS for random-guessing [20]. Based on these findings, our ES method provides insight into determining whether the data contain the information for classification. Furthermore, ES-K would help variable selection for classification.

4. Regression

4.1. Formulation of exhaustive search for linear regression

In this section, we consider regression as a machine-learning task and introduce exhaustive search for linear regression (ES-LiR). Let us specifically formulate ES-LiR. Using Eq. (9), we can write $f_{\mathbf{c}}(\mathbf{x}(\mathbf{c}); \mathbf{w}(\mathbf{c}))$ as

$$y = f_{\mathbf{c}}(\mathbf{x}(\mathbf{c}); \mathbf{w}(\mathbf{c})) = f(\mathbf{x}; \mathbf{w} \circ \mathbf{c}) = w_0 + \sum_{i=1}^{N} c_i w_i x_i. \tag{22}$$

We use CVE and BFE as the evaluation criteria of variable selection, as denoted in §2.3. Cross validation error is derived from CVE($\mathbf{c}$) using the error function $\epsilon(\mathbf{w}; \mathbf{Y}, \mathbf{X})$ described in Eqs. (10) and (6), respectively. Similarly, we introduce BFE in Subsubsection 2.3.2 for linear regression. Using Eq. (22), we can write $P(\mathbf{Y}|\mathbf{X}, \mathbf{w}(\mathbf{c}), \mathbf{c})$ from Eq. (16),

$$P(\mathbf{Y}|\mathbf{X}, \mathbf{w}(\mathbf{c}), \mathbf{c}) = \frac{1}{\det(2\pi\Sigma)^{1/2}} \exp\left(-\frac{1}{2}\Delta^{\mathrm{T}}\Sigma^{-1}\Delta\right), \quad \Delta = [\mathbf{Y} - \{\mathbf{1}w_0 + \mathbf{X}(\mathbf{c} \circ \mathbf{w})\}], \tag{23}$$

where $\mathbf{1}$ is a p-dimensional all-one vector, and Σ represents the covariance matrix of measurement noise, whose elements are given by $\Sigma_{ii} = \sigma^2$ and $\Sigma_{ij} = 0$ $(i \neq j)$.

Let us consider $P(\mathbf{w}(\mathbf{c})|\mathbf{c})$ in Eq. (16). We assume that w_i in the case of $c_i = 1$ is generated from the Gaussian distribution where the mean and variance are 0 and s^2, respectively, and w_0 in Eq. (22) is also generated from the same distribution.

Then, we can write BFE in Eq. (16) as

$$\mathrm{BFE}(\mathbf{c}) = -\log \int \frac{dw_0}{\sqrt{2\pi s^2}} \exp\left(-\frac{w_0^2}{2s^2}\right) \prod_{\{i|c_i=1\}} \frac{dw_i}{\sqrt{2\pi s^2}} \exp\left(-\frac{w_i^2}{2s^2}\right) P(\mathbf{Y}|\mathbf{X}, \mathbf{w}(\mathbf{c}), \mathbf{c}) \quad (24)$$

$$= \frac{p}{2}\log(2\pi) + \frac{1}{2}\log\det(\Sigma) + K\log s + \frac{1}{2}\mathbf{Y}^{\mathrm{T}}\Sigma^{-1}\mathbf{Y} - \frac{1}{2}\boldsymbol{\mu}^{\mathrm{T}}\Lambda^{-1}\boldsymbol{\mu} - \frac{1}{2}\det(\Lambda), \quad (25)$$

where we set $\Lambda = (\mathbf{X}_{\mathrm{I}}^{\mathrm{T}}\Sigma^{-1}\mathbf{X}_{\mathrm{I}} + \frac{1}{s^2}\mathrm{I})^{-1}$, $\boldsymbol{\mu} = \Lambda\mathbf{X}_{\mathrm{I}}^{\mathrm{T}}\Sigma^{-1}\mathbf{Y}$ and $\mathbf{X}_{\mathrm{I}}$ is a matrix composed of non-zero explanatory variables. We estimate the variance s^2 using the observed data as described in the next paragraph. With our ES method, the Bayesian free energy (BFE) is calculated for all combinations of explanatory variables, and the combination minimizing the BFE is taken as the optimal one.

4.2. Results

We applied ES-LiR to a synthetic dataset for sparse variable selection in linear regression. We derived the DoS for BFE($\mathbf{c}$) and CVE($\mathbf{c}$) and investigated the solution space. For the synthetic data, we assumed that the N dimensional coefficient vector $\mathbf{w}$ has two non-zero variables and set $w_1 = -0.65, w_2 = 1.2, w_i = 0 (i \neq 1, 2)$. The synthetic data y_μ were obtained as follows

$$y_\mu = w_0 + \sum_{i=1}^{N} w_i x_{i,\mu} + \epsilon_\mu. \quad (26)$$

Each element of $x_{i,\mu}$ $(\mu = 1, \ldots, p)$ is generated from the Gaussian distribution $\mathcal{N}(0, \sigma_w^2)$ where the mean and variance are 0 and $\sigma_w^2 = 1$, respectively. Each noise component ϵ_μ is assumed to follow $\mathcal{N}(0, \sigma_N^2)$, and we set $\sigma_N^2 = 0.1$, signal-to-noise ratio was set as $\sigma_w^2/\sigma_N^2 = 10$, and the data dimension and sample number as $N = 200$ and $p = 50$, respectively.

4.2.1. DoS and solution analysis using ES-LiR We applied ES-LiR to the synthetic data and investigated whether the method can extract the non-zero variables $\{1, 2\}$ and visualize the two dimensional DoS corresponding to CVE and BFE for studying the solution structure. Figure 7 shows the DoS for synthetic data obtained with ES-1, -2, and -3. The combination consisting of $\{1, 2\}$ had a remarkably lower CVE and BFE than the other combinations of two explanatory variables, as shown in Fig. 7(b). Then let us consider the results of ES-3. Figure 7(c) shows the structure consisting of combinations of two variables including $\{1, 2\}$ and another variable. By comparing CVE and BFE, we found that BFE leads to better understanding of the solution structure than CVE, as shown in Fig. 7. Similar results were obtained in recent research in which parts of the authors calculated the DoS for an astronomical dataset [19]. We can estimate the sparseness K_{opt} from the data by calculating the minimum values of BFE and CVE for all combinations of $K = 1, 2, \ldots, N$ explanatory variables and comparing the values [19].

5. Discussion and conclusion

In this paper, we proposed an exhaustive search with density-of-states estimation (ES-DoS) method for sparse variable selection. We applied ES-linear regression (ES-LiR) to synthetic data as an example of the regression problem. As an example of the classification problem, we also applied ES-logistic regression (ES-LoR) and ES-support vector machine (ES-SVM) to synthetic and real data, respectively. Practically, to apply our ES method to data analysis, we

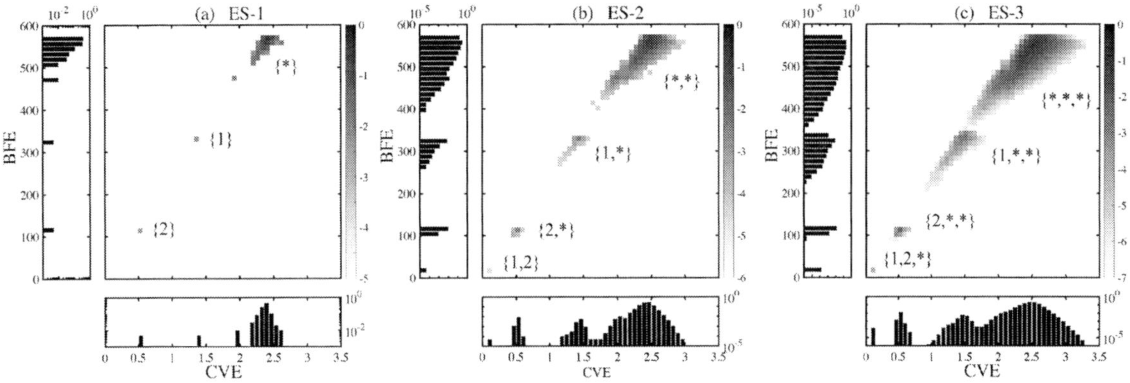

Figure 7. DoS for synthetic data obtained using ES-1, -2, -3. In each figure, horizontal and vertical histograms represent DoS corresponding to CVE and BFE, respectively, and central figure shows two-dimensional DoS. They are expressed in logarithmic scales. Combinations forming each cluster are written in $\{\cdot\}$.

used K-sparse ES (ES-K), which searches all possible subsets that have exactly K variables [19] and is important for understanding the solution structure, as shown in §3.2.

The most important aspects of our ES-DoS method is to extract not only the optimal solution [2] but also density of states (DoS). We discussed the importance of the DoS in terms of machine learning and data-driven science [1]. Generally, various approximate methods, such as LASSO [30], and Iterative thresholding [31], can be applied to the target measurement data. Using the state indicator of solutions obtained with these methods, we can map the solutions of various approximate methods onto the DoS. We thus can comprehensively discuss and evaluate the advantages and disadvantages of each method.

We then discussed the importance of the DoS for data-driven science [1]. Depending on the various scientific disciplines, each scientist forms a hypothesis and conducts experiments and measurements for hypothesis testing. Generally, the hypothesis is formulated as a feature selection to determine which variable essentially explains the data. We can map the scientist's hypothesis onto the DoS using the state indicator corresponding to the hypothesis. We thus can comprehensively discuss and evaluate the different hypotheses. From this standpoint, the solution obtained with an approximate method is regarded as one of the hypotheses.

The proposed method, ES-DoS method, opens the way for sparse variable selection in various fields, which promotes the high-dimensional data-driven science.

Acknowledgments

This work was supported by a Grant-in-Aid for Scientific Research on Innovative Areas (No. 25120009, 16H01555), Grant-in-Aid for Young Scientists (B) (No. 17K12735,17K12749) and Grants-in-Aid for JSPS Fellows (No. 15J07765) from Japan Society for the Promotion of Science (JSPS) and by JST CREST(JPMJCR1761) and PRESTO (JPMJPR15E8, JPMJPR17N2, JPMJPR1773).

References

[1] Igarashi Y, Nagata K, Kuwatani T, Omori T, Nakanishi-Ohno Y, and Okada M 2016 *J. Phys. Conf. Ser.* **699** 012001
[2] Beale E M L, Kendall M G, and Mann D W 1967 *Biometrika* **54** 357–366
[3] Miller A J 1990 *Subset selection in regression* Chapman and Hall

[4] George E I 2000 *J. Am. Stat. Assoc.* **95** 1304–1308
[5] Garside M J 1965 *J. R. Stat. Soc. Ser. C* **14** 196–200
[6] Kudô A 1963 *Memoirs of the Faculty of Science, Kyushu University. Series A, Mathematics* **17** 63–75
[7] Healy M J R 1963 *Comput. J.* **6** 57–61
[8] Efroymson M A 1960 *Mathematical Methods for Digital Computers* John Wiley and Sons 191–203
[9] Furnival G M and Wilson R W 1974 *Technometrics* **16** 499–511
[10] Metropolis N, Rosenbluth A W, Rosenbluth M N, Teller A H, and Teller E 1953 *J. Chem. Phys.* **21** 1087–1092
[11] George E I and McCulloch R E 1993 *J. Am. Stat. Assoc.* **88** 881–889
[12] Green P J 1995 *Biometrika* **82** 711-732
[13] Kuo L and Mallick B 1998 *Sankhyā Ser. B* **60** 65–81
[14] Kim S, Tadesse M G, and Vannucci M 2006 *Biometrika* **93** 877–893
[15] Lindley D V 1968 *J. R. Stat. Soc. Ser. B* **30** 31–66
[16] Mitchell T J and Beauchamp J J 1988 *J. Am. Stat. Assoc.* **83** 1023–1032
[17] Ishwaran H and Rao J S 2005 *Ann. Stat.* **33** 730–773
[18] Tarumi T and Kudô A 1974 *J. Japan. Statist. Soc.* **4** 47–56
[19] Igarashi Y, Takenaka H, Nakanishi-Ohno Y, Uemura M, Ikeda S, and Okada M 2017 *arXiv preprint* arXiv:1707.02050
[20] Nagata K, Kitazono J, Nakajima S, Eifuku S, Tamura R, and Okada M 2015 *IPSJ Online Trans.* **8** 25–32
[21] Ferrenberg A M and Swendsen R H 1989 *Phys. Rev. Lett.* **63** 1195–1198
[22] Hukushima K and Nemoto K 1996 *J. Phys. Soc. Jpn.* **65** 1604–1608
[23] Geyer C J 1991 in *Proceedings 23th Symp. Interface Comput. Sci. Stat* 156–163
[24] Dam H C 2016 private communications
[25] Akaike H 1973 in *2nd Int. Symp. Inf. Theory* 267–281
[26] Stone M 1977 *J. R. Stat. Soc. Ser. B* **39** 44–47
[27] Schwarz G 1978 *Ann. Stat.* **6** 461–464
[28] Eifuku S, De Souza W C, Tamura R, Nisijo H, and Ono T 2004 *J. Neurophysiol* **91**, 358–371
[29] Eifuku S, De Souza W C, Nakata R, Ono T, and Tamura R 2011 *PLoS One* **6**, e18913
[30] Tibshirani R 1996 *J.Royal Stat. Soc. B* **58**(1), 267-288
[31] Blumensath T, and Mike E D 2008 *J. Fourier Analys and Applications* **14**(5), 629-654

IOP Conf. Series: Journal of Physics: Conf. Series **1036** (2018) 012002 doi:10.1088/1742-6596/1036/1/012002

Estimating causal effects using triplet orderings: scale free graphs and dependence on the node order

Alexander K. Hartmann

Institut of Physics, University of Oldenburg, 26111 Oldenburg, Germany

E-mail: a.hartmann@uni-oldenburg.de

Gregory Nuel

LPMA, CNRS 7599, Université Pierre et Marie Curie, Paris, France

Abstract. For many data analysis tasks, obtaining the causal relationships between interacting objects is of crucial interest. Here, the case of modelling causal relationships via causal orderings is considered. Triplet ordering preferences are used to perform Monte Carlo sampling of the posterior causal orderings originating from the analysis of experiments involving observation as well as, usually few, interventions, like knockouts in case of gene expression. The performance of this sampling approach is compared to a previously used sampling via pairwise ordering preference as well as to the sampling of the full posterior distribution. This is performed for artificially generated causal, i.e. directed acyclic graphs (DAGs) with a scale-free structure, i.e. a power-law distribution for the out-degree. The sampling using the triplets ordering turns out to be superior to both other approaches, similar to our previous work, where the less-realistic case of Erdős-R é nyi random graphs was considered.

1. Introduction

When analysing large sets of data, on important task is to identify causal relationships [1] between the objects described by the data. One example is the analysis of gene regulatory networks, which have received a great deal of attention. In this context, Gaussian models like the Graphical Lasso [2] or approaches based on mutual information [3] are very popular for inferring gene regulation networks.

When time series of the data are available, the Granger causality [4] is a common approach in general. If time-resolved gene-expression data is available, methods like dynamic Bayesian networks [5] or ordinary differential equations [6] can be applied. Nevertheless, in many cases the temporal resolution is too coarse or it is not at all possible to measure time series. For this case, another popular approach, following the work of Pearl [1], focuses on causal Gaussian Bayesian networks. One can retrieve bounds on causal effects and thus partially determine causal relationships using only observational data [7]. To go beyond this, one can perform intervention calculus [8]. This means in real or numerical experiments, through suitable manipulations, one assigns to selected variables some fixed values. Therefore, they are not affected by the standard dynamics of the systems, while their state value is still available to their interaction partners. Examples are gene knockouts or gene knockdowns in gene regulatory networks. Or, for networks modelling infectious diseases, interventions could be implemented by medical treatments. In this paper we focus on estimating causal Bayesian networks in the presence of arbitrary mixtures of (non-time resolved) observational and interventional data [9, 10], i.e. wild-types and knockout/down experiments with possibly multiple interventions within each experiment.

As explained in Ref. [9] estimating the underlying DAG (Directed Acyclic Graph) structure of a causal Bayesian network is equivalent to finding of the so-called *causal ordering* between the genes of interest. In general, this causal ordering is unknown and belongs to a very large ordering space (p! possible orderings for p genes) which cannot be explored exhaustively. The solution suggested by Ref. [9] consists in sampling causal orderings in the posterior distribution using Markov chain Monte Carlo (MCMC) simulations.

At each MCMC step, a new causal ordering is sampled according to a proposal distribution (for example the Mallows distribution) and the maximum likelihood of the model must be computed given the new ordering to accept or reject the sampled ordering. Thanks to the closed formulas developed in Ref. [9], this likelihood maximisation can be done exactly and efficiently but requires a computational effort which still grows with the sixth power of the number p of interacting objects (e.g. genes). Thus, each single Monte Carlo step is computationally rather expensive.

Mathematically a proper MCMC is guaranteed to converge to the correct sampling, but only on diverging time scales. Given that for practical applications one only has a finite amount of computational resources available, only small networks can be treated in this way. For this reason, an approximation based solely on pairwise probabilities of ordering preference has recently been introduced [11]. This resulted in a considerable increase of efficiency, but led in many cases to less reliable parameter estimates.

Recently, we have extended [12] this approximation to triplet probabilities. We have shown that this results in a strongly increased accuracy with respect to the pairwise approach. Also we have shown that, when allocating a comparable amount of the numerical resources for the two algorithms, the triplet approach outperforms the sampling approach based on the full maximum likelihoods. Nevertheless, the past study was performed for Erdős-Rényi random graphs [13], which is the most simple graph model, showing no structure at all, resulting, e.g., in a Poisson degree distribution. For real systems the underlying graphs usually have a more complicated structure, [14] often resulting in a scale-free, i.e. power-law, degree distribution. For this reason, we have applied the pair- and triplet-based approaches to scale-free graphs. The most important results are that the approximations still outperform the full MCMC sampling and that the triplet-based approach is again better than the pair-based one. Also, we investigated which influence the out-degree of the manipulated node has on the amount of information about the causal structure obtained for the graph.

2. Model and Algorithms

First, we will present the graphical model which is behind this approach. Next, we show how model parameters are estimated by using likelihood calculation and how interventions are incorporated into the scheme. These two sections summarise the detailed descriptions of Ref. [9]. Next, we explain sampling of causal orderings, which we used to establish the causal structure. This allows us to calculate averages of the estimated quantities. Finally, we show how pair- and triplet-approximations are used to speed up the sampling procedure. For the convenience of the reader, this section contains a sufficient amount of details to allow for an implementation of the method and follows closely the presentation in Ref. [12].

2.1. Model

Here, we study directed graphs $G = (V, E)$ with p nodes $i \in V$. Pairs of nodes i, j are connected by directed edges $(i, j) \in E$, with weights $w_{i,j}$. The matrix $\mathbf{W} = (w_{i,j})$ contains all weights.

A nonzero weight describes a causal relationship. We assume that the graph is acyclic, i.e. a DAG. Without loss of generality, we can assume that the nodes are ordered according the causal relationships, i.e. $w_{i,j} > 0 \Rightarrow i < j$. This means within the following random process only nodes i can have causal effects on nodes j if $i < j$:

To each node $j = 1, \ldots, p$ a Gaussian random variable X_j is associated given by

$$X_j = m_j + \sum_{i<j} w_{i,j} X_i + \epsilon_j \quad \text{with} \quad \epsilon_j \sim N(0, \sigma_j^2)\,. \tag{1}$$

Thus, the equations are deterministic except the term ϵ_j which generates fluctuations of the variables, e.g., for fluctuations of gene expression and σ_j^2 determines the strength of fluctuations. In particular, the parameters $\mathbf{m} = (m_1, \ldots, m_p)$ and $\sigma = (\sigma_1, \ldots, \sigma_p)$ represent the mean values and the standard deviations if all interactions were absent. In the following an *experiment* corresponds to one realisation of the random process (1).

Furthermore, within the model it is possible to perform *interventions* on the nodes, which assign the variable of a node a fixed value instead of generating the value according to (1). In the DAG these values are used as inputs to the descendants when generating a realisation of the process, i.e. performing an experiment numerically [15].

2.2. Estimating Model Parameters

Given are N experimental data points $\mathbf{x}^k = (x_1^k, \ldots, x_p^k)$ $(1 \le k \le N)$ assumed to be generated according to (1). The set of nodes subject to interventions on experiment k is denoted by J_k, respectively ($J_k = \emptyset$ means no intervention for the k'th experiment). We denote by $K_j = \{k \,|\, j \notin J_k\}$ the experiments where there was no intervention on node j and by $N_j = |K_j|$ the number of times node j was not target of an intervention. The log-likelihood of the joint experimental outcome given the parameters has been derived in detail in Ref. [9]. Here we only state the final result for brevity:

$$\ell(\mathbf{m}, \sigma, \mathbf{W}) = -\frac{\log(2\pi)}{2} \sum_j N_j - \sum_j N_j \log(\sigma_i) - \frac{1}{2} \sum_j \frac{1}{\sigma_j^2} \sum_{k \in K_j} (x_j^k - \mathbf{x}^k \mathbf{W} \mathbf{e}_j^T - m_j)^2\,, \tag{2}$$

where $\mathbf{e}_j^T$ denotes the transpose of the unit vector which has a value 1 in position j and zero entries everywhere else. Note in order to write this equation in the standard form of the multinomial distribution, one uses the covariance matrix $\mathbf{\Sigma} = \mathbf{L}^T \mathrm{diag}(\boldsymbol{\sigma}^2) \mathbf{L}$, where $\mathbf{L} = (\mathbf{I} - \mathbf{W})^{-1}$ and $\mathbf{I} = \mathrm{diag}(\mathbf{1})$ [9]. The above stated form is more convenient, because it is already diagonal. We omit the dependence of ℓ on the data here for brevity of notation. For the given N measurements, the parameters $\hat{\mathbf{m}}, \hat{\sigma}, \hat{\mathbf{W}}$ leading to the maximum likelihood estimator (MLE)

$$\ell_{\max} = \ell(\hat{\mathbf{m}}, \hat{\sigma}, \hat{\mathbf{W}}) = \max_{\mathbf{m}, \sigma, \mathbf{W}} \ell(\mathbf{m}, \sigma, \mathbf{W}) \tag{3}$$

can be obtained [9] in a straightforward way by the following procedure: First one obtains for each experiment $k = 1, \ldots, N$ the measurements normalised with respect to the experiments where there was no intervention on node j, for each node j:

$$\mathbf{y}^{k,j} = \mathbf{x}^k - \frac{1}{N_j} \sum_{k' \in K_j} \mathbf{x}^{k'}\,. \tag{4}$$

Next one solves the linear system of size $p(p-1)/2$

$$\sum_{i'|i'<j} \hat{w}_{i',j} \sum_{k \in K_j} y_i^{k,j} y_{i'}^{k,j} = \sum_{k \in K_j} y_i^{k,j} y_j^{k,j} \quad \text{for}\, i < j,\ 1 \le i, j, \le N \tag{5}$$

to obtain estimates $\hat{w}_{i,j}$ of the weights for the MLE. Solving a linear system with $O(p^2)$ variables takes $O(p^6)$ steps. From this solution one obtains, still just following [9], estimates of the mean values

$$\hat{m}_j = \frac{1}{N_j} \sum_{k \in K_j} (x_j^k - \mathbf{x}^k \hat{\mathbf{W}} \mathbf{e}_j^T) \tag{6}$$

and of the variances

$$\hat{\sigma}_j = \frac{1}{N_j} \sum_{k \in K_j} (y_j^{k,j} - \mathbf{y}^{k,j} \hat{\mathbf{W}} \mathbf{e}_j^T)^2 . \tag{7}$$

2.3. Estimating the Posterior Distribution

So far, we have assumed that the causal ordering of the model is given by $\mathbf{o}_0 = (1, 2, \ldots, p)$. In experimental situations, if the data was actually generated according the DAG model, the ordering is most of the time unknown, i.e. all estimates will depend on the ordering: $\ell_{\max} = \ell_{\max}(\mathbf{o})$. For the general case, if the data was not generated according to a DAG model, the modelling must involve many orderings. Thus, in experiments and subsequent model estimation, one is actually interested in either the ordering which maximises the MLE, or, alternatively, in obtaining the posterior distribution involving all (or the dominant) orderings weighted by the corresponding ordering-dependent MLEs.

Both can be obtained in principle by iterating over all $p!$ possible causal orderings $\mathbf{o}$, i.e. permutations of the natural numbers $1, \ldots, p$. Each time one has to reorder the measurement data according this ordering, and to obtaine the MLE (3) via solving (4), (5), (6) and (7). Clearly, if p is too large, this enumeration is not feasible any more.

One alternative approach is to use a MCMC simulation, where orderings $\mathbf{o}(t)$ according the likelihood $\exp(\ell_{\max}(\mathbf{o}))$ are sampled, t denotes the number of steps. A convenient approach to achieve this is the *Metropolis algorithm*. Here, within each step, a *trial order* $\mathbf{o}'$ is generated. For the present study, we use local changes, i.e. an exchange of the order of two nodes with respect to the current ordering $\mathbf{o}(t)$. The trial ordering is *accepted*, i.e. $\mathbf{o}(t+1) = \mathbf{o}'$ with the probability

$$p_{\text{acc}} = \min\{1, \exp[\ell_{\max}(\mathbf{o}') - \ell_{\max}(\mathbf{o}(t))]\} . \tag{8}$$

Otherwise, the trial ordering is not accepted and the current ordering kept for the next time step, i.e. $\mathbf{o}(t+1) = \mathbf{o}(t)$. Note that for all simulations we performed (see below for details), the empirical acceptance rate of these locally generated trial orderings was below 0.5. The value of 0.5 is considered as a good choice by rule of thumb, balancing a desired high rate of changeing with a desired high acceptance rate. Therefore it would not make sense to consider trial orderings which differ from the current ordering by more than two exchanged positions, since this would increase the fluctuations and therefore decrease the acceptance rate even more.

This type of sampling guarantees, in principle, if the Markov chain is long enough, that the orderings are sampled according the desired posterior distribution. Note that for the computation of the change $\ell_{\max}(\mathbf{o}') - \ell_{\max}(\mathbf{o}(t))$ of the log-likelihood one has to recalculate the log-likelihood for the trial ordering $\mathbf{o}'$ from scratch. Thus, each MCMC Metropolis step takes $O(p^6)$ running time.

By starting with a random ordering $\mathbf{o}(0)$, performing a "long enough" MCMC sampling and by discarding the "initial" part (allowing for *equilibration*), a sample set S of orderings is obtained, which can be used to calculate averaged estimated parameters, see Section 2.4.

2.4. Calculation of Averaged Estimates

The aim is to study expectation values in ensembles defined by probabilities or likelihoods $P(\mathbf{o})$. Here we are interested in the true likelihoods $P(\mathbf{o}) \sim e^{\ell_{\max}(\mathbf{o})}$. Thus, for any measured quantity

$A(\mathbf{o})$, where the estimate depends on the assumed ordering $\mathbf{o}$, the expectation value is given by

$$\langle A \rangle \equiv \sum_{\mathbf{o}} A(\mathbf{o}) P(\mathbf{o}). \tag{9}$$

Note that the measured quantities of interest are usually estimates which are obtained from the maximum-likelihood calculation, e.g. the estimates of the weights obtained from (5) or estimates of the variances (7), or any other derived value.

If only a finite set S of samples is given, averages can be obtained, approximating the expectation values:

$$\hat{A} \equiv \frac{\sum_{\mathbf{o} \in S} A(\mathbf{o}) P(\mathbf{o})}{\sum_{\mathbf{o} \in S} P(\mathbf{o})} \tag{10}$$

These estimates are most accurate if the process used to generate the sample set follows the desired sampling $P(\mathbf{o}) \sim e^{\ell_{\max}(\mathbf{o})}$ as close as possible. Thus, the sample set S could be generated by a MCMC sampling according to the true probabilities $e^{\ell_{\max}}$, as outlined in the previous section. In this way automatically orderings with high contributions to (10) are preferentially generated. Note that since S is actually a mathematical *set*, there will be no multiple occurrences of orderings in S. If one allowed for multiple occurrence, then one would have to take simple arithmetic averages instead of weighted ones as in (10).

Anyway, here we work with sampling sets. The reason is that, alternatively, these sets can be obtained by sampling according to different probabilities, which only aim at approximating the true probabilities but which are computationally much cheaper to calculate. (This also means, we accept a sampling error, which would only disappear if the sampling set approaches the set of all possible configurations). One the other hand, if the size of the set is suitably restricted, we used always $|S| = 100$, the computationally expensive $O(p^6)$ full likelihood calculations have to be performed only for a small number of (here) 100 samples.

The approximate probabilities we have used are introduced in the following section.

2.5. *Pair and Triplet Probabilities*

Instead of sampling the full posterior distribution, in Ref. [11] it was proposed to perform an MCMC sampling from a different distribution, the Babington-Smith (BS) ordering distribution [16, 17]. It is based on pair preferences $\pi_{i,j}$ ($1 \leq i \neq j \leq p$) with $\pi_{i,j} \in [0,1]$ and $\pi_{i,j} + \pi_{j,i} = 1$. The meaning is that within the desired ordering distribution in any random ordering element i appears before j with this probability $\pi_{i,j}$. The pair preferences can be estimated from the experimental data with interventions by considering all possible two-node graphs $G_{i,j} \equiv (\{i,j\}, \{(i,j)\})$ with the nodes i and j and with exactly one directed edge (i,j). As above, for brevity of notation, we omit the dependence of the pair preferences and any derived quantities on the data here. Only the data values for the two nodes are considered. Note that in case of multiple interventions, we observed in tests, which are not contributing to the results shown here, that the overall performance of the sampling according pair preferences is somehow better if data points with interventions on other nodes than i, j are not included in the data set for the pair i, j, respectively. For each of the $p(p-1)$ directed two-node graphs the log-likelihood $\ell^{(2)}_{\max}(i,j)$ is obtained. The pair preferences are then given by

$$\pi_{i,j} = \frac{\exp(\ell^{(2)}_{\max}(i,j))}{\exp(\ell^{(2)}_{\max}(i,j)) + \exp(\ell^{(2)}_{\max}(j,i))}. \tag{11}$$

From the pair preferences, the BS probability of a full ordering $\mathbf{o}$ is obtained by

$$P(\mathbf{o}|\pi) \sim \prod_{i<j} \pi_{o_i,o_j} \tag{12}$$

with a suitable normalisation. The normalisation is not needed here, since, first, we only compare the (relative) values of (12) for different orderings. The corresponding log-likelihoods are denoted as

$$\ell^{\text{pair}} \equiv \ell^{\text{pair}}(\mathbf{o}) = \log \prod_{i<j} \pi_{o_i,o_j} \,. \tag{13}$$

Second, we performed MCMC sampling of orderings using the Metropolis algorithm according to (12) where also only relative likelihoods are needed. This was done in an equivalent way as above, only the true MLE is replaced by (13). Thus, starting again from a random ordering $\mathbf{o}(0)$, we generated trial orderings $\mathbf{o}'$ by exchanging the i-th and the j-th entry in the current ordering. The new orderings are accepted with the corresponding Metropolis probability. Note that one does not have to recalculate the BS probability from scratch, since the change in probability is easier to obtain. The Metropolis acceptance probability is given by

$$p_{\text{acc}}^{\text{pair}} = \min \left\{ 1, \frac{\pi_{o_j,o_i}}{\pi_{o_i,o_j}} \prod_{k|i<k<j} \frac{\pi_{o_j,o_k}\pi_{o_k,o_i}}{\pi_{o_i,o_k}\pi_{o_k,o_j}} \right\} . \tag{14}$$

This takes only $O(p)$ steps compared to the $O(p^2)$ steps which are needed for the calculation of the full probability. In particular it is much faster than computing the full likelihood which takes $O(p^6)$ steps.

Naturally, when sampling according to (12) the observed set of orderings will be different but somehow similar to when sampling according the true likelihood. The reason is that orderings with a high full likelihood induces in principle large pairwise probabilities for those pairs which are compatible with such an order. Nevertheless, the pairwise approximation cannot completely cover collective effects which involve the ordering of more than two nodes. Thus, the final estimates, like the weights, for the posterior distribution are obtained by keeping the n_{incl} samples with the highest probabilities (12) in the sample set S. For these orderings now the *true MLE* (3) is evaluated and used. This means, (10) is applied for any kind of estimation or averaging, i.e. these weights are now used in this final averaging step.

In [11] it was found that this sampling approach is in some case similar accurate as a full MCMC sampling as described in Sec. 2.3, but there were notable differences. In particular when the number p of nodes is growing, the orderings exhibiting the largest pair-based probabilities turned out to be more and more different from the orderings exhibiting large full likelihoods. This showed up in particular when calculating estimates of network parameters. Therefore it was proposed to maybe consider triplets instead of pairs.

Thus, it is the purpose of the present work to study this higher level approximation of the true posterior distribution. Similar to the above defined pair probabilities, we introduce triplet probabilities $\rho_{i,j,k} \in [0,1]$ such that $\rho_{i,j,k} + \rho_{i,k,j} + \rho_{j,i,k} + \rho_{j,k,i} + \rho_{k,i,j} + \rho_{k,j,i} = 1$ $(i < j < k)$. These probabilities can be estimated from the experimental data in a similar way as above, by considering all possible sub graphs $(\{i,j,k\}, \{(i,j), (i,k), (j,k)\})$ with three nodes and corresponding edges. For these sub graphs the corresponding MLE are obtained and suitably normalised, equivalent to (11) to yield the triplet probabilities $\rho_{i,j,k}$. They can be used to generalise the Babington-Smith probabilities of orderings to

$$P(\mathbf{o}|\rho) \sim \prod_{i<j<k} \rho_{o_i,o_j,o_k} \,. \tag{15}$$

Again, the normalisation is not needed here. The corresponding log-likelihood is denoted as

$$\ell^{\text{tripl}} = \log \prod_{i<j<k} \rho_{o_i,o_j,o_k} \,. \tag{16}$$

We perform an MCMC sampling of orderings according to these probabilities using the Metropolis algorithm and trial ordering generated via swapping of pairs of elements. Like for the case of sampling according the pair-based probabilities, these swapping pairs are chosen unbiased, i.e. each pair is chosen with the same probability. To guarantee a sampling according to (15), for the calculation of the acceptance probabilities only the change in probability of (15) has to be considered, which now takes $O(p^2)$ steps for such a swap.

Again, for all evaluations and estimates, the $n_{\text{incl}} = 100$ highest-probability samples with respect to the triplet probability are kept. For these samples the true likelihood is obtained and used for all averaging processes according to (10).

3. Results

We have performed simulations for scale-free random DAGs with p nodes. The graphs are generated similarly to the Barab á si-Albert model [14] with iterated preferential attachment of each node to m ($m = 3$ or $m = 4$ here) already existing nodes. Note that for the original undirected model, the graph generation starts with a complete graph with $m+1$ nodes. Here, to obtain a DAG, the initial graph consists of $m+1$ nodes such that node 1 is connected by directed edges $(1, j)$ to the other nodes $j = 2, \ldots, m+1$. Then, iteratively the nodes $j = m+2, \ldots, p$ are added to the graph. Each new node j is connected to m randomly chosen nodes among the already existing nodes $k = 1, \ldots, j-1$ by directed edges (k, j). The probability that the new node is connected to an existing node by an edge is proportional to the degree (sum of in-degree and out-degree) of the already existing node.

Each edge receives a weight which is drawn uniformly from the range $[-1, -0.4] \cup [0.4, 1]$. Thus, these edges should be distinguishable very well from the absent edges which correspond to weight 0. Finally, for each DAG instance, for each node j, mean values $m_j = 1/2$ are used and the variance values σ_j are drawn randomly uniformly in the interval $[0.01, 0.1]$. In Ref. [12] we have shown that the choice of m_j and σ_j does not play a crucial role. All simulations are performed for 1000 ($p = 20$) and 300 ($p = 50$) DAG instances generated independently in this way.

Next, for each DAG instance, a certain number of N measurements are performed, where the measurement vectors $\mathbf{x}^k$ ($k = 1, \ldots, N$) are generated according to (1). Typically, for a DAG of p nodes, we generated $N = 10p$ measurement vectors. Note that the scheme exhibited in Section 2.2 allows for multiple interventions. Nevertheless, since we are interested in comparing different sampling approaches here, we present for simplicity just single interventions which are systematically done during the first r ($r \leq N$) experiments of each set of experiments. We applied a systematic manner, such that for all nodes at least $\lfloor r/p \rfloor$ interventions are performed while for $r - p\lfloor r/p \rfloor$ nodes one intervention more, i.e. $\lceil r/p \rceil$ interventions are performed. This sums up to r interventions. For each intervention on node j, we fix $X_i = 0$, respectively, corresponding to a knockout. For the order in which the nodes are taken for interventions (experiment by experiment), we studied three different variants:

- *decreasing* out-degree
- *random* order
- *increasing* out-degree

This means for *decreasing* order, in experiment 1, the node with the highest out-degree will be set to zero, in experiment 2 the node with the second highest out-degree, and so on.

Thus, to measure the efficiency of an approach, we consider all edge weights $w_{i,j}$, where $w_{i,j}$ might be zero because it does not match the causal ordering, or because the causal interaction is just absent. This is done in the following way: From each sampling, we obtain averaged estimated

IOP Conf. Series: Journal of Physics: Conf. Series **1036** (2018) 012002 doi:10.1088/1742-6596/1036/1/012002

edge weights $\hat{w}_{i,j}$ $(i, j = 1, \ldots, p)$ according to (10). Now, we count the "bad" estimates of the edge weights by comparing with the true edge weight as follows:

$$\delta_{\text{bad}}(i,j) = \begin{cases} \Theta(|\hat{w}_{i,j}| - w_0) & \text{if } w_{i,j} = 0 \\ \Theta\left(\left|\frac{w_{i,j} - \hat{w}_{i,j}}{\hat{w}_{i,j}}\right| - w_1\right) & \text{if } w_{i,j} \neq 0 \end{cases} . \tag{17}$$

$\Theta(x)$ denotes the threshold function which is $\Theta(x) = 0$ for $x \leq 0$ and $\Theta(x) = 1$ for $x > 0$. Thus, for a weight which is zero in the original DAG used to generate the data, the averaged estimate is counted as bad if its absolute value exceeds a threshold value w_0. For an edge with nonzero weight of the original DAG the average estimate is counted as bad, if the relative deviation of the average estimated weight and the original weight exceeds threshold value w_1. We used $w_0 = 0.1$ and $w_1 = 0.5$. In general, details of the results might depend on the actual values of w_0 and w_1, but we verified that the principal trends, with respect to which sampling approach performs better, remain the same [12]. Also when using the threshold-independent area under the *Receiver-Operator Characteristics* (ROC), the principal nature of the results does not change [12]. Thus, we concentrate here on estimating the "bad" edges by iterating over all edges, i.e. we measured

$$n_{\text{bad}} = \frac{1}{p(p-1)} \sum_{i \neq j} \delta_{\text{bad}}(i,j) . \tag{18}$$

3.1. Comparing Different approaches

First, we compare the performance of the four different approaches, i.e. when using the original ordering, sampling orderings with MCMC based on the full likelihood, sampling with pair probabilities and sampling with triplet probabilities.

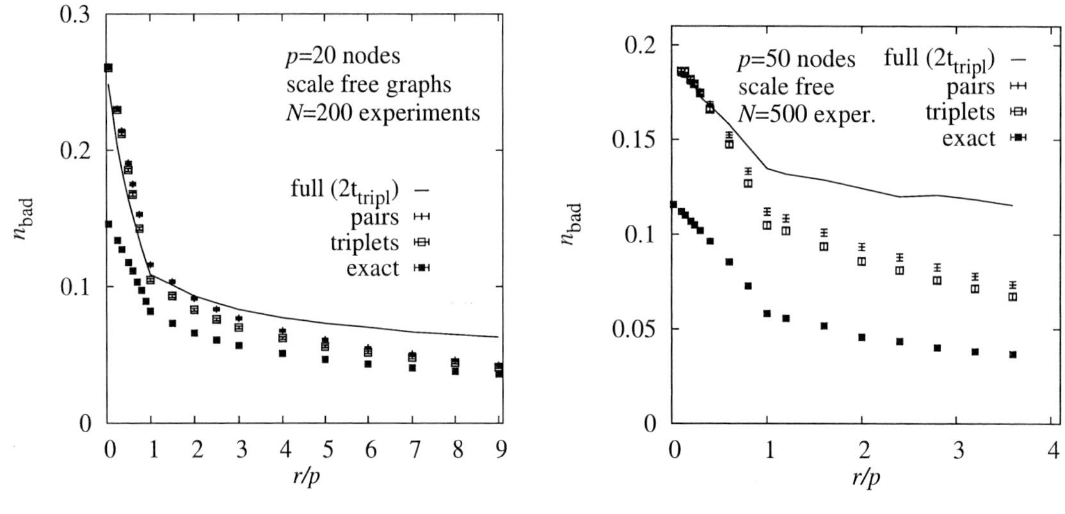

Figure 1. Average fraction n_{bad} of incorrectly estimated edge weights as a function of the number of interventions r per node. The results are obtained using four different sampling approaches using the true maximum likelihoods (**full**), the pair BS probabilities (**pair**), the triplet BS probabilities (**triplet**) and using just the exact ordering of nodes of the DAGs. The running time for the sampling using the true maximum likelihoods was restricted to two times the CPU time of the triplet sampling. (left) The data was generated for 1000 randomly generated scale free DAGs ($m = 3$) of size $p = 20$ nodes. (right) Results for 200 scale free DAGs ($m = 4$) of size $p = 50$.

In Fig 1 (left) the resulting average values for the fraction n_{bad} of incorrectly estimated edge weights is shown for scale-free graphs with $p = 20$ nodes as a function of the relative number

r/p of single-node interventions. The order in which the nodes are fixed was *random*. One can observe that with increasing number of interventions, the quality of the averaged weight estimate increases. Similar to the case of complete graphs [12], due to the limited number of MCMC steps performed, sampling using the true likelihoods is the worst approach, except for a very small number of interventions, where the estimates are bad anyway. For scale-free graphs the quality of the estimates is much better when using the triplet probabilities as compared to the pair probabilities. Still, one cannot reach the quality of the estimate which we obtained when using the single true ordering, which constitutes the lower limit of what is achievable using any algorithm.

We have repeated the same study for larger graphs of $p = 50$ nodes, here for $m = 4$. Again we have measured the fraction n_{bad} of incorrectly estimated edge weights. In Fig. 1 (right) the corresponding results are shown. The results for the triplet-based approach are better than for the pair-based method, and much better than the full MCMC sampling restricted to at most twice the numerical resources of the triplet-based simulation. Note that the difference between pair-based approach and triple-based approach is now larger than for the $p = 20$, $m = 3$ case. For $p = 50$, $m = 3$ (not shown) the differences are smaller than for $p = 50$, $m = 4$. This goes along with the previous results [12] that the difference between pair-based and triplet-based approaches grows with the edge-density of the graphs.

3.2. Influence of node degree

It is a natural assumption that the effect of an intervention on a node is larger for higher number of outgoing edges. For this reason, we have performed simulations for varying order (increasing/random/decreasing degree) of the nodes. Still, for each experiment exactly (at most) one node is fixed to zero. This means in particular that after p experiments, each node has been fixed for exactly one experiment, here the order should have no large influence.

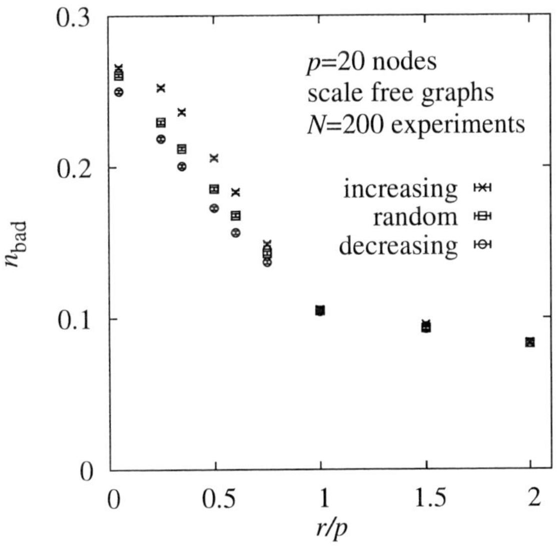

Figure 2. Average fraction n_{bad} of incorrectly estimated edge weights as a function of the number of interventions r per node. The data was generated for 1000 randomly generated scale free DAGs of size $p = 20$ nodes. The results are obtained using the triplet BS probabilities for different orderings of nodes: with increasing out-degree, random, and decreasing out-degree.

The results are shown in Fig. 2. For fixing the nodes with decreasing out-degree, the results for the number n_{bad} of incorrectly estimated edges are best, they are below the results for the two other approaches. When using the increasing out-degree order, the results are worst. This appears natural, because when fixing nodes with higher degree earlier, one learns the causal structure of a network quicker. Nevertheless for $r/p = 1$ the results agree within error bars for the three different orders. Also, for $r/p > 1$ the results are basically indistinguishable. The reason is that each node has been fixed already exactly or at least once. Therefore, the additional information one obtains from fixing a node again is small anyway. This means, the order does not play so much of a role. Note that for $r/p = 1$ the results should be statistically indistinguishable anyway, because here each node has been fixed exactly for one experiment. Furthermore, the differences between the three approaches are also not very big for $r/p < 1$. In particular the regime where n_{bad} decreases strongly seems to extend for all three cases up to $r/p = 1$. This means in particular, that fixing the nodes in a random order, which corresponds to real situations where the node degrees are not known beforehand, is no real drawback when determining causal structures.

4. Summary and Discussion

To summarise, we studied the estimation of causal orderings and corresponding parameters in sampled data using interventions. In particular, we compared pairwise Babington-Smith sampling, which was discussed before [11] with triplet-based sampling for scale-free graphs. This ensemble is more realistic for real data applications in comparison to Erdős-R é nyi random graphs for which these approaches have been compared in a previously published work [12]. All results show a better performance for the triplet sampling approach. When limiting the numerical effort to about two times the running time of the triplet sampling, a sampling using the full maximum likelihood turned out to be much worse than both pair- and triplet-based sampling. Nevertheless, for very sparse graphs, the performance of pair- and triplet-based sampling is more similar.

Therefore, the triplet-based approach and, to some extent, the pair-based approach appear to be well-balanced: They are computationally efficient enough such that long MCMC chains can be generated easily for systems large enough for practical applications. This would be impossible when using a sampling based on the full likelihood, except for very small systems. On the other hand, in combination with the final computation of the true maximum-likelihood estimators for a comparable small subset of "best" configurations, the triplet approach allows for accurate results, in some cases much — in other cases slightly — better than the pair-based approach.

For future work it will be interesting to extend the approaches to very large networks. There exist other approaches for the estimation of the causal structure of actually very large networks of thousands of nodes using ad hoc heuristic algorithms [18, 19] which are based, among others, on clustering approaches and work often on a coarse-grained level. Although the approaches presented here are based on generative models, allowing for probabilistic interpretations and allowing for detailed reconstruction of the underlying networks, they can be extended to much larger systems as well. This can be achieved for a given large set of nodes by considering many different subsystems (subgraphs) of medium size and treating them with the correct joint likelihoods. The resulting sub networks can be assembled to one large consensus network. Here, e.g. the "Iterative Sub-Network Component Analysis" approach [20] or similar approaches can be applied.

Furthermore, it could be interesting to study more thoroughly the point $r = p$ where most results exhibit a notable change of characteristics: For $r < p$ the progress when adding further interventions is much faster compared to $r > p$. It could be interesting whether this change corresponds to a kind of information-driven "learning" phase transition, similar to Hopfield neural networks where the memory does not work well beyond a certain number of patterns to be stored.

Finally, it would be beneficial to try other intervention strategies, e.g interventions of more than one node per experiment. Also adaptive schemes could be useful, where the choice of the next intervention node depends on the outcome of the previous interventions.

Acknowledgments
We thank Hendrik Schawe for critically reading the manuscript. A visit of GN to the University of Oldenburg, during which the project was set up, was supported by the Graduiertenkolleg 1885 "Molecular Basis of Sensory Biology" funded by the Deutsche Forschungsgemeinschaft (DFG). AKH is grateful to the Foundation Sciences Mathématiques de Paris (FSMP) for providing financial support for a three month (February to April 2016) visit at the LPTMA of the Université Pierre et Marie Curie during which this project was designed and started.

The simulations were performed on the *Carl* cluster of the University of Oldenburg jointly funded by the DFG (INST 184/157-1 FUGG) and the Ministry of Science and Culture (MWK) of the Lower Saxony State.

[1] Pearl J 2009 *Causality* (Cambridge: Cambridge University Press)
[2] Friedman J, Hastie T and Tibshirani R 2008 *Biostatistics* **9** 432–441
[3] Wang J, Chen B, Wang Y, Wang N, Garbey M, Tran-Son-Tay R and et al 2013 *Nuc. ac. res.* **41** e97
[4] Granger C W J 1969 *Econometrica: J. Econ. Soc.* **37**(4) 424–438
[5] Zou M and Conzen S D 2005 *Bioinf.* **21** 71–- 79
[6] Wu H, Lu T, Xue H and Liang H 2014 *J. Amer. Stat. Assoc.* **109** 700 - –716
[7] Maathuis M H, Colombo D, Kalisch M and Bühlmann P 2010 *Nature Methods* **7** 247–248
[8] Maathuis M H, Kalisch M and Bühlmann P 2009 *Annals Stat.* **37** 3133–3164
[9] Rau A, Jaffrézic F and Nuel G 2013 *BMC systems biology* **7** 1
[10] Hauser A and Bühlmann P 2015 *J. Royal Stat. Soc.: Series B* **77** 291–318
[11] Nuel G, A A R and Jaffrézic J 2014 *Quality. Techn. Quant. Manag.* **11** 23–37
[12] Hartmann A K and Nuel G 2017 *PLOS one* **12** e0170514
[13] Erdős P and Rényi A 1960 *Publ. Math. Inst. Hungar. Acad. Sci.* **5** 17–61
[14] Albert R and Barabási A L 2002 *Rev. Mod. Phys.* **74** 47
[15] Hartmann A K 2015 *Big Practical Guide to Computer Simulations* (Singapore: World Scientific)
[16] Critchlow D E, Fligner M A and Verducci J S 1991 *J. Math. Psych* **35** 294–318
[17] Joe H and Verducci J S 1993 *Cell* **102** 109–126
[18] Liang T L H and Li H 2011 *J. Amer. Stat. Assoc.* **106** 1242–1258
[19] Wang Y, Xu M, Wang Z, Tao M, Zhu J and et al L W 2011 *Brief. Bioinform.* **13** 162–174
[20] Jayavelu N D, Aasgaard L S and Bar N 2015 *BMC Bioinf.* **16** 366

Reflectance spectra of Asteroids and Meteorites: their classifications and statistical comparisons

Hideaki Miyamoto[1,2,3,4], Peng K Hong[1], Takafumi Niihara[2], Takeshi Kuritani[5], Kenji Fukumizu[6,7], Hideitsu Hino[8], Kenji Nagata[9], Shotaro Akaho[10,9,11], J Alexis P Rodriguez[12,13], Hemmi Ryodo[1], Seiji Sugita[3], and Masato Okada[4,9,14,15]

[1] Department of Systems Innovation, The University of Tokyo, 7-3-1 Hongo, Bunkyo-ku, Tokyo 113-8656, Japan
[2] The University Museum, The University of Tokyo, 7-3-1 Hongo, Bunkyo-ku, Tokyo 113-0033, Japan
[3] Department of Earth and Planetary Science, The University of Tokyo, 7-3-1 Hongo, Bunkyo-ku, Tokyo 113-0033, Japan
[4] Department of Complexity Science and Engineering, The University of Tokyo, 5-1-5 Kashiwanoha, Kashiwa, Chiba 277-8561, Japan
[5] Department of Natural History Sciences, Hokkaido University, N10W8 Kita-ku, Sapporo 060-0810, Japan
[6] Department of Mathematical Analysis and Statistical Inference, The Institute of Statistical Mathematics, 10-3 Midori-cho, Tachikawa, Tokyo 190-8562, Japan
[7] Department of Statistical Science, The Graduate University for Advanced Studies, Shonan Village, Hayama, Kanagawa 240-0193, Japan
[8] Department of Computer Science, University of Tsukuba, 1-1-1 Tenoudai, Tsukuba, Ibaraki 305-8573, Japan
[9] Artificial Intelligence Research Center, AIST, 2-3-26 Aomi, Koto-ku, Tokyo 135-0064, Japan
[10] Human Informatics Research Institute, AIST, 1-1-1 Umezono, Tsukuba, Ibaraki 305-8568, Japan
[11] Department of Computational Intelligence and Systems Science, Tokyo Institute of Technology, 4259 Nagatsuta-cho, Midori-ku, Yokohama, Kanagawa 226-8503, Japan
[12] Planetary Science Institute, 1700 East Fort Lowell, Suite 106, Tucson, AZ 85719-2395, USA
[13] NASA Ames Research Center, Mail Stop 239-20, Moffett Field, CA 94035, USA
[14] Department of Physics, The University of Tokyo, 7-3-1 Hongo, Bunkyo-ku, Tokyo 113-0033, Japan
[15] Materials research by Information Integration Initiative, NIMS, 1-2-1 Sengen, Tsukuba, Ibaraki 305-0047, Japan

E-mail: hm@sys.t.u-tokyo.ac.jp

Abstract. Asteroids have been observed both from the ground and through space missions for decades, which accumulated large amount of their observational data. These data are used to estimate the sizes, orbits, and even possible chemical compositions of asteroids. Even though the chemical composition is generally difficult to be accurately determined without a sample return or in-situ observation by a spacecraft, asteroids are classified based on their reflectance spectra, which are compared with those of meteorites, which are known to be mostly originated from asteroids. This scheme works reasonably well for some asteroid types, but others, mostly featureless ones in reflectance spectra, remained controversial due to the fact that the observational data of asteroids and measured data of meteorites are different in terms of the data coverage, precision and resolution. Our aim is to connect asteroids with meteorites based on sparse modelling in order to search for the optimal integration scheme for two different databases without relying on preliminary knowledge. For the above purpose, we develop large databases of asteroids and meteorites for easy application of sparse modelling. Through our analyses including principal component analysis, Bayesian spectral deconvolution and dimensionality reduction, we found that our data-driven approach can extract potential information without using empirical knowledge. Our methods show a new type of data handling scheme for asteroid and meteorite data, potentially having a significant contribution for future missions.

1. Introduction

Small bodies in the solar system, such as asteroids and comets, are considered as remnants of the early evolution of the solar system [1]. Unlike planets which have experienced large scale melting due to impact and/or radioactive heating, small bodies are considered to have experienced less thermal metamorphism, except for shock heating on the surfaces and initial radioactive heating. Therefore, even today small bodies would likely possess primordial materials which have been lost from planets and large satellites through the evolution of solar system. Thus, understanding the formational histories of asteroids may provide constraints on the physical and chemical conditions of the solar nebula and subsequent evolution [1]. Small bodies are also considered as sources of volatile materials, which provide water and organic materials to the Earth. Thus the distribution of small bodies in the early solar system provide a great influence on the formation of Earth's atmosphere-ocean system and on the origin of life. In order to understand asteroids' formation history, compositions of asteroids are the first clue to infer the initial conditions of asteroids. Asteroids' compositions can be measured with high accuracy and high precision through in-situ observations by spacecraft. Several spacecraft missions have been successfully conducted thus far to explore asteroids, such as Galileo [2], NEAR Shoemaker [3], Rosetta [4] and Dawn [5], as well as sample-return missions such as Stardust [6] and Hayabusa [7]. However, since more than three million asteroids are estimated to exist in the solar system, it is impossible to explore every asteroid with spacecraft. Thus inferring surface compositions of asteroids by ground-based telescopes is vital to complement spacecraft missions. In many cases, observable data of asteroids by ground-based telescopes are limited to orbital characteristics and reflectance spectra. Therefore, comparing asteroids' reflectance spectra with those of meteorites is the most essential approach to infer the composition of asteroids (Figure 1).

Majority of meteorites are considered to originate from asteroids. More than twenty thousand meteorites have been collected to date and their reflectance spectra have been measured. However, direct comparison of reflectance spectra between meteorites and asteroids is not simple because meteorites are classified into more than 70 classes, while asteroids are typically 30 classes. Furthermore, most of ordinary chondrites which are the most common meteorite type is spectrally different with typical S-type asteroids which usually have been assumed as parent bodies of ordinary chondrites [8]. Hayabusa spacecraft resolved this problem by returning surface materials from a S-type asteroid Itokawa, proving that the surface materials of Itokawa are consistent with LL-type ordinary chondrites [7]. This indicates that the spectral difference between ordinary chondrites and S-type asteroids would be originated from the effect of space weathering [9]. The lessons learned from the Hayabusa mission is that studies of high-precision measurement of minor elements and isotope analysis of meteorites can be connected with studies of ground-based observations of asteroids, which cover a wide spatial range with low-precision measurement. Thus ground observations of asteroids reinforced by meteoritical science would greatly improve our understanding of compositional distribution of asteroids.

Despite the success of sample-return missions represented by Hayabusa, there still remains a few major issues to connect asteroids with meteorites. One of the largest obstacles is that the classification scheme between asteroids and meteorites are fundamentally different. Asteroids are classified mainly based on the shape of their reflectance spectra and orbital parameters [10], while meteorites are classified by detailed petrology and mineralogy [11]. In addition, reflectance spectra of asteroids are greatly influenced by surface properties (e.g. roughness, particle size) and space weathering effect. Reflectance spectra of meteorites are also significantly influenced by various terrestrial weathering and contamination. Sampling bias of meteorites is another major problem because some of the meteorites, such as iron meteorites, can more frequently survive destruction by aerodynamic heating during entry. Such meteorites are also easy to be discovered on the ground because of their pronounced colors compared with other type of meteorites. From the perspective of data science, the above problems can be described as (1) choosing the optimal integration scheme for two databases with significantly different accuracies and certainties, and (2) extracting appropriate base functions to explain different kinds of physical models. Previous spectra analyses, for example, have focused on certain wavelength regions or intensity ratios based on preliminary knowledge given by petrology or meteoritics. By using sparse modelling, however, we could extract optimal base functions without relying on preliminary knowledge. Sparse modelling is also strongly required for automatic analyses of big data in the present situation, where amount of observation data is dramatically increasing. In this paper we report our attempt to connect asteroids with meteorites using statistical scheme based on sparse modelling.

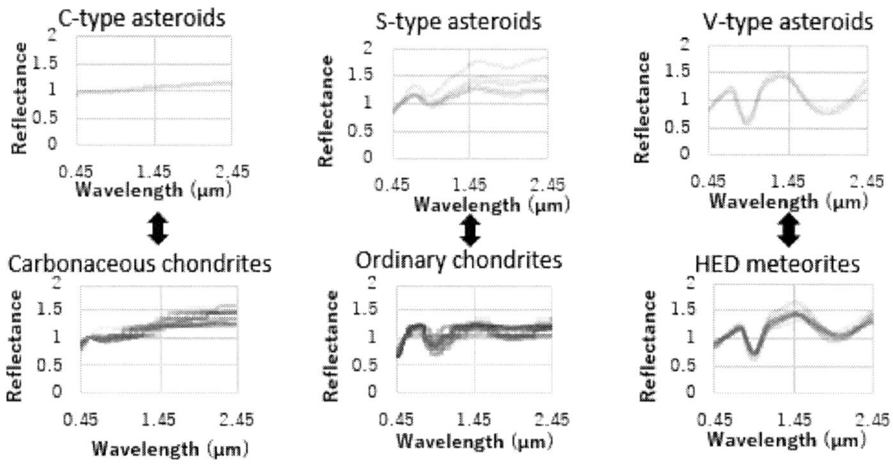

Figure 1. Typical reflectance spectra for asteroids (C-type, S-type, and V-type) and their presumably-corresponding meteorites (C chondrites, O chondrites, and HED).

2. Cluster analysis of the bulk elemental compositions of meteorites

Identifications of the corresponding meteorite types of the targeted bodies of asteroid missions are among the major goals of asteroid missions. Besides such scientific importance, rapid identification of the surface materials during the reconnaissance phase is also critically important. Modern spacecraft carry cameras/spectrometers in the visible to infrared wavelengths, which can identify surface materials. However, irradiation by cosmic rays and solar wind as well as bombardment by interplanetary dust particles modify the surfaces of airless bodies through processes known as space weathering. Impact events also mix materials on the surface of the body. These processes may flatten or change the absorption characteristics of reflectance spectra. Therefore, elemental compositions, which can be measured by X and gamma-ray spectrometers, may be useful for the above purpose. However, it has not been investigated extensively how well we can classify these planetary materials based on elemental composition alone.

We perform principal component and cluster analyses on 12 major and minor elements of the bulk compositions of 500 meteorites reported in the NIPR database [12]. Our analyses, including hierarchical cluster analysis, indicate that meteorites can be classified into about 10 groups solely by their bulk elemental compositions. We suggest that Si, Fe, Mg, Ca, and Na are the optimal set of elements, as this set has been used successfully to classify meteorites of the NIPR database with more than 94% accuracy. Principal components analysis indicates that elemental compositions of meteorites form 8 clusters in the three-dimensional space of the components. Three major principal components (PC1, PC2, and PC3) can be interpreted as degree of differentiations of the source body (i.e., primitive vs. differentiated), degree of thermal effects, and degree of chemical fractionation, respectively. Although the exact ranges of elements of each cluster could suffer from the systematic intra-laboratory error, realized through comparing our results with those of another elemental composition database, our new method shows promise in the classification of the surface materials of a small body into a known group of meteorites, having a significant importance in future reconnaissance.

3. Automatic deconvolution method for Modified Gaussian model using the exchange Monte Carlo method

Deconvolution analysis of reflectance spectra has been a useful method to infer mineral composition and crystal structure for major rock forming minerals, including olivine and pyroxene. Clinopyroxene is one of the most important mineral groups due to both its rich abundance on solid bodies in the solar system and distinguished absorption features. The band centers of visible to near-infrared spectra of clinopyroxene are known to vary due to total iron and calcium content. Visible/near infrared spectra of synthetic clinopyroxene are characterized based on a modified Gaussian Model (MGM) [13]. The numerical algorithm of the widely used MGM, however, utilizes the steepest descent method, which has a local minima problem [14]. With bad initial parameters, the steepest descent method converges into a local minimum, thus one needs to manually adjust initial parameters and calculate the model repeatedly to obtain the desired solution.

In order to avoid the local minimum problem, we analyzed visible/near infrared spectra of clinopyroxene using Bayesian spectral deconvolution with the exchange Monte Carlo method [15]. This method is an improved algorithm of the Markov chain Monte Carlo method, aimed to both avoid local minima traps and remove the arbitrariness originated from initial parameters [16]. We used the visible/near infrared spectra of clinopyroxene, measured at 5 nm intervals over the wavelength range of 0.3-2.6 μm [13]. We chose only sieved powder samples whose grain sizes are smaller than 45 μm with individual grains about 15-25 μm in size. We collected 31 clinopyroxene spectra from published database with wide ranging Ca, Mg, Fe compositions (8-52%, 0-52% and 3-90%, respectively). We found that: (1) 1 μm band shifts regularly to longer wavelengths with an increase of Ca content; (2) 1 μm band shifts to shorter wavelengths with an increase of Fe content, although the variance is larger than that of orthopyroxene; (3) for Ca contents $< 30\%$, 2 μm band shifts to longer wavelength as a function of Ca, while for Ca contents $> 30\%$, the center of 2 μm band remains almost constant at ~2.4 μm; and (4) the center of 2 μm band of augites (Ca $<$ ~20%) does not depend on Fe content significantly, while the center of 2 μm band of pigeonites (Ca $>$ ~20%) shifts to longer wavelengths as a function of Fe content. These results are consistent with the previous study [13], suggesting that the exchange Monte Carlo method can yield the same results obtained by conventional MGM analysis, which is important to remove issues of local minima and arbitrariness of initial parameters. The successful application of the exchange Monte Carlo method to a wide range of clinopyroxene would pave the way for further deconvolution analysis of reflectance spectra of mineral mixtures, such as olivine, orthopyroxene, and clinopyroxene.

4. Relationship between reflectance spectra of asteroids and meteorites

4.1. Albedo and reflectance spectra

Asteroids have been classified into several types based on principal component analysis of reflectance spectra [10]. Different types of meteorite samples, however, can yield similar reflectance spectra such as black ordinary chondrites and carbonaceous chondrites [17]. Thus the degeneracy of reflectance spectra may suggest that there is an application limit of the classification scheme based solely on reflectance spectra. Albedo data also have been accumulated, however, their relationship with reflectance spectra are not fully understood [18]. Our objective is to combine albedo with reflectance spectra in an attempt of improving asteroid classification.

We compiled visible/infrared reflectance spectra within the region of 0.45 to 2.45 μm based on published databases. All the spectra were sampled with cubic spline fits at a wavelength interval of 0.05 μm, resulting in 41 data points. Each spectrum was subtracted by its mean. Using the reduced spectrum, we defined the spectral type index R, which is the difference between correlation coefficients with average S and C-type spectra as follows:

$$R = \frac{\boldsymbol{t} \cdot \boldsymbol{r}_{\mathrm{C}}}{\|\boldsymbol{t}\| \|\boldsymbol{r}_{\mathrm{C}}\|} - \frac{\boldsymbol{t} \cdot \boldsymbol{r}_{\mathrm{S}}}{\|\boldsymbol{t}\| \|\boldsymbol{r}_{\mathrm{S}}\|}$$

where $\boldsymbol{t}$, $\boldsymbol{r}_c$, and $\boldsymbol{r}_s$ are the reduced target spectrum, average C-type, and average S-type spectra, respectively. Positive R values more closely approximate C-type spectra, while negative R values S-type. We also compiled asteroids' geometric albedo data mostly from Supplemental IRAS Minor Planet Survey [19].

The albedo-spectra map indicates that there is a general trend in the distribution of asteroid types (Figure 2). V-type, C-type and S-type asteroids are distinctly separated from each other on the albedo-spectra map. The variance of each cluster appears to increase in the order of V-type, S-type and C-type. The reflectance-spectra map for meteorites also shows a trend in the distribution of meteorite types. HED meteorites distribute at the top left, carbonaceous chondrites distribute at the bottom right, and ordinary chondrites distribute between HED meteorites and carbonaceous chondrites (Figure 2). It appears that geometric albedo or reflectance plays a significant role in the resulting spectral signature. There are many possible factors which could influence albedo, such as (1) mineral and/or elemental composition, (2) fragmented particle size, (3) space weathering and (4) crystal size. We consider, however, crystal size to be the primary factor in the resultant albedo-spectra map because: (1) based on the analyses of meteorites, there is no significant difference between ordinary and carbonaceous chondrites in terms of carbon content or modal composition, (2) smaller particles tend to yield higher albedo, though not enough to explain the significant albedo difference observed among asteroids, and (3) spectral darkening due to space weathering requires reduced Fe (noting that there is no clear evidence that C-type asteroids, which are significantly darker than S-type or V-type asteroids, have higher Fe abundances). With the above consideration, we discuss the crystal size in the context of thermal processes of asteroids to help explain albedo-spectra map.

Smaller crystals generally result in darker reflectance. V-type asteroids are believed to have experienced differentiation and magmatism, as evidenced by the analyses of HED meteorites. Thus the crystals would become coarser due to a slow cooling rate, resulting in brighter and more pyroxene-rich surface spectra. On the other hand, S- and C-types would be undifferentiated chondrite asteroids, believed to correspond to ordinary and carbonaceous chondrites, respectively. Those meteorites preserve chondrules generally having finer crystals due to rapid cooling in the presolar nebula [20]. This would result in dark featureless surface spectra of C-types. S-type asteroids have experienced moderate thermal metamorphism after their accretion, causing recrystallization which results in larger crystals and in more evident pyroxene signals than C-type.

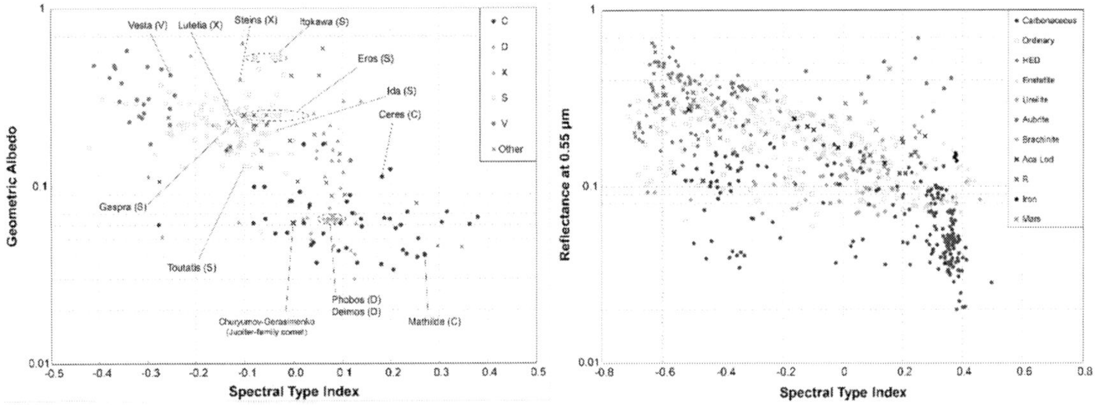

Figure 2. (left) Relationship between geometric albedo and visible/near infrared spectral type for asteroids. Also shown are the names of asteroids that have been explored by spacecraft. (right) Relationship between reflectance at 0.55 μm and visible/near infrared spectral type for meteorites.

4.2. Principal component analysis including 3 μm wavelength region

In order to better understand the compositions of asteroids, mineralogical relationship between asteroids and meteorites have been studied based on reflectance spectra obtained by ground- and space-based telescope. Their relationship, however, remains poorly constrained except for a S-type asteroid Itokawa and LL chondrites [7]. Although spectral similarities have been suggested between V-type asteroids and HED meteorites and between carbonaceous chondrites and C- and/or D-type asteroids, detailed relationship is not well constrained. The major obstacle to compare asteroids and meteorites is that the classification scheme between asteroids and meteorites are fundamentally different. Asteroids are classified mainly based on the shape of their reflectance spectra and orbital parameters [10], while meteorites are classified by detailed petrology and mineralogy [11]. Based on principal component analysis, Britt et al. (1992) [17] compare reflectance spectra of asteroids with those of meteorites. In this pioneer work, they pointed out some offsets remain between meteorites and asteroids, which might be due to the fact that they used eight color spectra within visible wavelength from 0.35 to 1.0 μm. Since characteristic absorptions are observed in the near-infrared range, including pyroxene (2 μm) and hydrated silicates (3 μm), using reflectance spectra with a wider wavelength range could result in a better spectral matching between asteroids and meteorites. In this study, we developed a database of reflectance spectra for asteroids and meteorites with wavelengths ranging from 0.3 to 4 μm and perform multivariate analysis.

We obtained reflectance spectra for meteorites and asteroids from RELAB [21] and the database of Planetary Spectroscopy at MIT [22], respectively. Asteroid spectra for 3 μm band are obtained from previous studies [e.g., 23]. All the spectra were sampled with cubic spline fits at a wavelength interval of 0.05 μm. Meteorite spectra are chosen based on the following criteria: (1) particulate sample, (2) phase angle is 30°, (3) sample is from valid/known meteorite, (4) not heated/laser-irradiated, inclusion or impact melt sample, (5) not moon sample or lunar meteorite. The developed database includes 534 meteorite spectra and 369 asteroid spectra. We performed principal component analysis on the database and measure how well each meteorite group and asteroid group is separated on the principal component space (Figure 3). Our analyses show that using spectra from 0.4 to 2.5 μm, accuracy of separation among ordinary chondrites, carbonaceous chondrites, HED meteorites is significantly improved compared with the case using spectra from 0.4 to 1.0 μm. In fact, cluster analysis (Ward's method) on asteroid spectra show that including 2.5 μm improve accuracy of classification from 65% to 75%. On the other hand, we found that the accuracy of separation is not significantly improved when using meteorite spectra from 0.4 to 4 μm compared with the case using spectra from 0.4 to 2.5 μm. This suggests that 3 μm band may not be useful for classification of some meteorites, because 3 μm band is widely observed among chondrites and achondrites. Our analysis also revealed that an inherent spectral difference exist between asteroids and meteorites. This may be attributed to space weathering effect. Or it may be originated from the difference of observation target, in which reflectance spectroscopy of asteroids measure surface compositions while reflectance spectroscopy of meteorites measure bulk compositions.

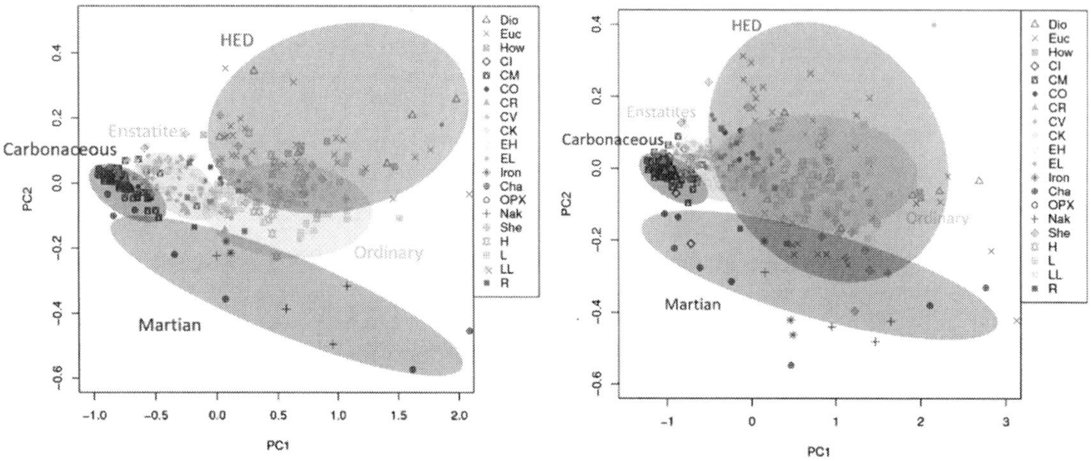

Figure 3. (left) Principal components of meteorites' spectra from 0.3 to 2.5 μm. (right) Principal components of meteorites' spectra from 0.3 to 4 μm.

4.3. Spectral relationship visualized by correlation distance and t-SNE

Statistical classifications of spectral types without detailed interpretation of spectral shapes can be useful to overview the variation and relationships within a spectral data set, even though there are known difficulties of comminution, melting, mixing, and space weathering. For example, as discussed in section 4.2, simple scheme such as principal component analysis cannot overcome the inherent spectral difference between asteroids and meteorites.

Thus we expand our analysis by applying to a wider and denser datasets of reflectance spectra for both meteorites and asteroids based on mathematically more advanced methods. We analyse sixteen kinds of distance of spectra including Partial Autocorrelation, Dynamic Time Warping, Pearson Correlation, and Euclidean distance. The distances are visualized by using six kinds of schemes including t-SNE (t-Stochastic Neighbor Embedding) [24]. We find that correlations of both meteorites and asteroids are generally good by this simple scheme. Preliminary results (Figure 4) indicate that (1) V-type asteroids generally match HED meteorites, (2) S-type asteroids locate near ordinary chondrites but they do not entirely match each other, which may reflect the effect of space weathering as discussed in the previous section (3) C-type asteroids match carbonaceous chondrites and they are separated into a few sub clusters. We consider that our unique approach is the first to quantitatively visualize the corresponding relationship between asteroid and meteorite spectra, which could obtain potential base functions to physically describe reflectance spectra in the future.

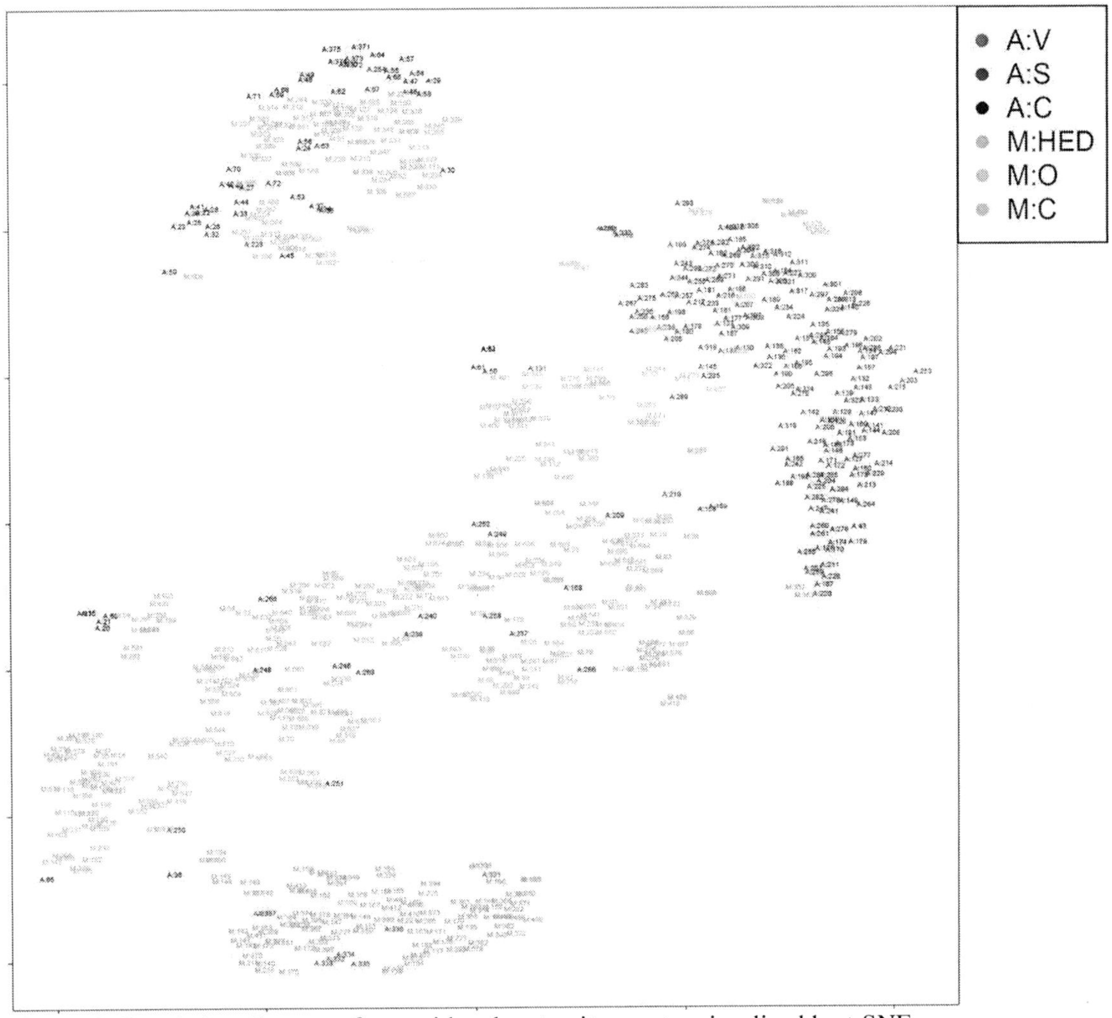

Figure 4. Correlation distance of asteroid and meteorite spectra visualized by t-SNE.

5. Summary

We have examined the relationship between asteroid and meteorite data by sparse modelling. These two kinds of data have significantly different spatial coverage and different precision and accuracy. By using sparse modelling, we aim to analyze asteroid and meteorite data without relying on preliminary knowledge and to perform automatic analysis of big data. We summarize our works as follows.

1. Building database of meteorites' elemental compositions (section 2)
2. Developing classification scheme for meteorites solely based on bulk elemental compositions (section 2)
3. Developing automatic deconvolution method of reflectance spectra for major rock-forming minerals based on the exchange Monte Carlo method (section 3)
4. Building database of reflectance spectra of asteroids and meteorites (section 4)
5. Visualization of albedo-spectra relationship of asteroids and meteorites (section 4.1)
6. Comparison between asteroids and meteorites spectra in PCA space using improved database (section 4.2)
7. Developing successful visualization scheme to compare asteroids' spectra with those of meteorites (section 4.3)

Since we have tested our method on relatively simple problems, we are now extending our analysis to more challenging problems. One is to develop a method to analyze dark featureless flat spectra, such as C-type asteroid and carbonaceous chondrites. These spectra are very important because they are often observed on small bodies which are expected to possess primordial materials. In addition, some of these bodies are the target for future space missions, including Hayabusa-2 and OSIRIS-Rex. Another project is to compare asteroids spectra with meteorites' spectra and elemental compositions. Through supervised learning, adding elemental compositions to the classification of reflectance spectra could improve the accuracy of comparison. Finally, spectral deconvolution without preliminary knowledge on mixtures of minerals is also an important, because an appropriate handling scheme for big data would be required for future remote sensing observations.

Acknowledgements
This work is supported in part by JSPS grant-in-aid 25120006 and TeNQ/Tokyo-dome.

References

[1] DeMeo F E and Carry B 2014 *Nature* **505** 629-634
[2] Veverka J, Belton M, Klaasen K and Chapman C 1994 *Icarus* **107** 2-17
[3] Veverka J *et al.* 2001 *Science* **292** 484-488
[4] Keller H U *et al.* 2010 *Science* **327** 190-193
[5] Jaumann R *et al.* 2012 *Science* **336** 687-690
[6] Zolensky M E *et al.* 2006 *Science* **314** 1735-1739
[7] Nakamura T *et al.* 2011 *Science* **333** 1113-1116
[8] Vernazza P, Binzel R P, Thomas C A, DeMeo F E, Bus S J, Rivkin A S and Tokunaga A T 2008 *Nature* **454** 858-860
[9] Hiroi T, Abe M, Kitazato K, Abe S, Clark B E, Sasaki S, Ishiguro M and Barnouin-Jha O S 2006 *Nature* **443** 56-58
[10] DeMeo F E, Binzel R P, Slivan S M and Bus S J 2009 *Icarus* **202** 160-180
[11] Weisberg M K, McCoy T J and Krot A N 2006 *Meteorites and the Early Solar System II* ed D S Lauretta and H Y McSween Jr (Tucson: University of Arizona Press) pp 19-52
[12] Miyamoto H, Niihara T, Kuritani T, Hong P K, Dohm J M and Sugita S 2016 *Meteorit. Planet. Sci.* **51** 906-919
[13] Klima R L, Dyan M D and Pieters C M 2011 *Meteorit. Planet. Sci.* **46**, 379-395
[14] Sunshine J M, Pieters C M and Pratt S F 1990 *J. Geophys. Res.* **95**, 9075-9087
[15] Hong P K, Miyamoto H, Niihara T, Sugita S, Nagata K, Dohm J M and Okada M 2016 *J. Geology Geophys.* **5**, 1000243
[16] Nagata K, Sugita S and Okada M 2012 *Neural Networks* **28**, 82-89
[17] Britt D T, Tholen D J, Bell J F and Pieters C M 1992 *Icarus* **99**, 153-166
[18] Bus S J and Binzel R P 2002 *Icarus* **158**, 146-177
[19] Tedesco E F, Noah P V, Noah M and Price S D 2002 *Astronomical J.* **123**, 1056-1085
[20] Yurimoto H and Wasson J T 2002 *Geochim. Cosmochim. Act.* **66**, 4355-4363
[21] Pieter C M and Hiroi T 2004 *LPSC* **XXXV**, 1720
[22] http://smass.mit.edu/home.html
[23] Rivkin A S, Howell E S, Lebofsky L A, Clark B E and Britt D T 2000 *Icarus* **145**, 351
[24] Van der maaten L and Hinton G 2008 *J. Machine Learning Res.* **9**, 2579-2605

IOP Conf. Series: Journal of Physics: Conf. Series **1036** (2018) 012004 doi:10.1088/1742-6596/1036/1/012004

Optimizing Change Detection for Planetary Remote Sensing Datasets

Reid Parsons[1,2] and Hideaki Miyamoto[3]

[1]Earth and Geographic Sciences Dept., Fitchburg State University, Fitchburg, MA, USA
[2]Department of Museum Collection Utilization Studies, University of Tokyo, Japan
[3]Department of Systems Innovation, School of Engineering, University of Tokyo, Japan

E-mail: `rparson4@fitchburgstate.edu`

Abstract. The fast-growing Remote Sensing dataset for Mars has challenged scientists in the community to make full use of the imagery collected from spacecraft. Cameras aboard long-lived orbiter missions capturing single-band, high-resolution images provide a special opportunity to monitor surface and atmospheric changes associated with geologic, weather and climate processes. A potential method for detecting changes is to employ principle component analysis (PCA) - an existing image processing method normally applied to multi-spectral imagery. When PCA is applied to a time-series stack of single-band images of the same location, subtle changes in pixel brightness are enhanced, and are represented as a series of transformed image components. PCA of a time-series image stack sorts these variations in the dataset according to their amplitudes. The most significant pixel variations are found in the lower principle components (PCs) while more subtle changes in pixel brightness are revealed in higher PCs. This method makes visual identification of subtle, ongoing geologic processes easier as is shown in a series of HiRISE images containing recurring slope lineae (RSL). As a final step, the change detection method proposed herein can estimate the likelihood of each of the original images containing the feature found in the PCA components. This is performed via a PCA-to-image coordinate rotation using the Eigen matrix calculated from the original PCA analysis. This proposed method is called "Change Detection via PCA of Stacked Time-series" (CDPCAST).

1. Introduction

Change detection is an analytical tool of Remote Sensing in which repeat observations of a particular location can be overlain, subtracted, or ratioed to detect differences between images. Many change detection techniques have been developed for multi-spectral sensors (cameras capturing several wavelength bands simultaneously) to monitor landcover and environmental changes on Earth (see, for example, [1, 2, 3]). Mars-observing spacecraft, however, commonly use a high-resolution, single band (capturing one range of light wavelength) a camera sensor resulting in grayscale images. Repeat images taken at the same location on Mars using these cameras as led to several discoveries of ongoing geologic processes operating at the sub-kilometer scale.

The arrival of the first high-resolution imaging sensors began with the Mars Orbital Camera ($\approx$ 2 m/pixel) in 1997 aboard Mars Global Surveyor and provided images which suggested liquid water was driving ongoing geologic activity in the formation of gullies [4]. Subsequent observations in a variety of locations made by the High Resolution Imaging Science Experiment (HiRISE, 0.25 m/pixel) aboard the Mars Reconnaissance Orbiter (2006 arrival) provided evidence for aqueous activity in forming RSL [5] which were previously interpreted as landslides [6]. RSL undergo a repetitive seasonal cycle in which dark, finger-like albedo changes originate near rock outcrops in early summer [7] and extend down steep, dust-covered slopes (see Figure 1). Eventually, the lineae stop growing, and slowly lighten in color, sometimes becoming brighter than their surroundings [6].

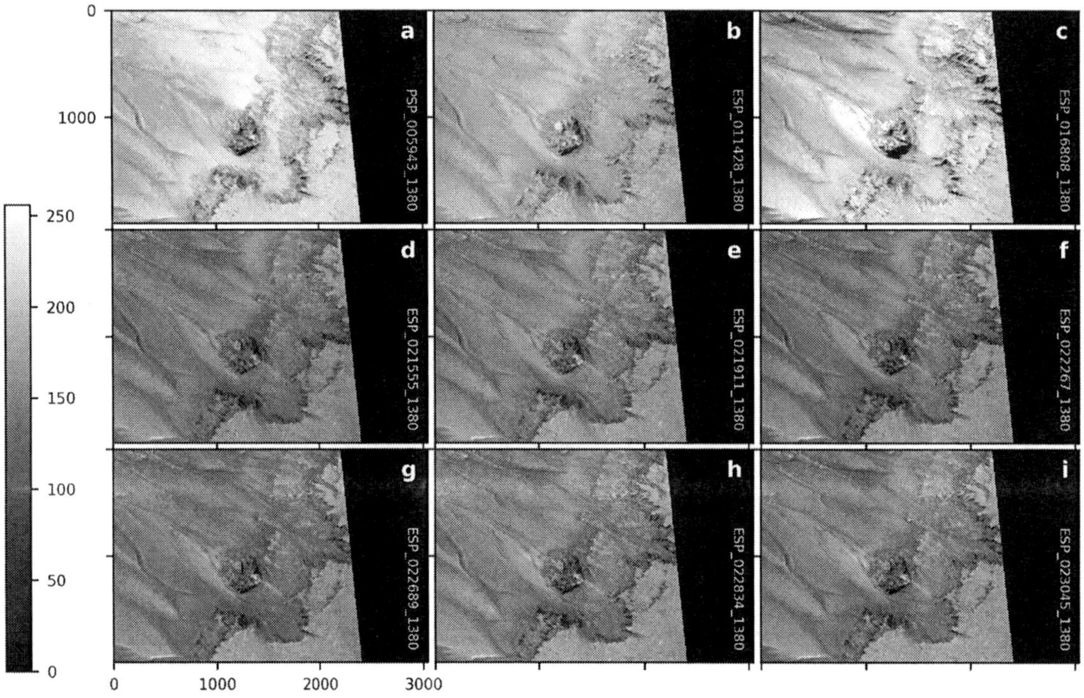

Figure 1. Uncalibrated HiRISE time-series image stack located on the southeastern wall of the 16 km diameter Palikir Crater (41.65°S, 202.71°E). Image IDs are given in yellow. The calendar date of acquisition are listed hereafter along with the Martian solar longitude (L_s) where $L_s = 180°$, $L_s = 270°$, and $L_s = 360°$corresponds to the spring equinox, summer solstice, and fall equinox in the southern hemisphere: a) 2007/11/02 ($L_s = 341°$) b) 2009/01/02 ($L_s = 185°$) c) 2010/02/26 ($L_s = 57°$) d) 2011/03/03 ($L_s = 247°$) e) 2011/03/30 ($L_s = 265°$) f) 2011/04/27 ($L_s = 282°$) g) 2011/05/30 ($L_s = 302°$) h) 2011/06/10 ($L_s = 309°$) i) 2011/06/27 ($L_s = 318°$).

RSL are one example of features which continue to alter the Martian surface, but other features include new, small impacts [8], aeolian dune migration [9], sublimation of polar ice [10], and gully formation [11]. Generally, the detection of these changes is done manually. The scientist or user of the remote sensing data viewed the data in GIS (Geographic Information Systems) software and visually identified changes by overlaying, and "flipping" between the two images. As planetary remote sensing data sets continue to grow, so does the need for a more efficient means of change detection which is suited toward high-resolution single-band imagery.

The potential advantage of using PCA as a means for detecting change, is that it considers the values of all pixels in all the images to determine the most statistically significant variations in pixel values. These variations are then isolated from one another as a hierarchical set of "principle components" of the image dataset. The user can quickly identify patterns or features of interest which are present in one or more of these components. This method, hereafter referred to as CDPCAST, provides a means for locating changes both spatially and temporally in repeat imagery. Nine HiRISE images collected from Palikir crater on Mars between November 2007 and June 2011 documenting RSL formation (Figure 1) will be used to illustrate the CDPCAST method.

2. Methods

2.1. Preprocessing

Existing methods for detecting changes in repeat imagery (including the one proposed here) is a step-wise process which begins with image preprocessing (Figure 2). Preprocessing involves image calibration (converting camera data into reflectance), orthorectification (removing geometric distortion associated with viewing angle and topography), and co-registration (aligning images of the same location so that they perfectly overlap). Although image co-registration has traditionally been a time-consuming process of manually selecting identical points in a pair of images, progress has been made in automating this procedure [12, 13].

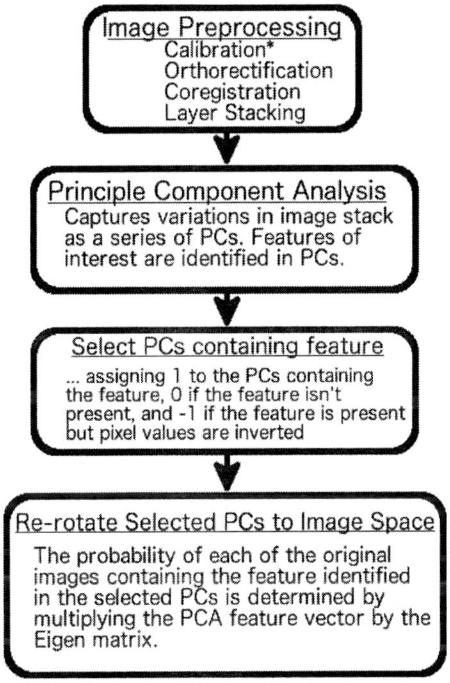

Figure 2. An outline of the CDPCAST method. The * indicates that image calibration (from digital numbers to reflectance) is optional. The example set of images shown in subsequent figures uses uncalibrated data.

The HiRISE images of Mars used here were not calibrated into reflectance data as it is not a required step in the change detection process. In this case, the same camera was used to collect all images so the relative change in brightness can be detected using the raw data. This example helps to illustrate potential broader application of this technique to other fields using repeat images and different camera data (e.g. medical imaging). In the example here, HiRISE images were orthorectified and co-registered by the HiRISE Operations Center and are provided to the public [14].

In addition to these initial steps, CDPCAST requires that each image in a time series be of the same extent and resolution and aligned (via coregistration) so that they can be "stacked" into a single, time-series image file. This final step was conducted using the open-source QGIS

Application. Figure 1 shows each of the preprocessed, uncalibrated HiRISE images in the final image stack. Note that the extent of these individual images has been cropped to a particular region of interest and region of overlap. The source image ID is given in yellow, axis labels indicate numbers of pixels, and the color bar indicates the pixel values. These images are arranged in order of time taken and each has a resolution of 0.25 m/pixel.

2.2. Principle Component Analysis

PCA is a common data reduction method employed in the field of remote sensing, but is most often applied to multi- or hyper-spectral images in order to distinguish between different materials in an image by accentuating differences in reflectivity at different wavelengths of light. In this standard application, the input image is a stack of images taken at many different wavelengths simultaneously. The CDPCAST method, in contrast, performs a PCA on a stack of images taken at the same wavelength at many different times. Although prior studies have employed PCA as a means of change detection on Earth using images taken at different times [15, 16, 17], the remote sensing community has focused more on alternative methods [1]. On planetary surfaces where geologic activity results in small features (requiring high-spatial resolution) which don't carry a strong spectral signal (dust-covered surfaces), PCA of single-band images stacked into a time-series could greatly improve change detection.

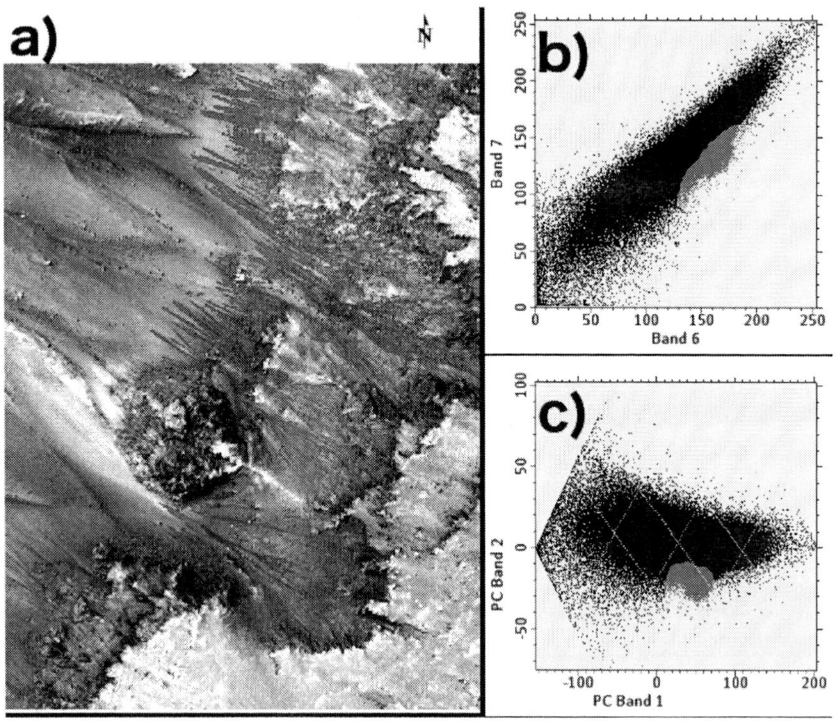

Figure 3. An illustration of PCA using only the sixth and seventh images from the time-series. a) The seventh image in the time-series documents the development of RSL (highlighted as elongated red regions). The RSL pixels in (a) were selected from the scatter plot shown in (b) which plots pixel values in the sixth image versus those in the seventh. This pixel highlighting is not part of the CDPCAST procedure for detecting changes, but is shown here to illustrate the PCA process and the magnitude of RSL pixel darkening (plots produced using ENVI® software). c) Rotation of the pixel cloud via PCA plots the data against axes which capture the maximum variance in the dataset.

A simple example of this process is depicted in Figure 3. The seventh image in the stack (Figure 1g) has well-developed, dark RSL features forming (highlighted in red in Figure 3a), but

IOP Conf. Series: Journal of Physics: Conf. Series **1036** (2018) 012004 doi:10.1088/1742-6596/1036/1/012004

those pixels were lighter in color in the previous image from this location (Figure 1f). Therefore, plotting the values of the sixth image versus those of the seventh image, gives a cloud of data points which are shown in Figure 3b. The upward trend along the line $y = x$ indicates close correlation of the pixel values in these two images, the exception being the points highlighted in red which correspond to the darker RSL pixels in image seven.

One way to extract differences between these two images involves re-plotting the pixel values on a new set of axes. One axis would closely follow the $y = x$ line in Figure 3b, but would actually follow the direction of maximum variance in the pixel cloud. The second axis runs perpendicular to the first highlighting changes in brightness between the two images. This rotated pixel cloud is shown in Figure 3c, with the thin lines running through the cloud indicating the orientation of the original axes.

PCA changes the values of the pixels in each image by plotting the pixel data on a set of axes which capture the maximum variance in the dataset. Each pixel in the original image stack can be treated as a 9-dimensional vector since there is a value for that pixel in each of the nine images. However, because much of the pixels remain unchanged between images, there is redundancy in the stacked image data that can be reduced by plotting the pixel data on a new set of axes that capture the most important, or PCs of the image stack.

In essence, the stacked image data can be treated as a matrix of numbers, S_{tp}, with rows indicating the number of images in the stack (t) and the number of columns indicating the number of individual pixels (p) in the image. Determining the orientation of the PC axes on which the data is to be re-plotted involves a series matrix calculations - the first of which is the covariance of S. The covariance matrix (C) is given by

$$C = SS^T \tag{1}$$

and has t columns and rows (a 9×9 matrix for our example data set) and describes how closely related each image in the stack is to every other image in the stack. An Eigen decomposition of C will give the direction (in image coordinate space) along which the maximum variation lies (as Eigen vectors) and the amplitude of the variation in the dataset in that direction (as Eigen values). The Eigen values of C are the the roots of the characteristic polynomial equation:

$$\det(\lambda I_9 - C) = 0 \tag{2}$$

where I_9 is the 9×9 identity matrix and will result in nine Eigen values (λ_i where i=[1,... ,$t = 9$]). Each of the nine Eigen values will have an associated unit length, 9-dimensional, Eigen vector ($\mathbf{e_i}$) given by

$$\mathbf{e_i} = N(\lambda_i I_9 - C) \tag{3}$$

where the right side of Equation (3) is the null space associated with λ_i. These Eigen vectors describe the axes of the PCs of the image data set. Thus, the Eigen vector associated with λ_1 (where the Eigen values have been sorted by their magnitude $\lambda_1 > \lambda_2$...) corresponds to the first PC and spans the maximum variance in the dataset. The value of the variance along $\mathbf{e_1}$ is given by $\sqrt{\lambda_1}$.

The key step in performing the transformation from image coordinates to PC coordinates it to construct the Eigen matrix (E):

$$E = \left[\mathbf{e_1}, \ldots, \mathbf{e_9}\right] \tag{4}$$

whose columns contain the Eigen vectors. These vectors point in the direction of the PC axes using the image data coordinates. Thus, to perform the transformation from image to PC coordinates, the inverse of E is multiplied by the image data as follows:

$$[S]_{PCA} = E^{-1}S \tag{5}$$

It naturally follows from Equation (5) that multiplying a pixel vector that has coordinates in PC space will convert back to the original image coordinates. This has relevance for locating the original images containing the features discovered in the PCA and is discussed in Section 3.2.

3. Results

Figure 4 shows the result of the PC transform (performed using ENVI® remote sensing image analysis software) on the stacked time-series. Because most of the variation in the images is captured in the first PC, it is colored according to the color bar given in Figure 1 with a pixel value range of 0 (black) to 256 (white). In subsequent components, the color bar shown on the lower-left of Figure 4 is used. Generally, variations covering a large spatial extent or are present in a large fraction of images that cause a big shift in pixel values will be captured in the lower PCA components. Higher components, on the other hand, will depict smaller amplitude variations occurring in fewer images (or over smaller spatial extents). Thus components 2 and 3 (Figure 4b and 4c, respectively) are dominated by variations caused by differences in lighting (sun direction and elevation angle) and possibly dust or frost deposition/erosion. RSL features are identifiable in PCA4 (Figure 4d) as elongated blue "fingers" and become more dominant and appear red in component 5 (Figure 4e). Subsequent components capture smaller variations caused either by changes in lighting, RSL shape (in PCA 8 and 9), or camera noise.

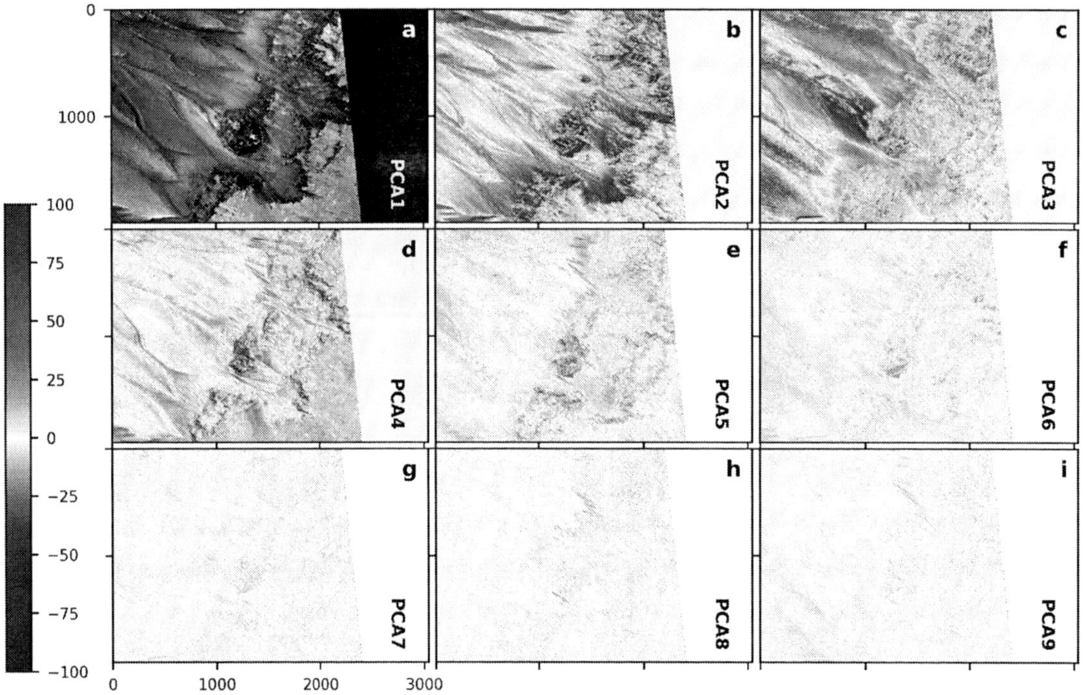

Figure 4. Nine components resulting from a PCA transformation of the image time-stack. PCA1 (a) uses the gray color scale used in Figure 1 due to its large variation in pixel values. The range of pixel values get progressively smaller in each PC due to reduction of data set variability. The higher components are capturing more subtle variations between the images in the stack which may be of geologic interest. In this case, we are interested in the RSL features seen in components 4, 5, 8, and 9 (d, e, h, and i).

The user can more easily identify features in the PCA components than the original images and can select those components to determine which images they correspond to. These selected components will be used in the next step in the CDPCAST method.

IOP Conf. Series: Journal of Physics: Conf. Series **1036** (2018) 012004 doi:10.1088/1742-6596/1036/1/012004

3.1. Constructing the PCA feature vector

Having identified PCs 5, 6, 8, and 9 as containing the feature of interest, the next objective is to identify which of the original images contain RSL. In Figure 3a, RSL were already determined to have formed in the seventh image (Figure 1g), but there may also be other images in the time-series which contain RSL. Future application of CDPCAST could potentially locate changes in much larger time-series involving tens of images without prior knowledge of which images contain changes.

Before the original images containing the feature can be identified, the user must first construct a "feature vector" whose elements correspond to each of the PCs as indicated in step 3 of Figure 2. Looking at each of the PC images in sequence, the user enters a zero into the feature vector, **f**, if the feature is not present. Because PCA1 is the average image of the dataset, it is neglected from this procedure as it is the reference from which changes are revealed in the subsequent PCs. Thus, $\mathbf{f_1}$ will always be set to zero. The first PC following PCA1 that contains the feature will be given a value of positive one. In the subsequent PCs, if the feature is present but the pixel values are reversed in sign compared to the first PC containing the feature, then a negative one is entered into the corresponding element of **f**.

3.2. Locating images containing feature

In practice, determining the elements of **f** can be ambiguous as will be discussed in Section 4. However, our goal is to get a qualitative measure of which images in the original stack are most likely to contain the feature identified in the PCs. Once the feature vector has been determined, we can quantify the potential (**p**) of an original image containing the feature via the following equation:

$$\mathbf{p} = E\mathbf{f}\sqrt{\lambda} \tag{6}$$

where the elements of **p** correspond to the sequence of images in the original stack. Equation (6) applies weights the feature vector (**f**) corresponding to the variance ($\sqrt{\lambda}$) of the PC containing the feature. The result is then transformed back into the original image coordinates using the Eigen matrix (step 4 of Figure 2). The result is a vector (**p**) whose elements indicate the relative influence each image had in producing the feature vector with larger, positive values indicating more influence than smaller, negative values. However, there is a potential for sign reversal depending on the orientation of the PCs that are unique to each dataset, so the user should also check whether negative or positive values indicate the presence of the feature. The actual calculation of **p** using the Eigen matrix, feature vector, and variances for the example HiRISE image data is given below:

$$\begin{bmatrix} 2.0 \\ -5.7 \\ -0.6 \\ -3.5 \\ -3.4 \\ -3.3 \\ 5.7 \\ 5.1 \\ 4.4 \end{bmatrix} = \begin{bmatrix} 0.41 & -0.07 & 0.89 & 0.2 & -0.01 & 0.02 & -0.03 & -0.02 & -0.03 \\ 0.36 & -0.04 & 0.01 & -0.78 & -0.34 & -0.33 & 0.14 & -0.03 & 0.03 \\ 0.41 & -0.83 & -0.27 & 0.09 & 0.21 & 0.13 & 0.01 & -0.06 & -0.02 \\ 0.29 & 0.25 & -0.09 & -0.26 & 0.34 & 0.2 & -0.73 & 0.17 & -0.26 \\ 0.31 & 0.29 & -0.06 & -0.13 & 0.33 & 0.49 & 0.52 & 0.28 & 0.32 \\ 0.3 & 0.3 & -0.14 & 0.14 & 0.36 & -0.3 & 0.05 & -0.74 & 0.14 \\ 0.29 & 0.18 & -0.19 & 0.18 & -0.59 & 0.48 & 0.06 & -0.29 & -0.38 \\ 0.3 & 0.17 & -0.18 & 0.32 & 0.07 & -0.5 & 0.26 & 0.45 & -0.47 \\ 0.3 & 0.09 & -0.19 & 0.32 & -0.37 & -0.15 & -0.32 & 0.24 & 0.66 \end{bmatrix} \begin{bmatrix} 0 \\ 0 \\ 0 \\ 1 \\ -1 \\ 0 \\ 0 \\ 1 \\ -1 \end{bmatrix} \begin{bmatrix} 220 \\ 21 \\ 16 \\ 9.8 \\ 6.3 \\ 4.2 \\ 2.9 \\ 2.7 \\ 2.5 \end{bmatrix} \tag{7}$$

4. Discussion

In order to assess the performance of the qualitative prediction made by Equations (6) and (7) regarding the presence of RSL, we take a closer look at the largest elements of **p** and the

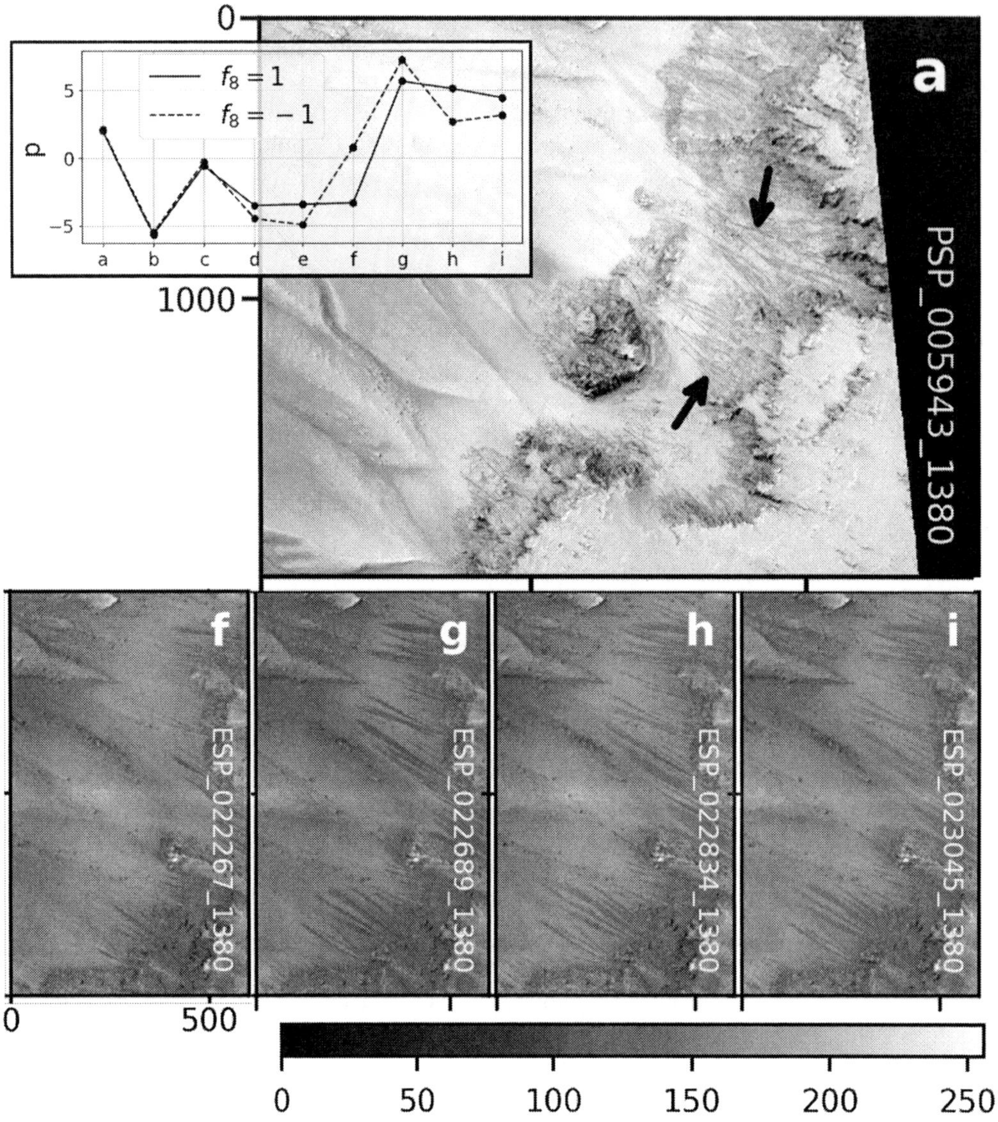

Figure 5. Images which contain RSL. Image (a) and cropped sections of images (f), (g), (h), and (i) from Figure 1 are reproduced here to highlight those images which contain RSL. Other images do not have RSL. The inset gives the "feature potential" value, **p**, for all images in the stack. The solid line gives **p** using $f_8 = 1$ as shown in Equation (7) which gives good results for all images except (f). The dashed line sets $f_8 = -1$ and results in better indication of which images contain RSL.

original images in our dataset. In Equation (7) the seventh element of **p** contains the highest value. Therefore the seventh image in the original stack is correctly predicted to contain RSL (Figure 3a or 5g). Elements eight and nine of **p** also have large values relative to the other elements of **p** and those images also contain RSL. A zoomed-in portion of these images (g, h, and i from Figure 1) are reproduced in Figure 5 to illustrate the presence of RSL.

The fourth largest value of **p** is $p_1 = 2.0$ and is associated with image "a" shown in Figures 1a and 5a. Although less distinct, incipient RSL can be identified downslope of the steep rim of Palikir Crater (arrows in Figure 5a). Using the four largest values of **p**, the method outlined here is successful in not only predicting the presence of RSL, but also providing an indication of the RSL distinctness. However, there is an image for which this method was not successful.

Image "f" in Figure 5 also exhibits incipient RSL, but has low predictive score ($p_6 = -3.3$). The reason for this can be identified by looking at the matrix multiplication shown in Equation (7). During the calculation of p_6, the sixth row of E is multiplied element-wise with the two vectors furthest to the right and each product is summed to give p_6. Looking at the sixth row of E, one notices that the eighth element most greatly affects the value of p_6. Stated differently, the sixth image in the original stack is captured mostly by the eighth PC. Taking a closer look at PCA8 (Figure 4h) and comparing it to the sixth image (Figure 1f or 5f) reveals that the incipient RSL locations in image six are colored red in PCA8 - suggesting that the value for f_8 should be negative instead of positive. This example illustrates the potential for ambiguity in defining $\mathbf{f}$ because there are also parts of PCA8 with blue colored RSL shapes.

The inset plot in Figure 5 indicates the value of the elements of $\mathbf{p}$ for images (a) through (i) from Figure 1 using a f_8 value of one (solid line) or negative one (dashed line). The $\mathbf{p}$ value for image "f" is most affected by this change. Given the observation of incipient RSL in image "f," a value of negative one for f_8 provides a better agreement with the presence of RSL. Alternatively, if f_8 were set to zero, the elements of $\mathbf{p}$ would take a value halfway between the two lines shown in the inset plot in Figure 5.

5. Conclusions

Determining which images in the time-series stack that are responsible for the variations found in the PCA analysis is the potentially unreliable part of the CDPCAST method, and requires some caution as shown in Section 4. The method is reliable in predicting which images contain the most prominent RSL, but images containing less distinct manifestations of the feature of interest may not be picked out by Equation (6). Therefore the values of the elements of $\mathbf{p}$ should not be interpreted as definitive predictions of a feature's presence in an image, but merely point the user toward the most likely candidates.

Optimizing change detection for the rapidly growing Martian remote sensing dataset will further our understanding of ongoing geologic processes and take full advantage of the spacecraft imagery that has been collected. The method proposed here applies an existing image processing technique (PCA) to stacked images in a time series to enhance the change detection processes. PCA is well-suited for change detection in high-resolution, single-band imagery of planetary surfaces when they are stacked into a multi-temporal image file. Although variations in lighting between images does cause significant variation between images, the higher PCA components help to identify lower-amplitude brightness changes which may result from ongoing geologic processes (Figure 4) which might otherwise be difficult to detect. Future development of the CDPCAST method together with automation of image preprocessing steps will greatly enhance the planetary science community's ability to monitor surface changes on terrestrial Solar System objects.

References

[1] Singh A 1989 *International Journal of Remote Sensing* **10** 989–1003
[2] Mas J F 1999 *International Journal of Remote Sensing* **20** 139–152
[3] Brewer C K, Winne J C, Redmond R L, Opitz D W and Mangrich M V 2005 *Photogrammetric Engineering & Remote Sensing* **71** 1311–1320
[4] Malin M C and Edgett K S 2000 *Science* **288** 2330–2335
[5] McEwen A S, Ojha L, Dundas C M, Mattson S S, Byrne S, Wray J J, Cull S C, Murchie S L, Thomas N and Gulick V C 2011 *Science* **333** 740–743 ISSN 0036-8075 (*Preprint* `http://science.sciencemag.org/content/333/6043/740.full.pdf`) URL `http://science.sciencemag.org/content/333/6043/740`
[6] Sullivan R, Thomas P, Veverka J, Malin M and Edgett K S 2001 *Journal of Geophysical Research: Planets* **106** 23607–23633 ISSN 2156-2202 URL `http://dx.doi.org/10.1029/2000JE001296`

[7] Ojha L, McEwen A, Dundas C, Byrne S, Mattson S, Wray J, Masse M and Schaefer E 2014 *Icarus* **231** 365 – 376 ISSN 0019-1035 URL http://www.sciencedirect.com/science/article/pii/S0019103513005393
[8] Daubar I, McEwen A, Byrne S, Kennedy M and Ivanov B 2013 *Icarus* **225** 506 – 516 ISSN 0019-1035 URL http://www.sciencedirect.com/science/article/pii/S0019103513001693
[9] Bourke M, Edgett K and Cantor B 2008 *Geomorphology* **94** 247–255
[10] Thomas P, Malin M, James P, Cantor B, Williams R and Gierasch P 2005 *Icarus* **174** 535 – 559 ISSN 0019-1035 mars Polar Science III URL http://www.sciencedirect.com/science/article/pii/S0019103504003203
[11] Diniega S, Byrne S, Dundas C M and McEwen A 2009 *AGU Fall Meeting Abstracts*
[12] Di K, Liu Y, Hu W, Yue Z and Liu Z 2014 *IOP Conference Series: Earth and Environmental Science* **17** 012015 URL http://stacks.iop.org/1755-1315/17/i=1/a=012015
[13] Scheffler D, Hollstein A, Diedrich H, Segl K and Hostert P 2017 *Remote Sensing* **9** ISSN 2072-4292 URL http://www.mdpi.com/2072-4292/9/7/676
[14] HiRISE Operations Center Gullies in palikir crater https://www.uahirise.org/dtm/dtm.php?ID=PSP_005943_1380 accessed: 2017-07-03
[15] Byrne G F, Crapper P F and Mayo K K 1980 *Remote Sensing of Environment* **10** 175–184
[16] Ingebritsen S E and Lyon R J P 1985 *International Journal of Remote Sensing* **6** 687–696
[17] Rokni K, Ahmad A, Selamat A and Hazini S 2014 *Remote Sensing* **6** 4173–4189

Proceedings of the International Meeting on "High-Dimensional Data-Driven Science" (HD^3-2017)
September 10–13, 2017, Kyoto, Japan

Taxonomy matching between asteroids and meteorites: supervised clustering approach

Yuki Saito*, Peng Hong†, Takafumi Niihara‡, Hideaki Miyamoto†, Kenji Fukumizu§*

*Department of Statistical Science, School of Multidisciplinary Sciences, Graduate University for Advanced Studies, 10-3 Midori-cho, Tachikawa, Tokyo 190-8562, Japan; †Department of Systems Innovation, School of Engineering, The University of Tokyo, 7-3-1 Hongo, Bunkyo-ku, Tokyo 113-8656 Japan; ‡University Museum, The University of Tokyo. 7-3-1 Hongo, Bunkyo-ku, Tokyo 113-0033, Japan; §The Institute of Statistical Mathematics, 10-3 Midori-cho, Tachikawa, Tokyo 190-8562, Japan

E-mail: saito.yuki@ism.ac.jp; hong@seed.um.u-tokyo.ac.jp; niihara@seed.um.u-tokyo.ac.jp; hm@sys.t.u-tokyo.ac.jp; fukumizu@ism.ac.jp

Abstract. We propose a data-driven method for investigating taxonomy relations between asteroids and meteorite. The method clusters data of reflectance spectra of asteroids with the guidance of relatively reliable taxonomy of meteorites. By comparing the clustering accuracy of asteroid between with and without guidance, we observe that the guidance of meteorites taxonomy improve the accuracy. This serves as an evidence that there is a common taxonomical structure between meteorites and asteroids, confirming a long-standing prediction.

1. Introduction

Linking meteorites to asteroidal bodies is an important subject toward better understanding of the Solar System. This is, however, not easy for various reasons such as space weathering. Some strong links between specific types are known so far. A link was confirmed between V-type asteroids and Howardite-Eucrite-Diogenite (HED) meteorites by the Dawn mission [2]. The link between S-type asteroids and ordinary chondrite meteorites was verified by the analysis of the sample taken from Itokawa by the Hayabusa mission [5]. A less certain link is known between C-type asteroids and carbonaceous chondrites. Other more uncertain links have been discussed also [3]. For investigating these links, there are some works that take data-driven approaches. Britt et al [1], Tholen et al [9], and Watabiki et al [10], for example, provided a map to overview the similarity in reflectance spectra between meteorites and asteroids by using principal component analysis.

This work proposes a data-driven method for linking the taxonomies between meteorites and asteroids. The proposed method classifies asteroid data with guidance of the known taxa of meteorites [11], and matches the taxa between the two domains of asteroids and meteorites. For this clustering and matching procedure, reflectance spectral data of asteroids are clustered so that the similarity between the cluster structures of the asteroids and meteorites is maximized. To represent the cluster structures, a nonlinear method, the so-called kernel method [7], is used to quantify the mutual relations among clusters. For the teaching data to guide the clustering of asteroid data, the reflectance spectral data and element composite data of meteorites are used. We call this analysis procedure *supervised clustering.*

As a first step, we confine our analysis to the three types in each of the asteroids and meteorites: C-, S-, and V-type for asteroids, and the carbonaceous chondrites, ordinary chondrites, and HED for meteorites. The reflectance spectral data are used for asteroids, while for meteorites each of the reflectance spectra and element composites is served for teaching data.

2. Method for supervised clustering

2.1. Data sets

The data set of reflectance spectra for asteroid contains 365 data from 0.45 to 2.45 μm in wavelength divided by 401 bins of equal size. The dataset includes also the labels for the taxonomical types including C-, S-, V-, which are given by domain experts. These labels are not used for clustering but only for evaluation. As pre-processing, DeMeo's method [4] is applied. The numbers of C-, S-, and V-type data are 19, 95, and 8, respectively; 122 in total.

For meteorites, the dataset of reflectance spectra contains 731 data, in which 221 carbonaceous chondrites, 245 ordinary chondrites, 108 HED meteorites, 574 in total, are used for the analysis. The preprocessing is the same as the asteroid data. The element composition dataset of meteorites consists of 481 data, consisting of 30 carbonaceous, 388 ordinary and 63 HED meteorites. The dataset includes 12 elements: Fe, Si, Al, Ca, Mg, Na, P, K, Mn, Ni, Cr, and Na_2O.

2.2. Supervised clustering

Our method for supervised clustering consists of two steps: [1st step] as initialization, the spectral clustering [6], which is a standard non-hierarchical clustering method, is applied to asteroid data, [2nd step] the meteorite data with taxonomic information is used to guide the clustering procedure for the asteroid data. We also refer to the standard clustering in the first step as *unsupervised clustering* in contrast with supervised clustering.

We review the spectral clustering in the sequel, since it serves as a basics of our supervised clustering. The spectral clustering operates on a similarity matrix, which contains as an element a similarity between two data points x and y. In this paper, the similarity is evaluated by Gaussian kernel function $\exp(-\|x-y\|^2/2\sigma^2)$ with bandwidth parameter σ. In the case of asteroid data, $N_A \times N_A$ data-similarity-matrix K_A is used, which contains similarity between each pair of spectral data, where N_A is the size of the data set. The algorithm computes eigenvectors of the normalized graph Laplacian, defined by

$$L_A := T_A^{-1/2}(T_A - K_A)T_A^{-1/2}, \tag{1}$$

where T_A is the diagonal matrix $T_{A,ii} = \sum_j K_{A,ij}$, representing the degree of the nodes in the graph represented by K_A. We used the algorithm proposed by Ng et al [6], which further applies K-means clustering after projecting the data onto the eigenspaces corresponding to the least eigenvalues.

The proposed method of supervised clustering is also based on similarity matrices. As an overview, the method first constructs a data-similarity-matrix for each of the asteroid and meteorite data, and then computes respective cluster-similarity-matrices for each domain to represent the cluster structures. A similarity measure between these two cluster-similarity-matrices is introduced for evaluating the similarity of taxonomies, serving as an objective function for clustering asteroid data and matching the clusters of the two domains.

For more details, we explain for the asteroid data. Given data-similarity-matrix K_A computed in the same way as spectral clustering, the cluster-similarity matrix is calculated by

$$S_A = D_A W_A^T K_A W_A D_A, \tag{2}$$

where W_A is $N_A \times C$ cluster assignment matrix with common cluster size C for the asteroid and meteorite domains, and D_A is the normalizer. More specifically, $W_{A,ia} = 1$ if and only if i-th data belongs to a-th cluster with the other components 0, and $C \times C$ matrix D_A is given by $D_{A,ab} = \delta_{ab}/N_A^a$ $(a = 1, \ldots, C)$, where N_A^a is the number of data in the a-th cluster C_A^a of the asteroid dataset. The (a, b)-component of the cluster-similarity-matrix is then equal to

$$S_{A,ab} = \frac{1}{N_A^a N_A^b} \sum_{i \in C_A^a, j \in C_A^b} K_{A,ij}, \tag{3}$$

which represents the average total similarities between the data in the two clusters. The cluster-similarity-matrix S_M for the meteorite dataset is obtained in a similar way. Note that different data-formats, dimensionalities, and data sizes can be handled by the proposed similarity representations; the meteorite and asteroid datasets can have different data formats such as reflectance spectra and element composition, as demonstrated in Section 3.

The proposed objective function for our supervised clustering is to maximize

$$\frac{Tr(S_A H S_M H)}{\|H D_A W_A^T W_A D_A H\|_F} = \frac{Tr(H W_A^T D_A K_A D_A W_A H S_M)}{\|H D_A W_A^T W_A D_A H\|_F}, \tag{4}$$

where H is the centering matrix defined by $H_{ij} = \delta_{ij} - 1/C$, and $\|M\|_F$ denotes the standard Frobenius (Euclidean) norm. To optimize W_A, a greedy search is used in a similar way to CLUHSIC [8]. Before the search, W_A is initialized by the spectral clustering, W_M is obtained by the labels given by experts, and the cluster order is also optimized so that Eq.(4) is maximized.

3. Clustering results

For evaluation, the *cluster purity* and *cluster matching accuracy* are computed based on the type information labeled by experts for the asteroid data. The cluster purity of a clustering is the portion of data that accord with a dominant type in a cluster. The cluster matching accuracy is the portion of the correct cluster assignments to the corresponding meteorite type. This is similar to classification accuracy. In the following results, the best bandwidth parameters σ_A (asteroid) and σ_M (meteorite) are chosen, but from the experimental results not shown here, similar results are obtained by a wide range of the parameters.

Figure 1 shows supervised clustering of the asteroid data set with guidance of meteorite reflectance spectra data ($\sigma_A = 3.0, \sigma_M = 1.0$) (left) together with the teaching data of meteorites (right) for reference. They are shown after projecting onto the three dimensional space given by the principal component analysis. The cluster purity is 99.1% at best ($\sigma_A = 2.5, \sigma_M = 1.0$); this means significant improvement from the the cluster purity of the unsupervised clustering (spectral clustering), which is 92.8% at best ($\sigma_A = 1.5$).

As for cluster matching accuracy, the supervised clustering gives 99.0%. In Figure 1, the teaching cluster labels (right) are shown by A (carbonaceous chondrite), O (ordinary chondrite), and H (HED), while in the left figure C, S, and V show the labels given by domain experts (not used for the supervised clustering). The colors indicate the matched clusters; e.g. the red cluster A carbonaceous chondrite in the right figure is matched to red cluster in the left figure.

Next, we show supervised clustering results with guidance of element composites dataset for meteorites. Figure 2 is made in a similar manner to Figure 1. The cluster purity is 98.8% ($\sigma_A = 2.5$, $\sigma_M = 1.5$), which is significantly improved from the unsupervised clustering. The cluster matching accuracy is 98.9% ($\sigma_A = 2.5, \sigma_M = 2.0$) at best. It is important to note that the proposed supervised clustering works for different data types: reflectance spectra and element composites.

Those analyses show that taxonomic guidance of meteorite data in fact improves making asteroidal clusters, implying that a common cluster structure exists between the meteorite and asteroid domains.

4. Conclusion

We have studied a common taxonomical structure in asteroids and meteorites using a data-driven approach. We have proposed supervised clustering to approach this problem: it is investigated whether the taxonomy information of meteorite improves clustering of asteroid data. The numerical studies have demonstrated that the taxonomical information of meteorites does improve the accuracy of clustering of asteroids. This serves as an evidence that there is a common taxonomical structure between these two domains.

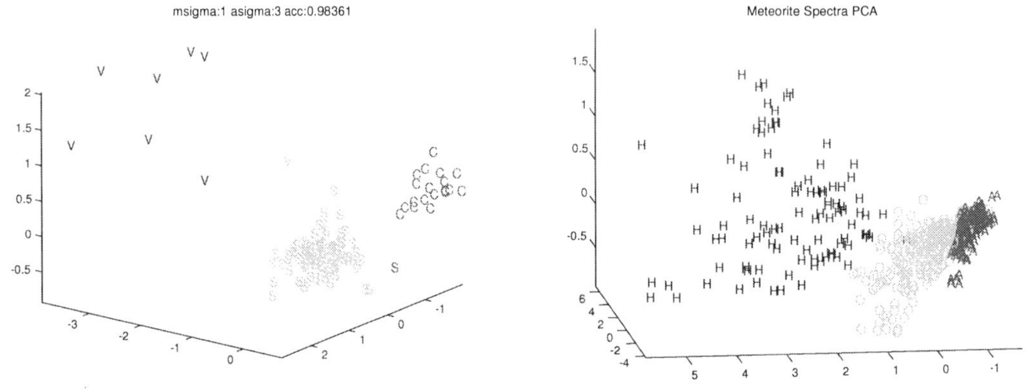

Figure 1. Clustering of asteroid with guidance of meteorite reflectance spectra. Three dimensional plots by PCA for the clustering results of asteroid data (left) and the teaching data of meteorite (right).

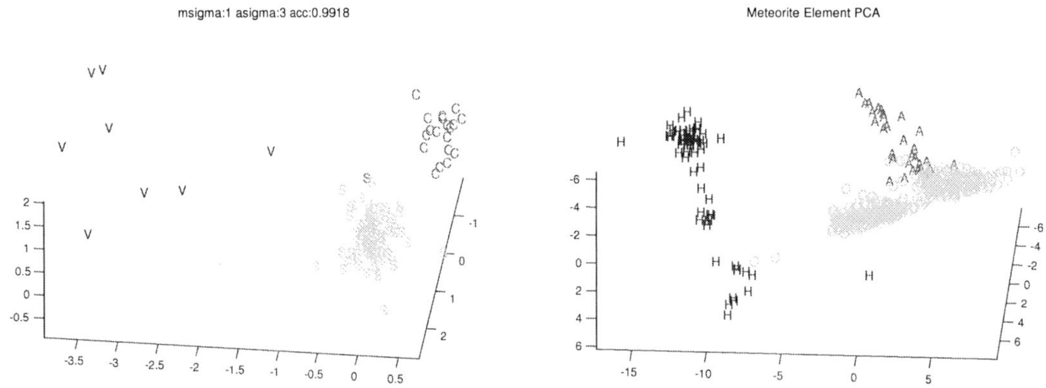

Figure 2. Clustering of asteroid with guidance of meteorite element compositions. Three dimensional plots by PCA for the clustering results of asteroid data (left) and the teaching data of meteorite (right).

Acknowledgments

The present study was supported by KAKENHI Innovative Areas 25120012.

References

[1] Britt D T, Tholen J D, Bell J F, and Pieters M C 1992. *Icarus* **99** 153.

[2] Buratti J B, Dalba A P, Hicks D M et al. 2013. *J. Geophys. Res. Planets* **118** 1991.
[3] Cloutis A E, Binzel P R, and Gaffey J M 2014 *Elements* **10** 25.
[4] DeMeo E F, Binzel P R, Slivan M S, and Bus J S 2009 *Icarus* **202** 160.
[5] Nakamura T, Noguchi T, Tanaka M, Zolensky E M et al. 2011 *Science* **333** 1113.
[6] Ng A, Jordan M, and Weiss Y 2002. *Advances in NIPS* **14** 849.
[7] Schölkopf B and Smola A J 2001. *Learning with Kernels.* MIT Press.
[8] Song L, Smola A, Gretton A, and Borgwardt M K 2007 *Proc. ICML2007* 815.
[9] Tholen J D 1984 *Doctoral thesis, Univ. Arizona.*
[10] Watabiki Y, Niihara T, Hong P, Miyamoto H, Saito Y, and Fukumizu K 2016 *Japan Geoscience Union Meeting* MTT27-03.
[11] Weisberg K M, McCoy J T and Krot N A 2006 *Meteorites and the Early Solar System II.*

IOP Conf. Series: Journal of Physics: Conf. Series **1036** (2018) 012006 doi:10.1088/1742-6596/1036/1/012006

Spatially continuous dynamic factor modeling with basis expansion using L_2 penalized likelihood

Takamitsu Araki and Shotaro Akaho

Human Informatics Research Institute, National Institute of Advanced Industrial Science and Technology, Central 2, 1-1-1 Umezono, Tsukuba, Ibaraki 305-8568, Japan

E-mail: tk-araki@aist.go.jp

Abstract. In spatio-temporal data analysis, dimension reduction is necessary to extract intrinsic structures and to avoid over-parametrization problems. The spatial dynamic factor model (SDFM) reduces dimension of the data by decomposing them into spatial and temporal variations. The spatial variation is represented by a few spatially structured vectors, called factor loading vectors. The SDFM cannot be directly applied when the data contain missing values and their observation sites vary with time. We extend the factor loading vector to a smooth continuous function obtained by basis expansion, where we call the extended model the spatially continuous dynamic factor model (SCDFM), and estimate the SCDFM using the maximum L_2 penalized likelihood method. We derive model selection criteria to select a regularization parameter and the number of factors. Applications to synthetic and real data show the effectiveness of our modeling strategy in terms of estimation accuracy and stability.

1. Introduction

In many scientific and industrial fields, spatio-temporal data, which depend on time and space, are often observed. The amount of spatio-temporal data are increasing as measurement devices continue to be developed, thereby increasing the importance of statistical modeling of the spatio-temporal data [1]. In spatio-temporal data analysis, when the data at all observation sites are directly modeled by multivariate time series models such as the vector auto-regression moving average model [2], over-parametrization happens and complex models that are difficult to interpret are obtained. In order to avoid over-parametrization, the space-time autoregressive moving average model [3], where many parameters are fixed by topographical information, has been proposed. However, this model is not flexible enough to capture spatio-temporal structures.

For this reason, dimension reduction for the spatio-temporal data is necessary to extract their inherent and essential structures. The spatial dynamic factor model (SDFM; [4, 5, 6]) is a dynamic factor model [7, 8] extended to capture spatial structures. The SDFM reduces dimension of the spatio-temporal data by estimating spatial and temporal variations of the data based on the Bayesian approach. The spatial variation is represented by a few spatially correlated vectors, called factor loading (FL) vectors, which are assumed to follow a Gaussian process. The temporal variation is represented by autoregressive processes of the factors, called factor processes, which are the stochastic processes corresponding to the FL vectors. The estimated FL vectors and factor processes reveal the spatial, temporal and spatio-temporal structures of the data.

Since the SDFM represents spatial variations only on the observation points in discrete form, it cannot be directly applied to a data set that contains missing values or whose observation sites vary with time, such as weather buoy data. The SDFM also has to augment FL vectors and estimate them after the estimation of the other parameters, in order to interpolate unobserved points.

The spatio-temporal random effects model (STREM; [9, 10]) represents spatial variations continuously by basis expansion, which is a linear combination of spatial basis functions. The STREM can be directly applied if the data set contains missing values or if the observation sites change temporally, and can directly interpolate unobserved points using the estimated model. However, the STREM does not compress the dimension of the spatio-temporal data, and the essential spatial, temporal and spatio-temporal structures are not extracted.

We extend the SDFM to a spatially continuous model that has continuous FL functions instead of FL vectors; we call this the *spatially continuous dynamic factor model* (SCDFM), where the FL functions are constructed by basis expansion. Although the maximum likelihood method has served as a standard method for various statistical models, it often yields unstable and unsmooth FL functions for the SCDFM. Thus, we estimate the parameters in the SCDFM by the maximum penalized likelihood method with an L_2 penalty on the coefficients in the FL functions, giving us more stable and smooth FL functions than the maximum likelihood method. We derive an Expectation-Maximization (EM) algorithm [11] to maximize the penalized likelihood stably. A crucial issue in modeling of the SCDFM is the choice of a regularization parameter and the number of factors. Since the maximum penalized likelihood estimation can also be considered within the Bayesian framework, we derive a model selection criterion from a Bayesian viewpoint [12] for evaluating the SCDFM estimated by the penalized method. In the FL functions estimated by the maximum penalized likelihood method, the number of parameters is not a suitable measure of model complexity, since the complexity also depends on the penalty term. Based on the concept of the effective number of parameters of a regression model [13, 14], we also derive the effective number of parameters of the estimated FL functions from a formula in the EM algorithm. By replacing the number of parameters of the FL functions in consistent AIC (CAIC; [15, 16]) with the derived effective number of parameters, we formally obtain an information criterion for the estimated SCDFM. We use these criteria to choose the regularization parameter and the number of factors.

The maximum likelihood estimation for the SCDFM and the proposed modeling procedures are applied to synthetic data and ozone concentration data, and their effectiveness in terms of estimation accuracy and stability is shown by comparing the SCDFMs estimated by the maximum likelihood method and our modeling procedures.

The rest of the paper is organized as follows. The SDFM is extended to the SCDFM in Section 2. The maximum penalized likelihood method with an L_2 penalty for the SCDFM and the EM algorithm are introduced in Section 3. In Section 4, the model selection criterion for evaluating the estimated SCDFM is derived from a Bayesian viewpoint. The effective number of parameters of the estimated FL function is derived, and the model selection criterion with the effective number of parameters is obtained based on the CAIC formula. In Section 5, the performance of our modeling methods is evaluated through applications to synthetic data and ozone concentration data. Finally, we present conclusions in Section 6.

2. Spatially continuous dynamic factor model

We extend the spatial dynamic factor model (SDFM; [4]) in order to allow to analyze directly the spatio-temporal data that have missing values or whose observation sites vary with time. The SDFM consists of an observation model and a system model. The observation model describes a relationship between the observation data and the factors via the factor loading (FL) vectors that represent spatial variations of the data. The system model describes the temporal evolution of the factors that represent temporal structures of the data. We extend the FL vectors in the observation model to the continuous FL functions expressed by the basis expansion, and we call the model consisting of the system model and the extended observation model the *spatially continuous dynamic factor model* (SCDFM).

The SDFM assumes that the data $\mathbf{y}_t = (y_{1t}, \ldots, y_{Nt})'$ are observed in locations $\bar{\mathbf{s}} = (\mathbf{s}_1, \ldots, \mathbf{s}_N)$ contained in the region $\mathbf{S}$, where $\mathbf{S} \subset \mathbb{R}^p$, for each time $t = 1, \ldots, T$. The observation model of the SDFM is described as

$$\mathbf{y}_t = \mathbf{\Lambda f}_t + \boldsymbol{\epsilon}_t, \quad \boldsymbol{\epsilon}_t \sim N(\mathbf{0}, \sigma_\epsilon^2 \mathbf{D}_N),$$

where $\mathbf{f}_t = (f_{1t}, \ldots, f_{dt})'$ is a vector of the factors at time t, $\mathbf{\Lambda} = (\boldsymbol{\lambda}_1, \ldots, \boldsymbol{\lambda}_d)$ is an FL matrix, where $\boldsymbol{\lambda}_j$ is an FL vector, D_N is a known N order positive definite matrix, and σ_ϵ^2 is a positive scale parameter.

In our study, we consider data whose observation sites or site number may vary with time, that is, the data $\mathbf{y}_t = (y_{1,t}, \ldots, y_{N_t,t})'$ are observed in $\mathbf{s}_t = (\mathbf{s}_{1,t}, \ldots, \mathbf{s}_{N_t,t})$, where $\mathbf{s}_{it} \in \mathbf{S}$. For example, this assumption is satisfied when missing values lie in the data. To cope with the time varying observation sites, we introduce continuous FL functions $\lambda_j(\mathbf{s})$ against the factor loading vectors $\boldsymbol{\lambda}_j$ for $j = 1, \ldots, d$. The FL functions are constructed by the basis expansion as

$$\lambda_j(\mathbf{s}) = \sum_{i=1}^{m} b_i(\mathbf{s})\alpha_{ij} = \mathbf{b}(\mathbf{s})'\boldsymbol{\alpha}_j, \quad \mathbf{s} \in \mathbf{S}, \quad j = 1, \ldots, d,$$

where $\mathbf{b}(\mathbf{s}) = (b_1(\mathbf{s}), \ldots, b_m(\mathbf{s}))'$ is a vector of known spatial basis functions, and $\boldsymbol{\alpha}_j = (\alpha_{1j}, \ldots, \alpha_{mj})'$ is a coefficient vector for the j-th FL function. Then the observation model is expressed as

$$y_{it} = \boldsymbol{\lambda}(\mathbf{s}_{it})'\mathbf{f}_t + \epsilon_{it} = \mathbf{b}(\mathbf{s}_{it})'\mathbf{A}\mathbf{f}_t + \epsilon_{it}, \quad i = 1, \ldots, N_t, \tag{1}$$

where $\boldsymbol{\lambda}(\mathbf{s}) = (\lambda_1(\mathbf{s}), \ldots, \lambda_d(\mathbf{s}))'$, and $\mathbf{A} = (\boldsymbol{\alpha}_1, \ldots, \boldsymbol{\alpha}_d)$. A vectorized form of the observation model is

$$\begin{aligned} \mathbf{y}_t &= \mathbf{\Lambda}(\mathbf{s}_t)\mathbf{f}_t + \boldsymbol{\epsilon}_t \\ &= \mathbf{B}_t\mathbf{A}\mathbf{f}_t + \boldsymbol{\epsilon}_t, \end{aligned} \quad \boldsymbol{\epsilon}_t \sim N(\mathbf{0}, \sigma_\epsilon^2 \mathbf{D}_{N_t}), \tag{2}$$

where $\mathbf{\Lambda}(\mathbf{s}_t) = (\boldsymbol{\lambda}(\mathbf{s}_{1t}), \ldots, \boldsymbol{\lambda}(\mathbf{s}_{N_t,t}))'$, $\mathbf{B}_t = (\mathbf{b}(\mathbf{s}_{1t}), \ldots, \mathbf{b}(\mathbf{s}_{N_t,t}))'$, $\boldsymbol{\epsilon}_t = (\epsilon_{1t}, \ldots, \epsilon_{N_t,t})'$, and $\mathbf{D}_{N_t}$ is a known N_t order positive definite matrix.

The system model that represents the dynamic evolution of the factors is

$$\mathbf{f}_t = \mathbf{\Gamma f}_{t-1} + \boldsymbol{\omega}_t, \quad \boldsymbol{\omega}_t \sim N(\mathbf{0}, \mathbf{\Sigma}_\omega), \tag{3}$$

where $\mathbf{\Gamma} = \text{diag}(\boldsymbol{\gamma})$, $\boldsymbol{\gamma} = (\gamma_1, \ldots, \gamma_d)'$, and $\mathbf{\Sigma}_\omega$ is a positive diagonal matrix.

The SCDFM consists of the extended observation model (2) and the system model (3).

The SCDFM encompasses the SDFM and the STREM without covariates as its special cases. When $d = m$, $\mathbf{A} = I_d$, and $\mathbf{\Gamma}$ is non-diagonal, the SCDFM is equivalent to the STREM without covariates, and when the observation locations are invariant, that is, $\mathbf{s}_t = (\mathbf{s}_1, \ldots, \mathbf{s}_N)$, and $b_i(\mathbf{s}) = \mathbf{1}_{\{\mathbf{s}_i\}}(\mathbf{s})$, where $\mathbf{1}_{\{\cdot\}}(\mathbf{s})$ is an indicator function, i.e., $\mathbf{B} = I_N$, the SCDFM is the same model as the SDFM.

2.1. Scaling indeterminacy

The SCDFM has a scaling indeterminacy in the coefficients of the FL functions (Appendix A), so that a constraint on $\mathbf{A}$ or that on $\mathbf{\Sigma}_\omega$ is necessary in the estimation of the SCDFM, e.g., $\|\boldsymbol{\alpha}_j\| = c$ for $j = 1, \ldots, d$, or $\mathbf{\Sigma}_\omega = cI_d$, where c is a constant. In this study, we impose the constraint on $\mathbf{\Sigma}_\omega$, $\mathbf{\Sigma}_\omega = cI_d$, because the estimation under the constraint on $\mathbf{\Sigma}_\omega$ can be conducted by just the maximization without any constraint while the estimation under the constraint on $\mathbf{A}$ requires the maximization with the constraints. After the estimation, the scales of the FL functions are adjusted to the same value for the purpose of model interpretation. Accordingly the system noise variances $\mathbf{\Sigma}_\omega$ and covariances and means of $\mathbf{f}_0$ are also adjusted to maintain the same likelihood value. Note that the SCDFM doesn't have a rotational indeterminacy that lies in a traditional factor analysis model, because the factors in the SCDFM independently evolve according to different first AR models (3) whose AR coefficients are all different values. The detail of the indeterminacy is described in Appendix A.

3. Maximum penalized likelihood method with L_2 penalty

In the estimation of the SCDFM, the maximum likelihood (ML) method often results in overfitting and yields unstable and unsmooth FL functions. We estimate the unknown parameters $\boldsymbol{\theta} = (\mathbf{A}, \gamma, \sigma_\epsilon^2)$ using the maximum penalized likelihood (MPL) method with L_2 penalty for $\mathbf{A}$. The penalty stabilizes the estimation of the FL functions and provides smooth FL functions.

The MPL method estimates the parameters of the SCDFM by maximizing the penalized log-likelihood function of the SCDFM. For the observation data $\mathbf{y}_{1:T} = (\mathbf{y}_1, \ldots, \mathbf{y}_T)$, the penalized log-likelihood function of the SCDFM is expressed as

$$l_{\rho_\alpha}(\boldsymbol{\theta}) = \frac{1}{T} \log p(\mathbf{y}_{1:T}|\boldsymbol{\theta}) - \frac{\rho_\alpha}{2} \sum_{j=1}^{d} \boldsymbol{\alpha}_j' K_\alpha^{(j)} \boldsymbol{\alpha}_j, \tag{4}$$

where p denotes a probability density function of $\mathbf{y}_{1:T}$ with $\boldsymbol{\theta}$, $\rho_\alpha (> 0)$ is a regularization parameter that controls the smoothness of the estimated FL functions, and $K_\alpha^{(j)}$ are known positive definite matrices. The elements of $K_\alpha^{(j)}$ depend on the types of the L_2 penalty for $\mathbf{A}$, which are $\sum_{j=1}^{d} \|\alpha_j\|^2$, $\sum_{i=1}^{m-1} \sum_{j=1}^{d} |\alpha_{ij} - \alpha_{i+1,j}|^2$, where b_i and b_{i+1} are adjacent, and so on. We maximize the penalized log-likelihood function using the expectation-maximization (EM) algorithm and the quasi-Newton method.

The EM algorithm is more stable and less sensitive to an initial value than gradient-based optimum algorithms such as a quasi-Newton method, but the parameters updated by the EM algorithm converge very slowly near local optimum solutions. Thus, after the EM algorithm is run a sufficient number of iterations, the quasi-Newton method is applied in order to maximize the penalized log-likelihood function faster.

The EM algorithm maximizes a Q-function with the L_2 penalty iteratively to maximize the penalized log-likelihood function. The Q-function at iteration n, $Q_{\rho_\alpha}(\boldsymbol{\theta}|\boldsymbol{\theta}^{(n-1)})$, is defined by $\frac{1}{T} E\left[\log p(\mathbf{y}_{1:T}, \mathbf{f}_{0:T}, \boldsymbol{\theta})|\mathbf{y}_{1:T}, \boldsymbol{\theta}^{(n-1)}\right] - \frac{\rho_\alpha}{2} \sum_{j=1}^{d} \boldsymbol{\alpha}_j' K_\alpha^{(j)} \boldsymbol{\alpha}_j$, where $E[\cdot|\mathbf{y}_{1:T}, \boldsymbol{\theta}^{(n-1)}]$ denotes a conditional expectation with respect to $\mathbf{f}_{0:T}$ given $\mathbf{y}_{1:T}$ with the parameters $\boldsymbol{\theta}^{(n-1)}$. From

$\frac{\partial Q_{\rho_\alpha}(\boldsymbol{\theta}|\boldsymbol{\theta}^{(n-1)})}{\partial\boldsymbol{\theta}} = \mathbf{0}$, the updates of $\boldsymbol{\theta}$ at n iteration are obtained as follows.

$$\boldsymbol{\alpha}_j^{(n)} = \left\{\sum_{t=1}^{T}\left\langle f_{jt}^2\right\rangle^{(n-1)}\mathbf{B}_t'\mathbf{D}_{N_t}^{-1}\mathbf{B}_t + \rho_\alpha T\sigma_\epsilon^{2(n-1)}K_\alpha^{(j)}\right\}^{-1}\times \sum_{t=1}^{T}\mathbf{B}_t'\mathbf{D}_{N_t}^{-1}\left(\langle f_{jt}\rangle^{(n-1)}\mathbf{y}_t - \mathbf{B}_t\mathbf{A}_{-j}^{(n-1)}\langle f_{jt}(\mathbf{f}_t)_{-j}\rangle^{(n-1)}\right), \quad (5)$$

$$\boldsymbol{\gamma}^{(n)} = \mathrm{diag}\left\{\sum_{t=1}^{T}\left\langle \mathbf{f}_{t-1}^2\right\rangle^{(n-1)}\right\}^{-1}\left(\sum_{t=1}^{T}\langle \mathbf{f}_t\circ\mathbf{f}_{t-1}\rangle^{(n-1)}\right), \quad (6)$$

$$\sigma_\epsilon^{2(n)} = \frac{1}{\sum_{t=1}^{T}N_t}\left\{\sum_{t=1}^{T}\mathbf{y}_t'\mathbf{D}_{N_t}^{-1}\mathbf{y}_t - 2\sum_{t=1}^{T}\langle\mathbf{f}_t\rangle^{(n-1)'}\mathbf{A}^{(n)'}\mathbf{B}_t'\mathbf{D}_{N_t}^{-1}\mathbf{y}_t + \mathrm{tr}\left(\sum_{t=1}^{T}\left\langle\mathbf{f}_t\mathbf{f}_t'\right\rangle^{(n-1)}\mathbf{A}^{(n)'}\mathbf{B}_t'\mathbf{D}_{N_t}^{-1}\mathbf{B}_t\mathbf{A}^{(n)}\right)\right\}, \quad (7)$$

where $\circ$ denotes an element-wise product, $\mathbf{A}_{-j}^{(n-1)} = (\boldsymbol{\alpha}_1^{(n)},\ldots,\boldsymbol{\alpha}_{j-1}^{(n)},\boldsymbol{\alpha}_{j+1}^{(n-1)},\ldots,\boldsymbol{\alpha}_d^{(n-1)})$, $(\mathbf{f}_t)_{-j}$ denotes the vector of the factors at time t except f_{jt}, and $\langle\cdot\rangle^{(n-1)} = E[\cdot|\mathbf{y}_{1:T},\boldsymbol{\theta}^{(n-1)}]$, which is calculated by the Kalman smoothing [17, 18]. After the EM algorithm is completed, the parameters $\mathbf{A}$ and $\boldsymbol{\gamma}$ are further updated by the quasi-Newton method, setting the EM solutions as its initial values.

To avoid the scaling indeterminacy, the system noise variances $\boldsymbol{\Sigma}_\omega = \sigma_\omega^2 I_d$ is fixed while the algorithms are running. However, if we update the system noise variance σ_ω^2 only for the initial iterations in the EM algorithm, the parameters converge faster. Thus, in the EM algorithm we update σ_ω^2 until M iterations, e.g., $M = 50$, and fix it from the $M+1$ iteration. The update of σ_ω^2 is described in Appendix B.

After optimizing the parameters by the quasi-Newton method, we scale the estimated coefficient vectors $\hat{\boldsymbol{\alpha}}_j$ to interpret the estimated model clearly. Accordingly the means and covariances of $\mathbf{f}_0$ and the system noise variances $\boldsymbol{\Sigma}_\omega$ are adjusted in order to keep the same likelihood value (Appendix A). That is,

$$\begin{aligned}&\hat{\mathbf{A}}^{\mathrm{scaled}} \leftarrow \hat{\mathbf{A}}H, \quad E[\mathbf{f}_0]^{\mathrm{scaled}} \leftarrow H^{-1}E[\mathbf{f}_0],\\ &\mathrm{Cov}[\mathbf{f}_0]^{\mathrm{scaled}} \leftarrow H^{-1}\mathrm{Cov}[\mathbf{f}_0](H')^{-1}, \boldsymbol{\Sigma}_\omega^{\mathrm{scaled}} \leftarrow H^{-1}\boldsymbol{\Sigma}_\omega(H')^{-1},\end{aligned} \quad (8)$$

where $H = \mathrm{diag}\{\boldsymbol{c}\}$, where $\boldsymbol{c}$ is a d dimensional vector. For example, $\boldsymbol{c} = (c/\|\hat{\lambda}_1\|_{L_2},\ldots,c/\|\hat{\lambda}_d\|_{L_2})$ provides the FL functions having the same L_2 norm, c.

4. Model selection criteria

In the MPL estimation with the L_2 penalty, it is crucial to select appropriately the number of factors d and the regularization parameter ρ_α, which controls the smoothness of the estimated FL functions. We derive two model selection criteria based on the Bayesian approach and the concept of the effective number of parameters.

4.1. Bayesian approach

From a Bayesian viewpoint, we consider the prior distribution for $\boldsymbol{\alpha} = (\boldsymbol{\alpha}_1',\ldots,\boldsymbol{\alpha}_d')'$ as a multivariate normal distribution given by

$$\pi(\boldsymbol{\alpha}|\rho_\alpha) = (2\pi)^{-(md)/2}T^{(md)/2}\rho_\alpha^{(md)/2}|\boldsymbol{K}_\alpha|^{1/2}\exp\left(-\frac{T\rho_\alpha}{2}\boldsymbol{\alpha}'\boldsymbol{K}_\alpha\boldsymbol{\alpha}\right), \quad (9)$$

IOP Conf. Series: Journal of Physics: Conf. Series **1036** (2018) 012006 doi:10.1088/1742-6596/1036/1/012006

where $\boldsymbol{K}_\alpha = \text{block diag}\{K_\alpha^{(1)}, \ldots, K_\alpha^{(d)}\}$. We also consider the prior distribution for $\boldsymbol{\gamma}$ and σ_ϵ^2 as a uniform improper prior distribution $\pi(\boldsymbol{\gamma}, \sigma_\epsilon^2) \propto c$. Then, the Bayesian estimation maximizing the posterior probability corresponds to the MPL estimation.

Considering the MPL estimation as the Bayesian estimation, Konishi et al [12] presented a model selection criterion, the generalized Bayesian information criterion (GBIC), for evaluating models estimated by the MPL method. Applying their result, we obtain a model selection criterion for evaluating the SCDFM estimated by the MPL method, as

$$\begin{aligned} \text{GBIC}(\rho_\alpha, d) = &-2\log p(\mathbf{y}_{1:T}|\hat{\boldsymbol{\theta}}) + T\rho_\alpha \hat{\boldsymbol{\alpha}}' \boldsymbol{K}_\alpha \hat{\boldsymbol{\alpha}} + d\log\left(\frac{T}{2\pi}\right) \\ &+ \log|Q_{\rho_\alpha}(\hat{\boldsymbol{\alpha}}, \hat{\boldsymbol{\gamma}})| - md\log(\rho_\alpha) - \log|\boldsymbol{K}_\alpha|, \end{aligned} \tag{10}$$

where

$$Q_{\rho_\alpha}(\boldsymbol{\alpha}, \boldsymbol{\gamma}) = -\frac{1}{T}\frac{\partial^2}{\partial(\boldsymbol{\alpha}', \boldsymbol{\gamma}')'\partial(\boldsymbol{\alpha}', \boldsymbol{\gamma}')}\log p(\mathbf{y}_{1:T}|\boldsymbol{\alpha}, \boldsymbol{\gamma}, \hat{\sigma}_\epsilon^2) + \begin{pmatrix} \rho_\alpha \boldsymbol{K}_\alpha & \mathbf{0}_{md,d} \\ \mathbf{0}_{d,md} & \mathbf{0}_{d,d} \end{pmatrix}, \tag{11}$$

where $\mathbf{0}_{m,n}$ denotes an $m \times n$ matrix all of whose elements are zero. The Hessian matrix of the log-likelihood of the SCDFM cannot be calculated analytically, so we calculate it numerically. To obtain stable numerical calculations of the Hessian matrix, we regard the scale parameter $\hat{\sigma}_\epsilon^2$ estimated by the MPL method as a hyper-parameter.

4.2. Effective number of parameters

The GBIC for the SCDFM requires the numerical calculation of the Hessian matrix, and the calculation becomes more unstable as the number of parameters increases. For evaluating stably the SCDFM estimated by the MPL method, we introduce AIC-formal criteria [15] including the derived effective number of parameters (ENP; [13, 14]) for $\hat{\mathbf{\Lambda}}$ instead of the number of parameters. The criteria don't require any complicated numerical calculation, and thus they are stable for the SCDFM even with a number of parameters.

The AIC-formal criteria for the SCDFM estimated by the ML method are $-2\log p(\mathbf{y}_{1:T}|\hat{\boldsymbol{\theta}}) + g(T)(md+d)$, where $g(T)$ is a function of T, and the number of parameters in the penalty term, $md + d$, represents a measure of model complexity. However, in the FL functions estimated by the MPL method $\hat{\lambda}_j$, the number of parameters m is not a suitable measure of model complexity, since the complexity of the functions depends on the penalty term.

The concept of the number of parameters was extended to the ENP by [13], [14], and others, and the ENP was shown to be a reasonable measure of model complexity especially in regression models. We derive the ENP for $\hat{\lambda}_j$, $\text{df}(\hat{\lambda}_j)$, from the EM update equation (5) based on the concept of the ENP for the ridge linear regression [13] as

$$\text{df}(\hat{\lambda}_j) = \text{tr}\left\{\left(\sum_{t=1}^{T}\langle f_{jt}^2\rangle \mathbf{B}_t'\mathbf{D}_{N_t}^{-1}\mathbf{B}_t + \rho_\alpha T\hat{\sigma}_\epsilon^2 K_\alpha^{(j)}\right)^{-1}\left(\sum_{t=1}^{T}\langle f_{jt}\rangle^2 \mathbf{B}_t'\mathbf{D}_{N_t}^{-1}\mathbf{B}_t\right)\right\},$$

where $\langle\cdot\rangle = E[\cdot|\mathbf{y}_{1:T}, \hat{\boldsymbol{\theta}}]$.

When we assume that the observation sites are temporally invariant, i.e., $N_t = N$ and $\mathbf{B}_t = \mathbf{B}$, $D_N = I_N$, and that $Var(f_{jt}) = 0$, for $t = 1, \ldots, T$, then $\text{df}(\hat{\lambda}_j) = \text{tr}\left\{\left(\mathbf{B}'\mathbf{B} + (\rho_\alpha T\hat{\sigma}_\epsilon^2)/(\sum_{t=1}^T c_{jt})K_\alpha^{(j)}\right)^{-1}\mathbf{B}'\mathbf{B}\right\}$, where $c_{jt} = \langle f_{jt}\rangle^2 = \left\langle f_{jt}^2\right\rangle$. This ENP is the same formula as that for the ridge regression model. Furthermore, when we suppose $\rho_\alpha = 0$, then $\text{df}(\hat{\boldsymbol{\alpha}}_j) = m$.

By replacing the number of parameters of the FL function λ_j in the AIC-formal criteria with the derived ENP for $\hat{\lambda}_j$, we formally obtain information criteria for the SCDFM estimated by the MPL method as $-2\log p(\mathbf{y}_{1:T}|\hat{\boldsymbol{\theta}}) + g(T)(\sum_{j=1}^{d} \mathrm{df}(\hat{\lambda}_j) + d)$. In this study, we use the weight of the penalty in the consistent AIC (CAIC; [16]), i.e., $g(T) = \log(T) + 1$, and we call the criterion *modified* CAIC ($\mathrm{CAIC_M}$) hereafter.

We choose the regularization parameter ρ_α and the number of factors d minimizing the GBIC or $\mathrm{CAIC_M}$. Through all the MPL estimations during selecting ρ_α and d, we use the system noise variances of the same value, i.e., $\boldsymbol{\Sigma}_\omega = cI_d$, where c is a constant value for all the estimations.

5. Applications

In this section, we applied the proposed modeling procedure and the ML estimation for the SCDFM to synthetic datasets and ozone concentration data, and compared their results.

5.1. Synthetic data

The synthetic data set was generated from the SCDFM having three factors and corresponding FL functions.

The true FL functions cover the observation region, $\mathbf{S} = [0, 10] \times [0, 10]$, and they are constructed by the basis expansion of the 5×5 Gaussian basis functions, $\exp\left(-\frac{||\mathbf{s}-\boldsymbol{\mu}_i||^2}{2\nu_i^2}\right)$, where the centers of the basis functions, $\boldsymbol{\mu}_i$, are allocated on lattice-shaped points (the blue points in Fig. 1(a)), and $\nu_i^2 = 2.5$ for $i = 1, \ldots, 25$. The true FL functions $\lambda_j^*(\mathbf{s})$ are shown in Fig. 1(a), and L_2 norms of them are all 10, i.e., $||\lambda_j^*||_{L_2} = 10, j = 1, 2, 3$.

The temporal evolution of the first factor is described as $f_{1t} = T(t) + \zeta_t$, where $\zeta_t = 0.6\zeta_{t-1} + \omega_{1t}$ and $T(t)$ is a linear trend described as $T(t) = t/100 + 4$. The other factors evolve as $f_{jt} = \gamma_j^* f_{j(t-1)} + \omega_{jt}$ for $j = 2, 3$, where $(\gamma_2^*, \gamma_3^*) = (0.55, -0.2)$. The initial values $\zeta_0, f_{2,0}, f_{3,0} \sim N(\mathbf{0}, \boldsymbol{\Sigma}_\omega^*)$, and the system noises $\omega_{1t}, \omega_{2t}, \omega_{3t} \overset{i.i.d.}{\sim} N(\mathbf{0}, \boldsymbol{\Sigma}_\omega^*)$, where $\boldsymbol{\Sigma}_\omega^* = \mathrm{diag}(0.32^2, 0.17^2, 0.1^2)$. Following these models, the factors $\mathbf{f}_t^*$ were simulated for $t = 1, \ldots, T = 50$, as shown in Fig. 2.

Two synthetic datasets were generated on two point sets of 25 observation sites, (i) the centers of the basis functions, and (ii) uniformly random-sampled positions from S. Each of the datasets was generated on each of these point sets from the observation model in the SCDFM, Eq. (1), (2), with the simulated factor processes $\mathbf{f}_{1:T}^*$, the true FL functions $\lambda_j^*(\mathbf{s})$ and the observation noises $\boldsymbol{\epsilon}_t \overset{i.i.d.}{\sim} N(\mathbf{0}, 0.2^3 I_{25})$. We assumed that 20% of the observations were missing values every time point.

We estimated the SCDFM using the MPL method and ML method from each of the datasets. The model includes the same basis functions as those constructing the true FL functions, and the initial factors are $\mathbf{f}_0 \sim N(\mathbf{0}, 10I_d)$. In the MPL estimation, the EM algorithm described in Section 3 was run for 1000 iterations, and then the quasi-Newton method was conducted to maximize the penalized log-likelihood with respect to $\mathbf{A}$ and $\boldsymbol{\gamma}$, where the initial values were the EM solutions. The regularization parameter and the number of factors were selected by the GBIC and $\mathrm{CAIC_M}$. The matrices in the penalty term are $K_\alpha^{(j)} = I_m$. The ML estimation was conducted using the same procedure as the MPL estimation, where the EM algorithm was run with $\rho_\alpha = 10^{-10}$. The number of factors was selected by the AIC and the BIC [19].

The estimated coefficients of the FL functions, $\hat{\boldsymbol{\alpha}}_j$, were scaled so that the L_2 norms of all the estimated FL functions were equal to 10 that is the L_2 norm of the true FL functions. For each of the estimated FL functions, its number and sign were adjusted to the number of the true FL function and the sign of the estimated FL function that minimize the L_2 distance between these functions, respectively, i.e., for $a\hat{\lambda}_i$, $a = \pm 1$, $(i, a)^{(\mathrm{new})} \leftarrow \arg\min_{j,a} ||\lambda_j^* - a\hat{\lambda}_i||_{L_2}$.

Through all the estimations in this experiment, the number of factors was selected as the true number 3, by both the GBIC and $\mathrm{CAIC_M}$ in the MPL estimation, and both the AIC and

BIC in the ML estimation. In the dataset observed on (i), the GBIC and $\mathrm{CAIC_M}$ each selected $\rho_\alpha = 0.25, 0.05$ within the three-factor model. In the dataset observed on (ii), both the GBIC and $\mathrm{CAIC_M}$ selected $\rho_\alpha = 0.25$ within the three-factor model. In the SCDFM that has many factors, the determinant of the numerical Hessian matrix in the GBIC often took a negative value, and thus the GBIC was not obtained.

The estimated FL functions, $\hat{\lambda}_j(\cdot)$, are plotted in Fig. 1(b) and (c), where the FL function estimated by the MPL method with ρ_α selected by the $\mathrm{CAIC_M}$ in (i) is not plotted since it is visually indistinguishable from the FL function estimated with the GBIC. The L_2 errors of the estimated FL functions, $||\lambda_j^* - \hat{\lambda}_j||_{L_2}$, are described in Table 1. For the observation point set (i), the MPL method estimated the FL functions slightly more accurately than the ML method (Table 1), but the FL functions estimated by the ML and MPL methods could capture the global structures of the true FL functions (Fig. 1(b)). On the other hand, for the observation point set (ii), the MPL method estimated the FL functions much more accurately than the ML method (Table 1), and the FL functions estimated by the MPL method could capture the global structures of the true FL functions. However, the FL functions by the ML method could not capture the true structures at all (Fig. 1(c)).

Observation point set	(i)			(ii)		
	λ_1	λ_2	λ_3	λ_1	λ_2	λ_3
MPL method with GBIC	0.058	3.917	2.290	0.242	3.072	5.046
MPL method with $\mathrm{CAIC_M}$	0.057	4.416	2.371	same as above		
ML method	0.056	4.818	2.412	12.388	13.714	14.089

Table 1: The L_2 errors of the FL functions estimated by the MPL method with the GBIC and $\mathrm{CAIC_M}$, and by the ML method.

We estimated the factors by posterior means of the factors calculated by the Kalman smoothing, i.e., $\hat{\mathbf{f}}_t \equiv E[\mathbf{f}_t|\mathbf{y}_{1:T}, \hat{\boldsymbol{\theta}}]$. The estimated factor processes $\hat{\mathbf{f}}_{1:T}$ are plotted in Fig. 2, where the estimated factor processes by the MPL method with the $\mathrm{CAIC_M}$ in (i) are not plotted since they are visually indistinguishable from those estimated with the GBIC. Root mean squared errors (RMSEs) of the estimated factor processes, $\sqrt{\frac{1}{50}\sum_{t=1}^{50}(\hat{f}_{jt} - f_{jt}^*)^2}$, are described in Table 2. For the observation point set (i), the factor processes estimated by the ML and MPL methods could capture the simulated factor processes accurately (Fig. 2(a)), and the factor processes estimated by the MPL method were slightly more accurate than those by the ML method (Table 2). On the other hand, for the observation point set (ii), the factor processes estimated by the MPL method could capture the temporal variations of the simulated factor processes accurately (Fig. 2(c)), but those by the ML method could not capture them at all (Fig. 2(b)). Hence the factor processes estimated by the the MPL method were much more accurate than those by the ML method (Table 2). This is because the MPL method estimated the corresponding FL functions much more accurately.

These results show that the MPL estimation with the GBIC and $\mathrm{CAIC_M}$ of the SCDFM is more stable and accurate than the ML estimation in different observation conditions.

5.2. Ozone concentration data

The ozone (O_3) is an air pollutant near the ground, and the O_3 concentration data were observed at 26 sites on the ground in New York state (Fig. 3) . The data were observed daily for 46 days (Fig. 4), and contain 15 missing values. The dataset is available in the "spTimer"-package of R software [20]. The data were also log-transformed to normalize them. The coordinate of

IOP Conf. Series: Journal of Physics: Conf. Series **1036** (2018) 012006 doi:10.1088/1742-6596/1036/1/012006

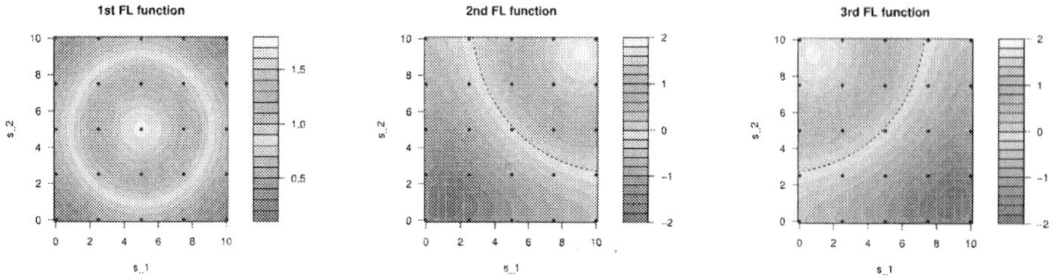

(a) True FL functions.

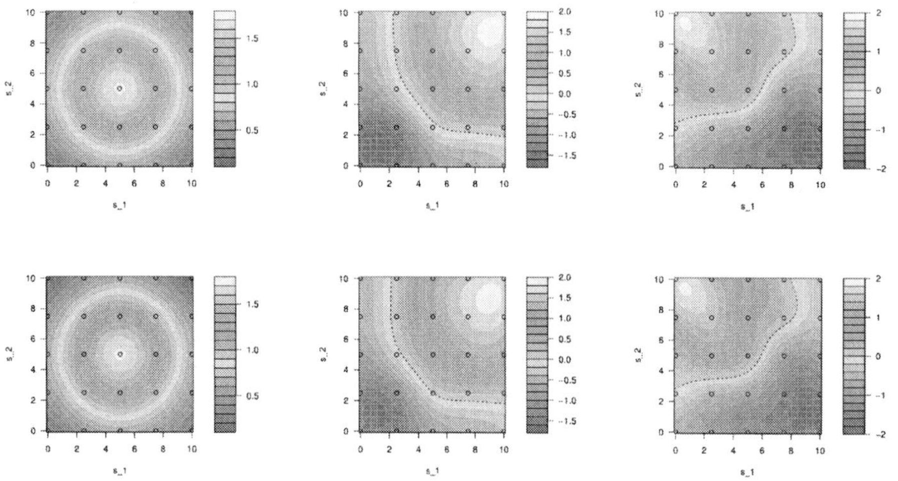

(b) Observation point set (i). Upper row: MPL method with ρ_α selected by the GBIC; lower row: ML method

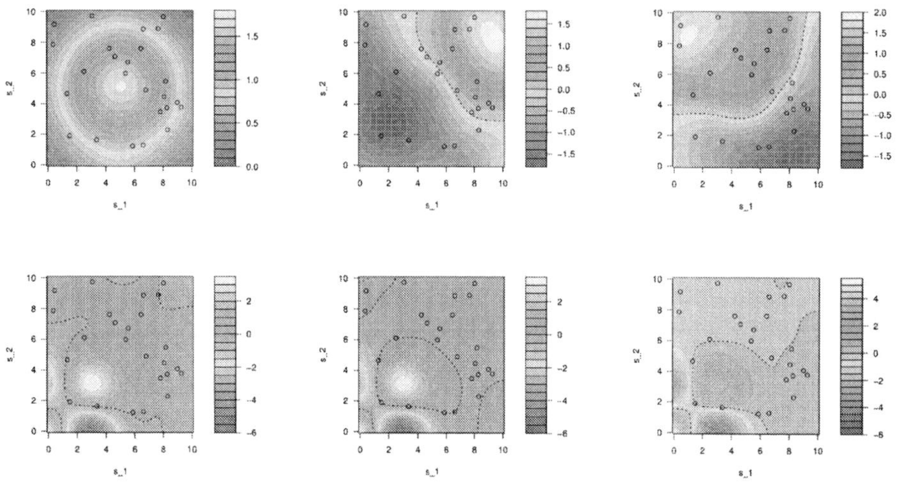

(c) Observation point set (ii). Upper row: MPL method with ρ_α selected by both the GBIC and CAIC_M; lower row: ML method

Figure 1: True and estimated FL functions. Figure (a) shows the true functions λ_j^*, where blue filled circles denote the centers of the Gaussian basis functions. Figure (b) and (c) show the FL functions estimated from the data observed in (i) and (ii), respectively, where the black open circles denotes the observation points. The FL functions estimated by the MPL and ML methods are plotted in the upper and lower rows, respectively, in Figure (b) and (c). Dotted lines denote contour lines of 0.

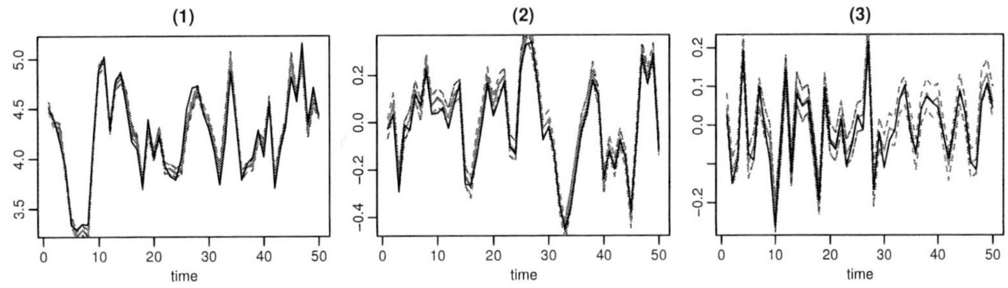

(a) Observation point set (i); simulated factor processes and factor processes estimated by the ML and MPL methods.

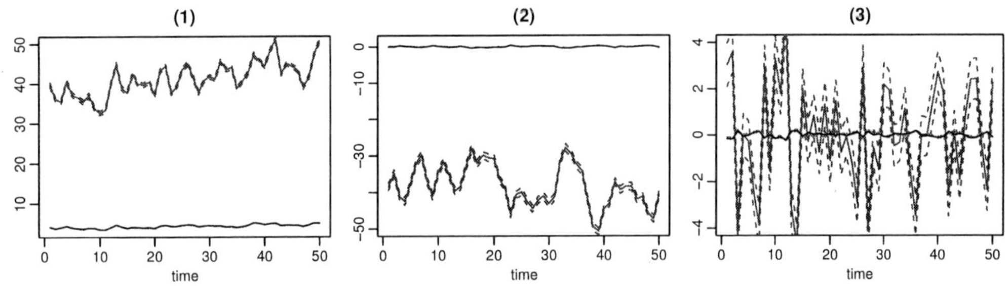

(b) Observation point set (ii); simulated factor processes and factor processes estimated by the ML and MPL methods.

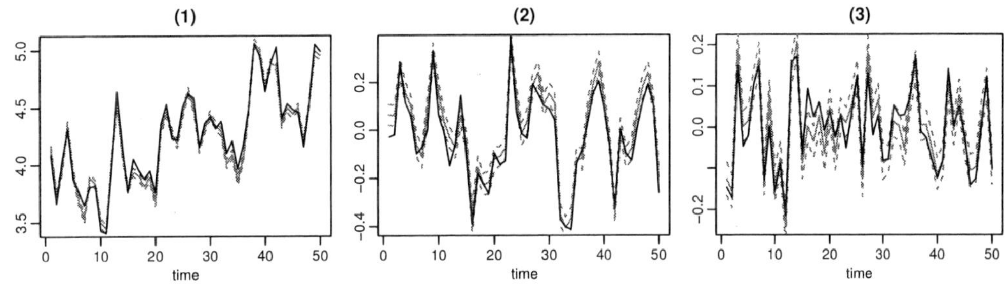

(c) Observation point set (ii); simulated factor processes and factor processes estimated by the MPL method.

Figure 2: The simulated factor processes (black line), the factor processes estimated by the ML method (blue line), and those estimated by the MPL method (red line). Figure (a) shows the estimated factor processes from the data observed in (i). Figure (b) and (c) show the estimated factor processes from the data observed in (ii). In the MPL estimation, Figure (a) shows the estimation result with the GBIC, and Figure (b) nd (c) show that with both the GBIC and CAIC_{M}. In Figure (c), the factor processes estimated by the ML method are omitted, because the temporal variations of the factor processes simulated and estimated by the MPL method cannot be recognized if those by the ML method are included as Figure (b). Dashed lines denote $\hat{f}_{jt} \pm 2 \times$ posterior standard deviations of f_{jt}.

Observation point set	(i)			(ii)		
	f_{1t}	f_{2t}	f_{3t}	f_{1t}	f_{2t}	f_{3t}
MPL method with GBIC	0.0809	0.0316	0.0201	0.0674	0.0517	0.0355
MPL method with $CAIC_M$	0.0902	0.0321	0.0199	same as above		
ML method	0.0982	0.0327	0.0199	37.179	38.6170	2.4177

Table 2: The RMSEs of the factor processes estimated by the MPL method with the GBIC and $CAIC_M$, and by the ML method.

the locations was converted from latitude and longitude to orthogonal coordinate, measured in kilometer.

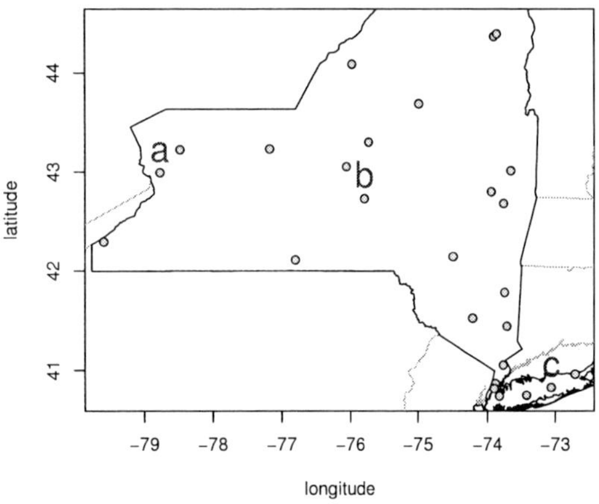

Figure 3: The map around New York state, the boundary of which is denoted by black lines. Green points show the observation sites of the O_3 concentrations.

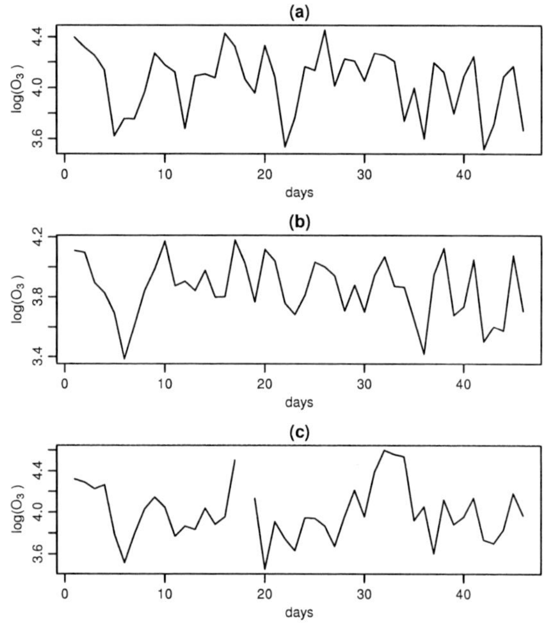

Figure 4: Figures (a), (b) and (c) show the logarithms of the data observed at the sites "a", "b" and "c" in Fig. 3, respectively. The blank at the 18 th day in the site "c" shows a missing value.

We used the Gaussian basis functions whose number was that of the observation sites, 26. The centers of the basis functions are the observation sites, and each of their spatial extents, ν_i, is the mean of the distances between the center of the basis function, $\boldsymbol{\mu}_i$, and the centers of its three nearest-neighbor basis functions. We estimated the SCDFM using the MPL and ML methods from the dataset through the same setup and procedure conducted in Section 5.1.

The estimated coefficients of the FL functions, $\hat{\alpha}_j$, were scaled so that the average of absolute values of the FL function on the observation sites was 0.5 for each of the estimated FL functions. The order and signs of the FL functions and related parameters were also adjusted with the likelihood value fixed in order to facilitate a comparison between the FL functions estimated by the ML method and those by the MPL method.

The number of factors was selected as five from three to seven by both the $\mathrm{CAIC_M}$ in the MPL estimation and the BIC in the ML estimation. When the number of factors was more than four, the determinant of the numerical Hessian matrix in the GBIC in the MPL estimation often took a negative value, and thus the GBIC was not obtained. The GBIC in the MPL estimation and the AIC in the ML estimation selected a seven-factor model that was obviously redundant since the model had quite similarly shaped FL functions. This shows that the GBIC and AIC could not select an appropriate model. Thus, hereafter we discuss the models selected by the BIC and the $\mathrm{CAIC_M}$.

The estimated FL functions and factor processes are plotted in Fig. 5 and 6, respectively. In both the ML and MPL estimations, the first FL function takes positive values all over the observation region (Fig. 5(a)(1), (b)(1)), so that the temporal variation of the corresponding first factor represents common temporal variation over the region. The first factor process also represents the nonstationary grand mean component of the temporal variations of the data (Fig. 6 (1)), and thus the time averages of the other factors are almost 0 (Fig. 6 (2)-(5)). The signs of the other FL functions vary by site, and the difference in the signs is considered to represent directions of migrations of the O_3. For example, the second FL function estimated by the MPL method takes positive values in the west area and negative values in the east (Fig. 5(b)(2)). Since the rise in the second factor increases the second component $f_{2t}\hat{\lambda}_2$ on the west area and decreases that on the east area, the rise in the second factor is considered to represent a migration of the O_3 from the east area to the west. We can find out the spatio-temporal structure of the migration of the O_3 from the estimated factor processes and the structures of the corresponding estimated FL functions.

The FL functions estimated by the MPL method have smooth and clear structures, but those estimated by the ML method have rough and unclear structures, which cannot provide a clear interpretation of the migration of the O_3. Especially, it is not easy to interpret a flux of the O_3 from the second FL function estimated by the ML method, since the 0-contour line of the FL function is winding, where the line decides the direction of O_3 migration in the second component.

6. Conclusions

We proposed the spatially continuous dynamic factor model (SCDFM) by extending the FL vectors of the spatial dynamic factor model to the continuous FL functions. The SCDFM can be directly applied to the spatio-temporal data whose observation sites and whose number vary with time, where the SDFM cannot be directly applied. To estimate smooth and clear FL functions of the SCDFM, we presented the maximum penalized likelihood method with the L_2 penalty for the FL functions. To maximize the penalized likelihood stably, we derived the EM algorithm. For the appropriate choice of the regularization parameter and the number of factors, we presented model selection criteria derived from a Bayesian view point and a concept of the effective number of parameters.

Through the applications to synthetic data and ozone concentration data, we showed that the proposed modeling procedure estimates the SCDFM accurately and stably, and especially provides the smooth and clear FL functions.

Acknowledgments

This study was supported by MEXT KAKENHI Grant Number 25120011.

Appendix A. Indeterminacy

We consider the SCDFM described in (2) and (3). Suppose that the matrix of the coefficients is $\tilde{\mathbf{A}} = \mathbf{A}H$, and correspondingly the initial factors and the system noises at all times are $\tilde{\mathbf{f}}_0 = H^{-1}\mathbf{f}_0$ and $\tilde{\omega}_t = H^{-1}\omega_t$, respectively, where H is a diagonal matrix or elementary matrix

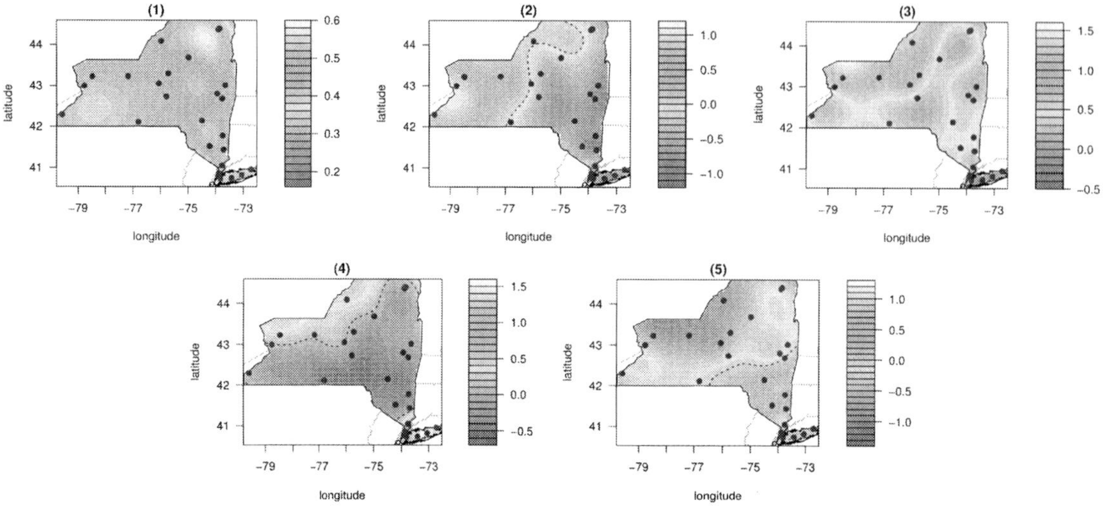

(a) ML method

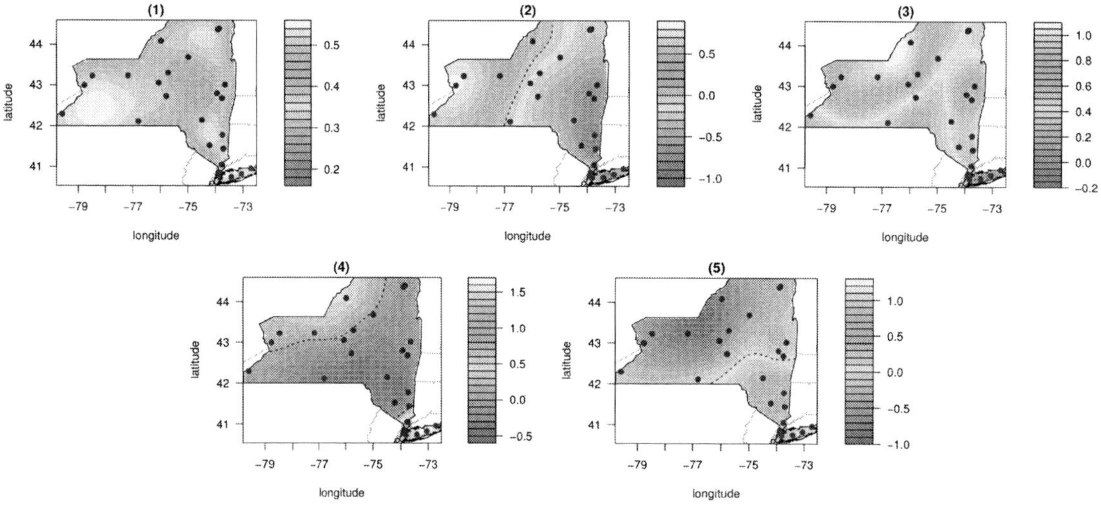

(b) MPL method

Figure 5: The FL functions estimated by the ML method with the BIC (a) and the MPL method with the $\mathrm{CAIC_M}$ (b). Blue circles denote the observation points, the centers of the Gaussian basis functions, and dotted lines denote contour lines of 0.

that exchanges rows or columns. Then the factors $\tilde{\mathbf{f}}_t$ are independent each other and expressed as $\tilde{\mathbf{f}}_t = H^{-1}\mathbf{f}_t$, for $t = 1, \ldots, T$. The observation $\tilde{\mathbf{y}}_t$ for the model with $\tilde{\mathbf{A}}$, $\tilde{\mathbf{f}}_0$ and $\tilde{\boldsymbol{\omega}}_t$ is $\tilde{\mathbf{y}}_t = \tilde{\mathbf{A}}\tilde{\mathbf{f}}_t + \epsilon_t = \mathbf{A}\mathbf{f}_t + \epsilon_t = \mathbf{y}_t$. Thus, the model converted by H with $\tilde{\mathbf{A}}$, $\tilde{\mathbf{f}}_0$ and $\tilde{\boldsymbol{\omega}}_t$ and the original model with $\mathbf{A}$, $\mathbf{f}_0$ and $\boldsymbol{\omega}_t$ generate the same probability density of the observations. That is, the likelihood for any data takes the same value for both the model converted by H and the original model.

Particularly we calculate the log-likelihood of the SCDFM before converted and converted

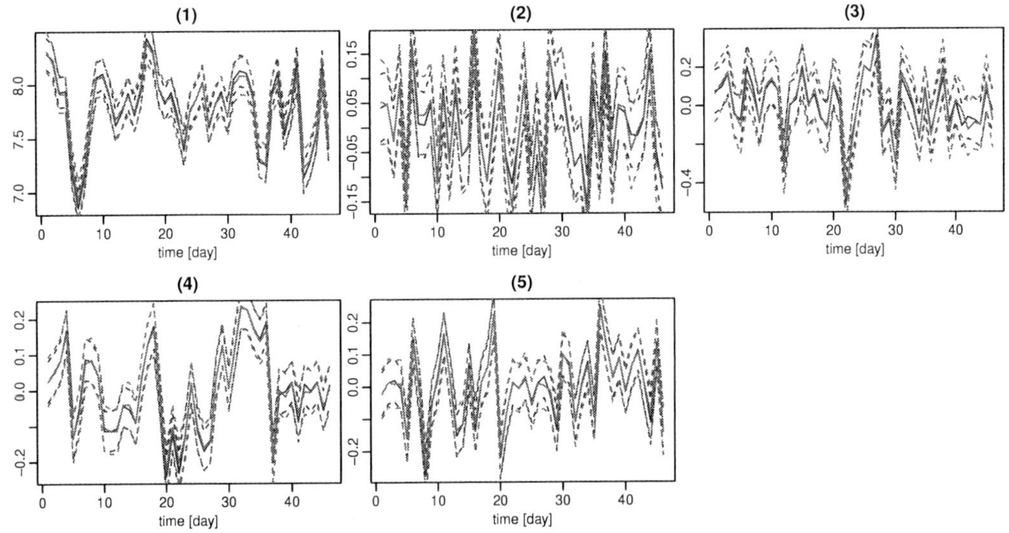

Figure 6: The factor processes estimated by the ML method (blue line) and the MPL method with the CAIC_M (red line). Dashed lines denote $\hat{f}_{jt} \pm 2 \times$ posterior standard deviations of f_{jt}.

by H hereafter. The log-likelihood of the SCDFM before converted is expressed as

$$\begin{aligned}\log p(\mathbf{y}_{1:T}|\mathbf{A}, \mathbf{\Gamma}, \sigma_\epsilon^2) &= \sum_{t=1}^{T} \log p(\mathbf{y}_t|\mathbf{y}_{1:t-1}, \mathbf{A}, \gamma, \sigma_\epsilon^2) \\ &= -\frac{1}{2}\left\{(\sum_{t=1}^{T} N_t)\log 2\pi + \sum_{t=1}^{T}\log|\mathbf{R}_{t|t-1}| \right. \\ &\quad \left. + \sum_{t=1}^{T}(\mathbf{y}_t - \mathbf{y}_{t|t-1})'\mathbf{R}_{t|t-1}^{-1}(\mathbf{y}_t - \mathbf{y}_{t|t-1})\right\},\end{aligned} \tag{A.1}$$

where $\mathbf{y}_{1:t} = (\mathbf{y}_1, \ldots, \mathbf{y}_t)$, $p(\mathbf{y}_1|\mathbf{y}_{1:0}, \mathbf{A}, \gamma, \sigma_\epsilon^2) = p(\mathbf{y}_1|\mathbf{A}, \gamma, \sigma_\epsilon^2)$, $\mathbf{y}_{t|t-1} = E[\mathbf{y}_t|\mathbf{y}_{1:t-1}]$ and $\mathbf{R}_{t|t-1} = \text{Cov}[\mathbf{y}_t|\mathbf{y}_{1:t-1}]$. The conditional means and variances, $\mathbf{y}_{t|t-1}$ and $\mathbf{R}_{t|t-1}$ are expressed as

$$\begin{aligned}\mathbf{y}_{t|t-1} &= \mathbf{B}_t\mathbf{A}\mathbf{f}_{t|t-1}, \\ \mathbf{R}_{t|t-1} &= \mathbf{B}_t\mathbf{A}\mathbf{V}_{t|t-1}(\mathbf{B}_t\mathbf{A})' + \sigma_\epsilon^2\mathbf{D}_{N_t},\end{aligned}$$

where $\mathbf{f}_{t|t-1} = E[\mathbf{f}_t|\mathbf{y}_{1:t-1}]$, $\mathbf{V}_{t|t-1} = \text{Cov}[\mathbf{f}_t|\mathbf{y}_{1:t-1}]$.

Suppose that $\mathbf{A}$ is replaced by $\tilde{\mathbf{A}} = \mathbf{A}H$, and correspondingly f_0 and ω_t are replaced by $\tilde{\mathbf{f}}_0 = H^{-1}\mathbf{f}_0$ and $\tilde{\omega}_t = H^{-1}\omega_t$, respectively. Then $\tilde{\mathbf{f}}_{t|t-1} = E[\tilde{\mathbf{f}}_t|\tilde{\mathbf{y}}_{1:t-1}, H]$ and $\tilde{\mathbf{V}}_{t|t-1} = \text{Cov}[\tilde{\mathbf{f}}_t|\tilde{\mathbf{y}}_{1:t-1}, H]$ are

$$\begin{aligned}\tilde{\mathbf{f}}_{t|t-1} &= H^{-1}\mathbf{f}_{t|t-1}, \\ \tilde{\mathbf{V}}_{t|t-1} &= H^{-1}\mathbf{\Gamma}\mathbf{V}_{t-1|t-1}\mathbf{\Gamma}'(H')^{-1} + H^{-1}\mathbf{\Sigma}_\omega(H')^{-1}, \\ &= H^{-1}\mathbf{V}_{t|t-1}(H')^{-1},\end{aligned}$$

and $\tilde{\mathbf{R}}_{t|t-1} = \mathrm{Cov}[\tilde{\mathbf{y}}_t|\tilde{\mathbf{y}}_{1:t-1}, H]$ and $\tilde{\mathbf{y}}_{t|t-1} = E[\tilde{\mathbf{y}}_t|\tilde{\mathbf{y}}_{1:t-1}, H]$

$$\begin{aligned}\tilde{\mathbf{y}}_{t|t-1} &= \mathbf{B}_t\mathbf{A}HH^{-1}\mathbf{f}_{t|t-1} = \mathbf{y}_{t|t-1}, \\ \tilde{\mathbf{R}}_{t|t-1} &= \mathbf{B}_t\mathbf{A}HH^{-1}\mathbf{V}_{t|t-1}(H')^{-1}H'(\mathbf{B}_t\mathbf{A})' + \sigma_\epsilon^2\mathbf{D}_{N_t} = \mathbf{R}_{t|t-1}.\end{aligned} \tag{A.2}$$

From (A.1) and (A.2), the probabilistic densities of $\mathbf{y}_{1:T}$ and $\tilde{\mathbf{y}}_{1:T}$ are identical, i.e., the likelihoods for the models of $\mathbf{y}_{1:T}$ and $\tilde{\mathbf{y}}_{1:T}$ are equal.

Appendix B. EM update of $\mathbf{\Sigma}_\omega$

Suppose $\mathbf{\Sigma}_\omega = \sigma_\omega^2 I_d$, then the update of σ_ω^2 in the EM algorithm (Section 3) is

$$\sigma_\omega^{2(n)} = \frac{1}{Td}\left\{\sum_{t=1}^{T}\langle \mathbf{f}_t'\mathbf{f}_t\rangle^{(n-1)} - 2\mathrm{tr}\left\{\left(\sum_{t=1}^{T}\langle \mathbf{f}_t\mathbf{f}_{t-1}'\rangle^{(n-1)}\right)\mathbf{\Gamma}\right\}\right. \\ \left. +\mathrm{tr}\left\{\left(\sum_{t=1}^{T}\langle \mathbf{f}_{t-1}\mathbf{f}_{t-1}'\rangle^{(n-1)}\right)\mathbf{\Gamma}'\mathbf{\Gamma}\right\}\right\}.$$

[1] Cressie N and Wikle C K 2011 *Statistics for spatio-temporal data* (John Wiley & Sons)
[2] Reinsel G C 2003 *Elements of multivariate time series analysis* (Springer Science & Business Media)
[3] Cliff A and Ord J K 1975 *Transactions of the Institute of British Geographers* 119–128
[4] Lopes Hedibert F, Salazar E and Gamerman D 2008 *Bayesian Analysis* **3** 759–792
[5] Strickland C, Simpson D, Turner I, Denham R and Mengersen K 2011 *Journal of the Royal Statistical Society: Series C (Applied Statistics)* **60** 109–124
[6] Ippoliti L, Valentini P and Gamerman D 2012 *Journal of the Royal Statistical Society: Series C (Applied Statistics)* **61** 175–200
[7] Geweke J 1977 *Latent variables in socio-economic models* **1**
[8] Stock J H and Watson M 2011 *Oxford Handbook on Economic Forecasting*
[9] Cressie N, Shi T and Kang E L 2010 *Journal of Computational and Graphical Statistics* **19** 724–745
[10] Katzfuss M and Cressie N 2011 *Journal of Time Series Analysis* **32** 430–446
[11] Dempster A P, Laird N M and Rubin D B 1977 *Journal of the Royal Statistical Society. Series B (Methodological)* **39** 1–38
[12] Konishi S, Ando T and Imoto S 2004 *Biometrika* **91** 27–43
[13] Hastie T J and Tibshirani R J 1990 *Generalized additive models* vol 43 (CRC press)
[14] Moody J E 1991 The effective number of parameters: An analysis of generalization and regularization in nonlinear learning systems *Advances in Neural Information Processing Systems (NIPS)* vol 4 pp 847–854
[15] Akaike H 1973 Information theory and an extension of the maximum likelihood principle *Second International Symposium on Information Theory* (Akademinai Kiado) pp 267–281
[16] Bozdogan H 1987 *Psychometrika* **52** 345–370
[17] Ghahramani Z and Hinton G E 1996 Parameter estimation for linear dynamical systems Tech. rep. Technical Report CRG-TR-96-2, University of Totronto, Dept. of Computer Science
[18] Durbin J and Koopman S J 2012 *Time Series Analysis by State Space Methods* (Oxford university press)
[19] Schwarz G 1978 *The annals of statistics* **6** 461–464
[20] Bakar K S and Sahu S K 2015 *Journal of Statistical Software* **63** 1–32

IOP Conf. Series: Journal of Physics: Conf. Series **1036** (2018) 012007 doi:10.1088/1742-6596/1036/1/012007

Selective isotope labeling strategy and computational interpretation of spectra for protein NMR analyses

Takuma Kasai[1,2,4] **and Takanori Kigawa**[1,3,4]

[1] RIKEN Quantitative Biology Center, 1-7-22 Suehiro-cho, Tsurumi-ku, Yokohama, Kanagawa 230-0045, Japan
[2] PRESTO, JST, 4-1-8 Honcho, Kawaguchi, Saitama 332-0012, Japan
[3] School of Computing, Tokyo Institute of Technology, 4259 Nagatsuta-cho, Midori-ku, Yokohama, Kanagawa 226-8503, Japan

E-mails: takuma.kasai@riken.jp, kigawa@riken.jp

Abstract. Signal assignment is a mandatory step for the site-specific analyses of proteins by NMR. In challenging cases, such as low solubility target proteins or in-cell NMR, amino-acid selective stable isotope labeling facilitates the main-chain assignment, since it can determine the amino-acid type of each signal, thus providing complementary information to that obtained by standard three-dimensional triple resonance experiments. Inspired by information science methodologies, we developed a combinatorial selective labeling strategy, named stable isotope encoding (SiCode), to enable amino-acid typing even from noisy NMR spectra, with small numbers of labeled samples. The high noise-resistance of SiCode is achieved by both optimization of the isotope labeling pattern according to the information distance between amino acids and model-based computational retrieval of the amino-acid information from spectra, using replica exchange Monte Carlo computation. We demonstrate amino-acid typing with simulated low signal-to-noise-ratio spectra, with sample concentrations as low as micromolar order. Since the main-chain signal assignment is often a bottleneck process for various amide-signal-based NMR analyses of challenging target proteins, SiCode will facilitate and accelerate such analyses.

1. Introduction

Nuclear magnetic resonance (NMR) is a common method for analyses of protein structures, dynamics, and interactions at atomic resolution. The protein sample is usually stable-isotope labeled, to obtain rich information about the targets using the NMR-detectable nuclei, ^{1}H, ^{13}C, and ^{15}N, and to perform various measurements that rely on stable isotope enrichment. Two-dimensional ^{1}H-^{15}N spectroscopy is one of the basic methods for protein analyses. It uses amide moieties as probes to obtain residue-specific information about the target protein. For example, it includes chemical shift perturbation, transferred cross saturation, ^{15}N relaxation, and paramagnetic effects. To perform such analyses, the assignment of the amide signals is required beforehand. The sequential assignment technique with triple resonance experiments is the standard procedure to perform the assignment [1, 2]. However, under challenging conditions, this is difficult to accomplish because some of the triple resonance experiments are insensitive. For example, an insufficient protein concentration caused by low

[4] To whom any correspondence should be addressed.

solubility or productivity of the protein leads to a low signal-to-noise ratio (SNR) of the spectra. The SNR is further reduced for large proteins, aggregated proteins, and/or proteins in crowded conditions, such as cellular environments, due to the relaxation enhancement of the magnetizations.

Amino-acid selective stable isotope labeling provides useful information for assignments even under such challenging conditions [3], since it enables amino-acid typing of each amide signal. For single selective labeling, the amino-acid typing of the residue can be performed by acquiring the standard ^{1}H-^{15}N correlation spectrum, such as ^{15}N-HSQC, with 19 labeled samples corresponding to all amino acids except for proline. For dual selective labeling, each sample is both ^{15}N-labeled with a specific amino acid and ^{13}C-labeled with another specific amino acid [4]. With the two-dimensional HN(CO) measurement, which is the most sensitive method among the triple resonance experiments, one can perform amino-acid typing of the preceding residue (residue i-1) as well as the residue including the amide (residue i), since the HN(CO) signal is only observed if both the carbonyl carbon at residue i-1 and the amide nitrogen at residue i are isotopically labeled. This greatly narrows down the assignment possibilities. For instance, if an amino-acid pair that appears only once in the amino acid sequence is labeled, then only a single peak is observed, and thus the corresponding amino-acid residue is assigned without further NMR experiments [5]. However, for the amino-acid typing of all of the signals, dual selective labeling requires tens or hundreds of labeled samples, corresponding to the number of amino acid pairs appearing in the sequence.

To reduce the required number of labeled samples, various combinatorial selective labeling methods have been developed [6], in which each amino acid corresponds to a combination of labeled samples, rather than a single sample. The methods were designed for various objectives, including the discrimination of 16 amino acids with only 5 samples [7]; the discrimination of amino acids that have close chemical shifts at the alpha and beta carbons [8, 9]; the design of a labeling pattern specialized for a specific amino-acid sequence to achieve maximal unique assignments [10-14]; the use of the alpha carbon as well as the carbonyl carbon to discriminate amino acids with various combinations of triple resonance experiments [15, 16]; the avoidance of signal overlap considering amino-acid frequency [17-19]; the labeling of a specific amino-acid set for the assignment of the transmembrane region of membrane proteins [16, 20, 21]; and the selective inverse labeling, or selective unlabeling, which is compatible with various protein production methods, including *in vivo* expression [9, 22].

In this paper, we describe a selective labeling strategy, facilitated by information science and computational analysis techniques, which enables amino-acid typing with a small number of labeled samples or with low-SNR spectra.

2. Stable isotope encoding (SiCode)

We developed a selective labeling strategy, named stable isotope encoding (SiCode), based on the unique point-of-view that the amino-acid information is encoded in labeling patterns and decoded from observed NMR spectra, by analogy to digital communications (figure 1a) [23]. To reduce the required number of labeled samples based on this point-of-view, the information content (entropy) per labeled sample should be increased. Therefore we introduced fractional stable isotope labeling to achieve ternary labeling levels; namely, 50%, 75%, or 100% for ^{15}N and 0%, 50%, or 100% for ^{13}C, while the other combinatorial labeling methods used binary labeling levels. An example of the labeling pattern is shown in figure 1b. Using this labeling scheme, we successfully discriminated all 19 non-proline amino acids with as few as 3 labeled samples [23].

The main-chain assignment of easy targets, such as small, highly-soluble, globular, and mono-dispersed proteins, can be achieved solely with the sequential assignment method. As mentioned, various selective labeling methods are used when the sequential assignment is difficult to accomplish. Therefore, selective labeling methods are needed to deal with the low-SNR spectra of challenging targets. To achieve the high noise-tolerance of SiCode, we improved both the encoding and decoding processes, thanks to SiCode's inherent affinity to information science methodologies.

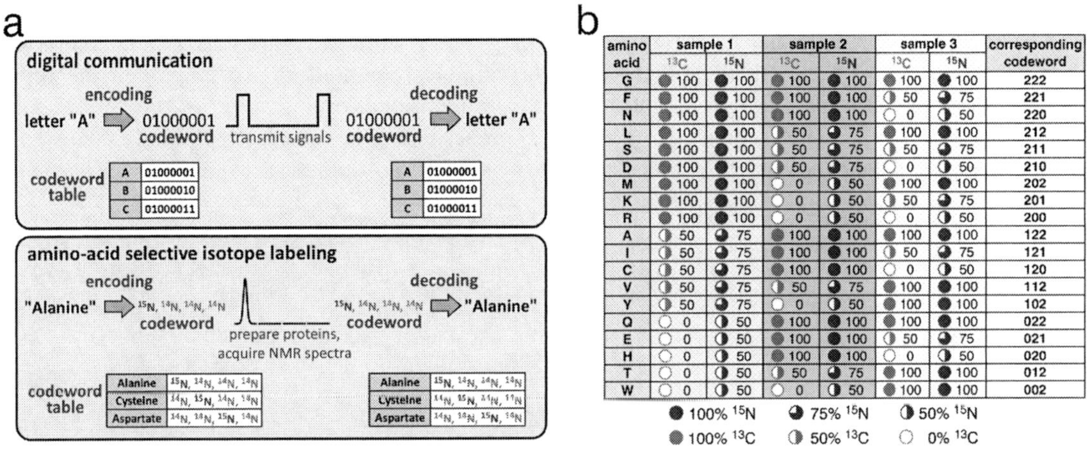

Figure 1. Stable isotope encoding (SiCode). (a) The basic concept of SiCode. In digital communication, a sender who wishes to send a message, in this case the letter “A”, converts the message to binary digits with a predefined codeword table. The binary digits are transmitted to a receiver, who converts them back to a letter using the identical codeword table. Amino-acid selective isotope labeling can be regarded as a similar process. In this case, the message is an amino acid type, and the codeword is an isotope labeling pattern. A receiver can retrieve the amino-acid information from the acquired NMR spectra. (b) An example of the labeling pattern of SiCode, which enables the discrimination of 19 amino acids with only 3 samples. The ^{13}C and ^{15}N labeling ratios are shown as percentages. The ternary digits that correspond to each amino acid are shown in the right column. Figures are reprinted from ref. [23].

For encoding, we introduced an optimization of the isotope labeling pattern [23]. In general, a larger information distance between two amino acids reduces misjudgements between the amino acids under noisy circumstances. Thus, according to the coding theory, we defined the minimal amino acid distance among all combinations of amino acids. The minimal distance should be maximized to obtain a “good” labeling pattern, which enables accurate amino acid discrimination even with noisy NMR spectra. We used the Euclidian distance as the information distance, because the thermal Gaussian noise is dominant for low-SNR NMR spectra [23]. We achieved the global optimization with the simulated annealing computation for the given number of amino-acids and the given number of labeled samples [23]. For example, figure 2a shows the optimized labeling pattern to discriminate all 19 non-proline amino acids with 3 labeled samples. This pattern is identical to the pattern shown in figure 1b, indicating that the pattern using the ternary three digits was the best for 19 amino acids. However, as shown in figure 2b, depending on the combination of the number of amino acids and the number of labeled samples, we can see various fractional labeling levels, such as 79.0% or 50.4%, in the optimized labeling patterns. To achieve such complicated labeling patterns, we utilized a cell-free protein synthesis system with suppressed amino-acid metabolism [24-26].

For decoding, we introduced a model-based analysis to obtain information from the low-SNR spectra [27]. Previously, we analyzed spectra based on the peak intensity; namely, at first we obtain the peak intensities from the spectra, then back-calculate the labeling ratio, and finally convert it to the amino-acid information by finding the nearest one in the labeling pattern table [23]. Obviously, this approach depends on the accuracy of the peak intensity, which may be reduced when the SNR of the spectra is low and/or the signal overlaps with another signal. In contrast, the model-based approach does not begin with obtaining the peak intensity. In this approach, we first make a list of all possible models; i.e., amino acid pairs representing residues i-1 and i, from the known amino-acid sequence of the target protein. We can calculate the intensity pattern for each model, using the known labeling pattern. Subsequently, to obtain the amino-acid information, we fit the intensity patterns to the

IOP Conf. Series: Journal of Physics: Conf. Series **1036** (2018) 012007 doi:10.1088/1742-6596/1036/1/012007

observed spectra and select the best-fitting model with the minimal residual error. As we use the model function with explanatory variables, including the continuous variables (signal amplitude, chemical shifts, and peak widths) and the discrete variable (the model), we employed the replica exchange Monte Carlo (REMC) computation to achieve the global optimization of the variables in a high-dimensional parameter space [27]. This approach can be used for simultaneous peak deconvolution and analysis, by fitting the spectra to the sum of multiple model functions [27].

a

	sample 1		sample 2		sample 3	
AA	^{13}C	^{15}N	^{13}C	^{15}N	^{13}C	^{15}N
1	100.0	100.0	100.0	100.0	100.0	100.0
2	100.0	100.0	100.0	100.0	50.0	75.0
3	100.0	100.0	100.0	100.0	0.0	50.0
4	100.0	100.0	50.0	75.0	100.0	100.0
5	100.0	100.0	50.0	75.0	50.0	75.0
6	100.0	100.0	50.0	75.0	0.0	50.0
7	100.0	100.0	0.0	50.0	100.0	100.0
8	100.0	100.0	0.0	50.0	50.0	75.0
9	100.0	100.0	0.0	50.0	0.0	50.0
10	50.0	75.0	100.0	100.0	100.0	100.0
11	50.0	75.0	100.0	100.0	50.0	75.0
12	50.0	75.0	100.0	100.0	0.0	50.0
13	50.0	75.0	50.0	75.0	100.0	100.0
14	50.0	75.0	0.0	50.0	100.0	100.0
15	0.0	50.0	100.0	100.0	100.0	100.0
16	0.0	50.0	100.0	100.0	50.0	75.0
17	0.0	50.0	100.0	100.0	0.0	50.0
18	0.0	50.0	50.0	75.0	100.0	100.0
19	0.0	50.0	0.0	50.0	100.0	100.0

b

		sample 1		sample 2		sample 3		sample 4	
AA		^{13}C	^{15}N	^{13}C	^{15}N	^{13}C	^{15}N	^{13}C	^{15}N
1	G	100.0	100.0	100.0	100.0	100.0	100.0	100.0	100.0
2	K	100.0	100.0	100.0	100.0	100.0	100.0	0.0	50.0
3	Q	100.0	100.0	79.0	89.5	0.0	50.0	100.0	100.0
4	L	0.0	50.0	100.0	100.0	79.0	89.5	100.0	100.0
5	I	79.0	89.5	0.0	50.0	100.0	100.0	100.0	100.0
6	T	100.0	100.0	37.0	68.5	50.4	75.2	51.8	75.9
7	M	50.4	75.2	100.0	100.0	37.0	68.5	51.8	75.9
8	Y	37.0	68.5	50.4	75.2	100.0	100.0	51.8	75.9
9	E	100.0	100.0	81.2	90.6	0.0	50.0	0.0	50.0
10	A	0.0	50.0	100.0	100.0	81.2	90.6	0.0	50.0
11	R	81.2	90.6	0.0	50.0	100.0	100.0	0.0	50.0
12	S	81.2	90.6	0.0	50.0	0.0	50.0	100.0	100.0
13	D	0.0	50.0	81.2	90.6	0.0	50.0	100.0	100.0
14	N	0.0	50.0	0.0	50.0	81.2	90.6	100.0	100.0
15	V	100.0	100.0	0.0	50.0	0.0	50.0	0.0	50.0
16	F	0.0	50.0	100.0	100.0	0.0	50.0	0.0	50.0
17	P	0.0	0.0	0.0	0.0	100.0	100.0	0.0	0.0
18	H	0.0	50.0	0.0	50.0	0.0	50.0	100.0	100.0

Figure 2. Optimized labeling pattern for (a) 19 amino-acids (AAs) with 3 labeled samples and (b) 18 amino-acids with 4 labeled samples. The ^{13}C and ^{15}N labeling ratios are shown as percentages. Amino acids are shown as numbers, because there are numerous possibilities for the assignment of amino acids to codewords, although there are some technical considerations [23]. The amino acid assignments for the Ubiquitin mutant protein used in this work are shown in the second column of (b).

3. Evaluation with experimental data and low-SNR simulation data

3.1. High-SNR experimental data

At first, we tried amino-acid typing with the high-SNR spectra of 1 mM Ubiquitin mutant protein. We adopted the labeling pattern for 18 amino acids with the 4 samples shown in figure 2b, because the protein lacks cysteine and tryptophan residues. Ubiquitin mutant protein is 84 amino-acids long and contains 3 proline residues. Four main-chain signals (S-5 in the N-terminal linker region, F4, E24, and G53) were not observed under our experimental conditions. In total, 76 main-chain signals were observed. We picked 94 signals from the HSQC spectrum, which included 75 main-chain amide signals and 19 side-chain amide signals (arginine, asparagine, and glutamine). One of the picked main-chain signals was actually two overlapped signals, G10 and T22. Excluding this overlapped signal, correct amino-acid typing was achieved for all of the other 74 signals at both residues i-1 and i (table 1). Assuming that we could determine the overlapping by other experiments, such as three-dimensional triple resonance experiments, we successfully deconvoluted and correctly obtained the amino-acid information (table 1).

3.2. Low-SNR simulation data

To simulate low-SNR data, such as those obtained from low-solubility proteins, we added artificial noise to the experimental data. The noise amplitude was adjusted so that the final noise level in the spectra was 100-fold (for simulation 1) or 200-fold (for simulation 2), as compared to the original experimental spectra. Therefore, these data sets emulate measurements with 10 μM and 5 μM protein solutions for simulations 1 and 2, respectively. For simulation 1, the accuracy of the amino-acid typing was the same as that obtained with the experimental data (table 1), suggesting that SiCode can deal with proteins with solubilities as low as 10 μM with sufficient reliability. For simulation 2, the accuracy dropped down close to 90%, regardless of the overlapped signal exclusion (table 1 and figure 3).

Table 1. Accuracy of amino-acid typing for the main-chain signals of Ubiquitin mutant protein[a].

Signals	Criteria of correctness	Experimental (1 mM)	Simulation 1 (10 μM equivalent)	Simulation 2 (5 μM equivalent)
Excluding overlap[b]	residues i-1, i[d]	74 (100%)	74 (100%)	69 (93%)
	residue i[e]	74 (100%)	74 (100%)	70 (95%)
Including overlap[c]	residues i-1, i[d]	76 (100%)	76 (100%)	69 (91%)
	residue i[e]	76 (100%)	76 (100%)	70 (92%)

[a] The numbers of correct signals are shown. The percentages of correct signals are shown in parentheses.
[b] Target signals are 74 observed main-chain signals, excluding the overlapped G10 and T22.
[c] Target signals are all of the 76 observed main-chain signals.
[d] Signals with correct amino-acid types at both residues i-1 and i are counted.
[e] Signals with the correct amino-acid type at residue i are counted.

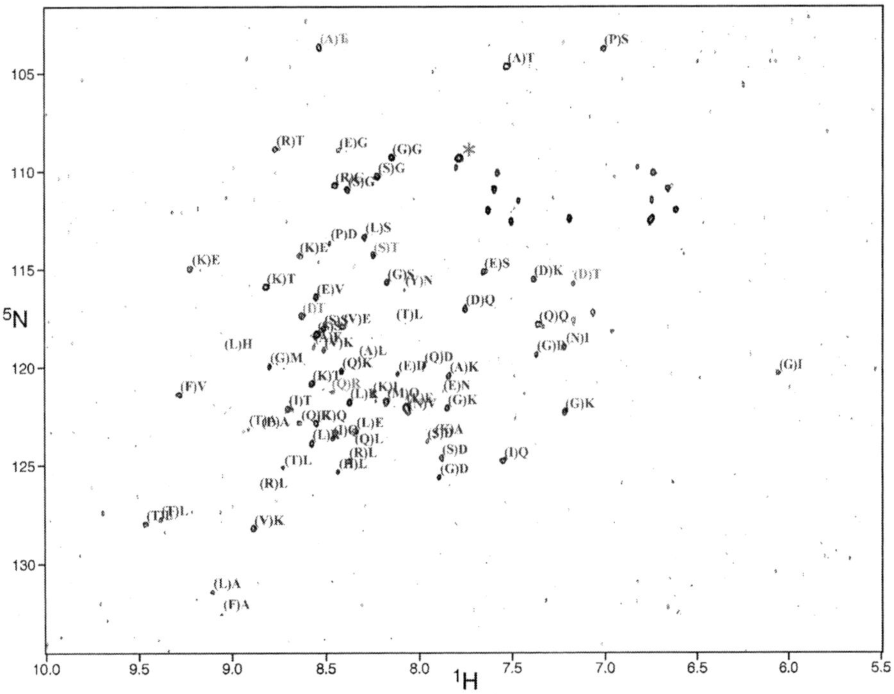

Figure 3. Amino acid typing with the "simulation 2" data. The results are shown on the ^{15}N-HSQC spectra of sample 1. Letters inside and outside the parentheses indicate the decoded amino acid information of residues i-1 and i, respectively. Blue letters indicate correct typing at both residues i-1 and i. Magenta letters indicate correct at residue i, but incorrect at residue i-1. Red letters indicate incorrect at residue i. The red asterisk denotes the overlapped signals of G10 and T22.

Figure 4 shows some examples of model fitting. Since the (N)I61 signal (the amide signal of the residue I61, which is preceded by N60) from the experimental 1 mM data has a sufficiently high SNR (figure 4a), the correct decoding is easy. In figure 4b, the same signal from "simulation 2" data (equivalent to 5 μM) is obscured by noise. In this case, we are unable to obtain the precise signal intensity from each spectrum. However, we successfully obtained the correct amino-acid information by the model-based fitting, thanks to the optimized labeling pattern to secure sufficient information distances between amino acids. The high-SNR (D)Y59 signal in the experimental spectra (figure 4c) is similarly buried in the noise in the "simulation 2" data (figure 4d). In this case, unfortunately, the best fit model to explain the spectra is (D)T, rather than the correct (D)Y. At the sampling steps of the lowest temperature replica of REMC, 45.4% (D)T, 37.2 % (D)Y, and 17.4% (D)G were sampled, showing that the correct model was the second likeliest candidate. For "simulation 2", the decoding was not perfect but still maintained high accuracy (over 90%). The sampled frequency of the model candidates also includes information that is helpful for the assignments to some extent, depending on the purpose of using SiCode.

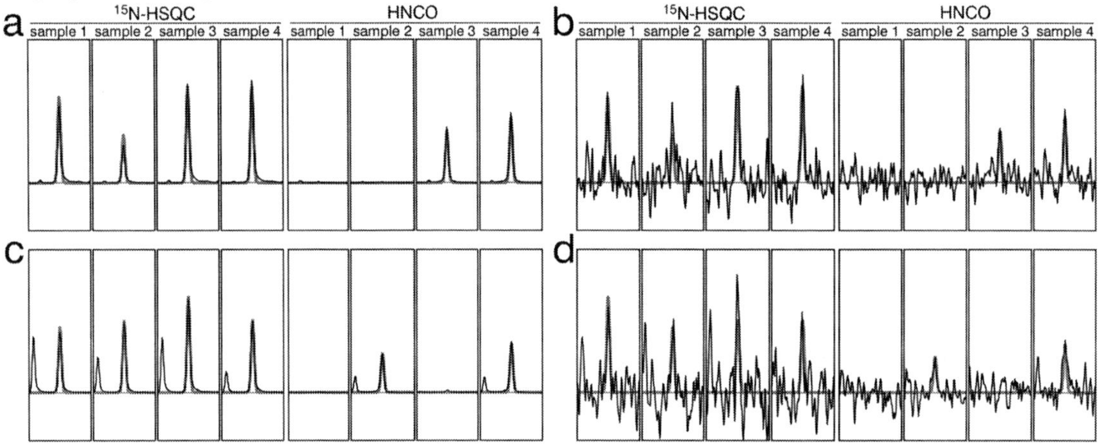

Figure 4. Spectral fitting for model-based decoding. Cross sections of the I61 (a and b) and Y59 (c and d) signals from the experimental spectra (a and c) and the "simulation 2" spectra (b and d) are shown. Black and red lines are observed spectra and fitted functions, respectively.

4. Conclusion

In this report, we described a selective stable isotope labeling strategy, SiCode, which is based on the concept that the selective labeling can be regarded as "encoding and decoding" processes. Improvements in both the encoding and decoding processes, inspired by information science methodologies, increased the robustness of amino-acid discrimination against noise. We demonstrated the noise tolerance of SiCode with the simulated low-SNR spectra corresponding to that of a low-solubility (5 to 10 μM) target protein. The total NMR measurement time for this demonstration was 56 hours (2.3 days). As we expect that N-fold lower concentration samples with N^2-fold longer experiment times will result in the same SNR, amino acid typing of even a several micromolar protein solution can be accomplished in a reasonable amount of time. Increasing the sample number above 4 may further improve the accuracy of amino acid discrimination, even with a fixed total NMR measurement time [23]. Therefore, we conclude that SiCode will facilitate the NMR analyses of low-solubility proteins.

In addition to low-solubility proteins, in-cell NMR is another applicable target of SiCode. In in-cell NMR, the SNR is much worse than that in *in vitro* NMR, due to the insufficient cellular protein concentration, the relaxation enhancement caused by macromolecular crowding, the heterogeneity of the sample, and the limited measurement time to ensure cell viability. These are the reasons why the successful reports of main-chain signal assignment solely with in-cell NMR spectra have been limited to a few small proteins so far [28-31]. The sequential assignment method using three-dimensional triple resonance measurements offers the information of connectivity to the neighboring residue and of the amino-acid type. However by the relaxation enhancement, the connectivity information is often lost due to insensitivity of some spectra. The information of amino-acid type is not complete because it relies on the chemical shift values of α and β carbons, which are degenerated in some amino acids. In contrast, the amino acid information of SiCode is derived not from the chemical shift values but from the isotope labeling pattern. In consequence, the information from the three-dimensional tripe resonance measurements and SiCode are complementary. Therefore we are sure that SiCode will also facilitate analyses of in-cell NMR spectra by the amino-acid type information.

In contrast to the other combinatorial selective labeling methods, SiCode spectra require computational interpretation, because the amino acid information represented by the signal intensities is difficult to analyze visually. Such computation actually enabled the interpretation of low-SNR spectra, by accumulating information about coincidences between the observed intensities and the model intensities among all of the samples. With various computational methodologies, SiCode will expand the applicable targets and situations of protein NMR analyses.

5. Materials and methods

The amino-acid sequence of the human Ubiquitin mutant protein used in this work is the same as residues 1 to 76 of UniProt P62979, except for three mutations (L8A, I44A, and V70A) and additional residues (GSSGSSG at the N-terminus and D at the C-terminus). The residue numbers of the N-terminal additional residues are indicated with the non-positive numbers in the main text. Four samples were prepared according to the pattern shown in figure 2b, with the cell-free protein synthesis system [23-26]. To achieve the desired labeling ratios, we used mixtures of different isotope-enriched amino acids. For example, amino acid I (isoleucine) of sample 1 showed on figure 2b should be 79.0% ^{13}C- and 89.5% ^{15}N-labeled. In this case we used a mixture of ^{13}C/^{15}N-enriched isoleucine, ^{15}N-enriched isoleucine, and naturally abundant isoleucine in the ratio of 79.0%, 10.5%, and 10.5%. In the cell-free protein synthesis reaction, these three types of isoleucine are randomly incorporated to the protein, resulting in the desired isotopically labeled sample. The synthesized proteins were purified by chromatography on a HisTrap affinity column (GE Healthcare) and a HiTrap SP cation exchange column (GE Healthcare). The purified proteins were concentrated to 1 mM solutions in NMR buffer (50 mM sodium phosphate buffer, pH 6.0, 10% ^2H_2O) with VIVASPIN 2 MWCO 5,000 concentrators (Sartorius). Two-dimensional ^{15}N-HSQC and HN(CO) spectra were measured with a 700 MHz Avance III spectrometer (Bruker) equipped with a TCI cryoprobe (Bruker) at 298 K. The spectral width and the time domain data points of the ^{15}N dimension were 33.0 ppm and 128 complex points, respectively, for both experiments. The numbers of scans were 8 and 32 for ^{15}N-HSQC and HN(CO), respectively. To simulate low-SNR spectra, random Gaussian white noise was added to each data point in the raw time-domain data. The noise level of the data was estimated from regions without signals in the processed frequency-domain spectra [27]. The amino-acid information was decoded from the spectra, as described previously [23, 27]. For evaluation purposes, the main-chain assignment was independently performed with the standard triple resonance experiments, using a uniformly ^{13}C/^{15}N-labeled sample.

Acknowledgements
We thank the lab members at RIKEN for assistance with sample preparation and for valuable comments. We also thank Dr. Kenji Nagata, Prof. Masato Okada, Prof. Toshiyuki Tanaka, and Prof. Shiro Ikeda for helpful discussions. This work was financially supported in part by JSPS KAKENHI (16K18508, 26650027, 25120003), JST PRESTO (JPMJPR1674), and JST CREST (JPMJCR13M3).

References

[1] Ikura M Kay L E and Bax A 1990 *Biochemistry* **29** 4659–67
[2] Kay L E, Ikura M, Tschudin R and Bax A 1990 *J. Magn. Reson.* **89** 496–514

[3] Hembram D S, Haremaki T, Hamatsu J, Inoue J, Kamoshida H, Ikeya T, Mishima M, Mikawa T, Hayashi N, Shirakawa M and Ito Y 2013 *Biochem. Biophys. Res. Commun.* **438** 653-9
[4] Kainosho M and Tsuji T 1982 *Biochemistry* **21** 6273–9
[5] Yabuki T, Kigawa T, Dohmae N, Takio K, Terada T, Ito Y, Laue E D, Cooper J A, Kainosho M and Yokoyama S 1998 *J. Biomol. NMR* **11** 295–306
[6] Jaipuria G, Krishnarjuna B, Mondal S, Dubey A and Atreya H S 2012 *Adv. Exp. Med. Biol.* **992** 95–118
[7] Parker M J, Aulton-Jones M, Hounslow A M and Craven C J 2004 *J. Am. Chem. Soc.* **126** 5020–1
[8] Shi J, Pelton J G, Cho H S and Wemmer D E 2004 *J. Biomol. NMR* **28** 235–47
[9] Krishnarjuna B, Jaipuria G, Thakur A, D'Silva P and Atreya H S 2011 *J. Biomol. NMR* **49** 39–51
[10] Hefke F, Bagaria A, Reckel S, Ullrich S J, Dötsch V, Glaubitz C and Güntert P 2011 *J. Biomol. NMR* **49** 75–84
[11] Maslennikov I and Choe S 2013 *Curr. Opin. Struct. Biol.* **23** 555–62
[12] Maslennikov I, Klammt C, Hwang E, Kefala G, Okamura M, Esquivies L, Mörs K, Glaubitz C, Kwiatkowski W, Jeon Y H and Choe S 2010 *Proc. Natl. Acad. Sci. USA* **107** 10902–7
[13] Staunton D, Schlinkert R, Zanetti G, Colebrook S A and Campbell I D 2006 *Magn. Reson. Chem.* **44** S2–9
[14] Xun Y, Tremouilhac P, Carraher C, Gelhaus C, Ozawa K, Otting G, Dixon N E, Leippe M, Grotzinger J, Dingley A J and Kralicek A V 2009 *Protein Expr. Purif.* **68** 22–7
[15] Hein C, Löhr F, Schwarz D and Dötsch V 2017 *Biopolymers* **107** e23013
[16] Löhr F, Reckel S, Karbyshev M, Connolly P J, Abdul-Manan N, Bernhard F, Moore J M and Dötsch V 2012 *J. Biomol. NMR* **52** 197–210
[17] Sweredoski M J, Donovan K J, Nguyen B D, Shaka A J and Baldi P 2007 *Bioinformatics* **23** 2829–35
[18] Ozawa K, Wu P S, Dixon N E and Otting G 2006 *FEBS J.* **273** 4154–9
[19] Wu P S, Ozawa K, Jergic S, Su X C, Dixon N E and Otting G 2006 *J. Biomol. NMR* **34** 13–21
[20] Sobhanifar S, Reckel S, Junge F, Schwarz D, Kai L, Karbyshev M, Löhr F, Bernhard F and Dötsch V 2010 *J. Biomol. NMR* **46** 33–43
[21] Trbovic N, Klammt C, Koglin A, Löhr F, Bernhard F and Dötsch V 2005 *J. Am. Chem. Soc.* **127** 13504–5
[22] Hiroaki H, Umetsu Y, Nabeshima Y, Hoshi M and Kohda D 2011 *J. Struct. Funct. Genomics* **12** 167–74
[23] Kasai T, Koshiba S, Yokoyama J and Kigawa T 2015 *J. Biomol. NMR* **63** 213–21
[24] Kigawa T, Yabuki T, Yoshida Y, Tsutsui M, Ito Y, Shibata T and Yokoyama S 1999 *FEBS Lett.* **442** 15–9
[25] Kigawa T, Yabuki T, Matsuda N, Matsuda T, Nakajima R, Tanaka A and Yokoyama S 2004 *J. Struct. Funct. Genomics* **5** 63–8
[26] Yokoyama J, Matsuda T, Koshiba S, Tochio N and Kigawa T 2011 *Anal. Biochem.* **411** 223-9
[27] Kasai T, Nagata K, Okada M and Kigawa T 2016 *J. Phys. Conf. Ser.* **699** 012003
[28] Reardon P N and Spicer L D 2005 *J. Am. Chem. Soc.* **127** 10848–9
[29] Sakakibara D, Sasaki A, Ikeya T, Hamatsu J, Hanashima T, Mishima M, Yoshimasu M, Hayashi N, Mikawa T, Wälchli M, Smith B O, Shirakawa M, Güntert P and Ito Y 2009 *Nature* **458** 102–5
[30] Hamatsu J, O'Donovan D, Tanaka T, Shirai T, Hourai Y, Mikawa T, Ikeya T, Mishima M, Boucher W, Smith B O, Laue E D, Shirakawa M and Ito Y 2013 *J. Am. Chem. Soc.* **135** 1688–91
[31] Ikeya T, Hanashima T, Hosoya S, Shimazaki M, Ikeda S, Mishima M, Güntert P and Ito Y 2016 *Sci. Rep.* **6** 38312

IOP Conf. Series: Journal of Physics: Conf. Series **1036** (2018) 012008 doi:10.1088/1742-6596/1036/1/012008

Fully automated assignment of methyl resonances of a 36 kDa protein dimer from sparse NOESY data

I Pritišanac*[1§], JM Würz[1§] & P Güntert*[1,2,3]

[1]Institute of Biophysical Chemistry, Center for Biomolecular Magnetic Resonance, Goethe University Frankfurt am Main, 60438 Frankfurt am Main, Germany

[2]Laboratory of Physical Chemistry, ETH Zürich, 8093 Zürich, Switzerland

[3]Graduate School of Science, Tokyo Metropolitan University, Hachioji, Tokyo 192-0397, Japan

*E-mail: pritisanac@em.uni-frankfurt.de, guentert@em.uni-frankfurt.de

Abstract. High-resolution solution-state NMR spectroscopy studies of large proteins typically require uniform deuteration of the system and selective protonation and isotope labelling of methyl groups. Under such circumstances, the assignment of methyl resonances presents a considerable experimental challenge and automation of the process using computational algorithms has been actively sought. Through-space connectivities between the labelled methyl groups can be established through nuclear Overhauser enhancement spectroscopy (NOESY). If a high-resolution structure of the system is available, the sparse connectivity restraints derived from this information enable structure-based methyl resonance assignment. Here, we outline a protocol for full automation of the methyl resonance assignment process using the CYANA software package. We tested the protocol on three-dimensional (3D) $^{13}C/^{13}C$-separated NOESY spectra of a dimer of regulatory chains of aspartate transcarbamoylase (ATCase-r_2). We used CYPICK to detect NOE signals, followed by automatic resonance assignment with FLYA. On this dataset, FLYA generated highly similar results using either automatically or manually generated peak lists, confidently assigning ~60% of the methyl groups with high accuracy (95 ± 2% correctness). We compared this performance to two alternative automatic methyl assignment protocols, MAP-XSII and FLAMEnGO2.0, both of which, similarly to FLYA, support unassigned NOESY peak lists as input.

[§]contributed equally

1. Introduction

High-resolution nuclear magnetic resonance (NMR) spectroscopy studies of biological macromolecules occupy a central role in biophysical chemistry, structural biology, drug design, structural genomics, and interactomics [1]. Hydrogen (^{1}H, or proton) is highly abundant in biological macromolecules and, owing to its high gyromagnetic ratio, is the most sensitive NMR probe of molecular structure and dynamics. Studies of complex biomolecules are aided by isotopic labelling that introduces NMR-active atomic nuclei to a system [2]. For instance, in studies of small and medium-size protein molecules, uniform labeling with carbon-13 (^{13}C) and nitrogen-15 (^{15}N) isotopes replaces NMR-inactive ^{12}C and ^{14}N nuclei. The *J*-coupling interactions between covalently linked ^{1}H, ^{13}C and ^{15}N nuclei are routinely exploited to establish atomic connectivity along the protein backbone and side chains [3,4].

A high density of NMR-active nuclei, especially protons, in biomolecular samples is a source of nuclear relaxation, which leads to rapid signal decays. Consequently, line broadening is a hallmark of high molecular weight (>30 kDa) protein NMR spectra [5]. To simultaneously optimize resolution and sensitivity for larger molecules, partial or complete deuteration (^{2}H) of samples can be employed, followed by back-introduction of solvent exchangeable ^{1}H nuclei [6]. However, as the molecular mass of proteins or protein complexes increases, additional considerations must be given to isotopic labelling. For instance, deuteration can be combined with ^{1}H, ^{13}C-labelling of only some amino acids or side chains. A particularly successful strategy is selective protonation of ^{13}C-labeled methyl groups ($^{13}CH_3$) in otherwise ^{12}C-labeled, uniformly deuterated proteins. Cost-effective and robust biosynthetic strategies have been established for selective or simultaneous labelling of all methyl-containing amino acids in *Escherichia coli* [7,8], and of isoleucine-δ1 methyls in a eukaryote, *Pichia pastoris* [9]. One additional advantage of methyl groups is their high abundance in protein sequences, as six out of 20 amino acids have at least one methyl group in their side chain. Moreover, methyls are found both in the core and, to a lesser extent, at the surface of proteins, making them particularly useful probes of protein structure and dynamics [10,11].

Cross-correlated relaxation effects in $^{13}CH_3$ spin systems in the macromolecular limit can be exploited in heteronuclear multiple quantum correlation (HMQC) spectroscopy, offering considerable improvements in spectral resolution. Hence, the measurement of [^{1}H,^{13}C]-HMQC spectra is referred to as methyl transverse relaxation-optimized spectroscopy (methyl-TROSY) [12]. Combined with the three-fold degeneracy of methyl ^{1}H spins, spectra can be recorded with intense and sharp resonances that are well dispersed in two-dimensional ^{1}H–^{13}C correlation plots. Taken together, the combination of isotopic labelling and advantageous spectroscopic properties of methyl groups now allows atomic-level insight into the structures and dynamics of large proteins and protein oligomers up to 1 MDa in molecular mass [13–17].

The assignment of methyl resonances is an essential prerequisite for the interpretation of structural and dynamical information derived from methyl-TROSY studies. Resonance assignment refers to the attribution of ^{1}H-^{13}C correlations from [^{1}H,^{13}C]-HMQC spectra to individual methyl-bearing residues in the protein sequence. The assignment step in NMR studies can be particularly laborious and time-consuming, especially with increasing complexity of the investigated systems [17,18]. The aforementioned line-broadening and extensive overlap of backbone chemical shifts in spectra of large proteins prevent standard through-bond NMR assignment strategies. Instead, a typically employed experimental strategy for large proteins is fragmentation of the system into smaller components for which through-bond NMR spectra of sufficient quality can be obtained [14,15,19]. Assignment of methyls is then obtained using standard experiments that correlate NMR backbone and side-chain resonances [4]. This strategy can be supplemented with site-directed mutagenesis of individual methyl-bearing residues [14,17]. Although highly reliable, these approaches are laborious and come with the disadvantage that any changes introduced to the protein for assignment purposes may result in considerable shifts of correlations in [^{1}H,^{13}C]-HMQC spectra. This may compromise the reliable transfer of assignments between the spectra of protein fragments/mutants and of the intact protein.

Alternatively, if a high-resolution structure or a homology model of the protein is available, automatic, structure-based NMR resonance assignment strategies can be considered [20–25]. In a methyl-TROSY study, isotopically labelled $^{13}CH_3$ groups are the only available source of connectivity information, and therefore, automatic assignment strategies can only make use of highly sparse and ambiguous experimental information. Currently available automatic methyl assignment approaches make use of short- and long-range distance restraints, which are obtained by nuclear Overhauser enhancement spectroscopy (NOESY) and/or measurements of paramagnetic relaxation enhancements (PREs). Both of these restraints can provide inter-methyl distance information, with PREs yielding inter-methyl distances up to 25-35 Å [20], depending on the paramagnetic tag employed, and NOEs reported on distances up to 12 Å in highly deuterated systems [26]. Importantly, NOEs do not require mutagenesis to introduce a paramagnetic tag and can reveal a network of connectivity between spatially proximal methyl groups [25].

Presently available automatic methyl resonance assignment approaches utilize these through-space restraints in different ways. An approach introduced by Venditti *et al.* [20] uses experimental ^{1}H-methyl PRE rates as primary restraints. A Metropolis Monte Carlo algorithm is employed to find assignment solutions that minimize the difference between measured PRE rates and those expected from a high-resolution protein structure. Measured chemical shifts and NOEs provide additional restraints to refine assignments. Two other programs, Methyl Assignment Prediction from X-ray Structures (MAP-XS) [21,22] and Fuzzy Logic Assignment of Methyl Groups (FLAMEnGO) [23,24] also employ Metropolis Monte Carlo optimization, but rely primarily on methyl-methyl NOEs as restraints. Both programs evaluate matches between structure-based predictions and experimentally measured NOEs and methyl chemical shifts. Additional restraint-specific terms are defined for optional incorporation of PREs, pseudocontact shifts (PCS), residual dipolar couplings (RDC), and TOCSY data in the search [21–24]. Most recently, Methyl Assignment by Graph MAtching (MAGMA) was introduced, which compares graphs derived from methyl-methyl NOEs and a protein structure using a combination of exact algorithms for graph-subgraph monomorphism and maximal common edge subgraph problems [25]. To treat the high sparsity of input restraints, MAGMA exhaustively samples all theoretically possible methyl assignment solutions that maximize the number of explained methyl-methyl NOEs. Relying solely on NOE restraints, MAGMA could achieve perfectly accurate methyl resonance assignments on a range of protein targets of various molecular sizes and shapes [25].

All currently available automatic methyl assignment strategies rely on the identification of methyl-methyl NOE cross-peaks by manual analysis of 3D or 4D ^{13}C-resolved NOESY spectra. Peak picking produces NOESY peak lists, in which each peak position is defined by proton (^{1}H) and carbon (^{13}C) frequencies of the two spatially proximal methyl groups that give rise to an NOE. In the case of a 4D ^{13}C-NOESY peak list, ^{1}H and ^{13}C frequencies from both NOE-contributing methyls are known ($^{1}H_1$,$^{13}C_1$,$^{1}H_2$,$^{13}C_2$), whereas for a 3D ^{13}C-resolved NOESY list, either the ^{13}C or ^{1}H frequency of one of the methyls is unknown (e.g. $^{1}H_1$,$^{13}C_1$,$^{13}C_2$; Fig. 1). Manually prepared peak lists are typically curated by an experienced spectroscopist before being fed to the automatic assignment programs, which introduces a user bias to the process.

A missing link in the full automation of the methyl assignment using currently available protocols [22,24,25] is the automatic detection of methyl-methyl NOEs. Recently, a contour-based automatic signal detection protocol (CYPICK) was introduced and compared to other approaches [27]. CYPICK analyses the NMR spectrum in the form of two-dimensional contour plots, mimicking the manual approach taken by a spectroscopist as far as possible. The human visual inspection of the spectrum is replaced by the analysis of certain geometric contour line properties, i.e. local extremality, approximate circularity (after appropriate scaling of spectral axes), and convexity. CYPICK has been tested on several soluble proteins up to a molecular mass of ~20 kDa, and was shown to work optimally with the FLYA automatic resonance assignment protocol, performing particularly favourably on ^{13}C-edited and ^{15}N-edited NOESY data, as compared to other existing algorithms [27]. FLYA is a generic automatic resonance assignment protocol incorporated in CYANA that supports the input of NMR peak lists from a variety of through-bond (*J*-coupling), or through-space (NOESY) experiments for the assignment of protein backbone and side-chain resonances [28,29]. To optimize the match between expected and experimentally observed peaks, FLYA employs an evolutionary algorithm that operates on a population of assignment solutions (individuals) governed by two scoring functions that consider the fitness of the assignment of individual atoms (local) and of the complete assignment solution (global). FLYA supports structure-based resonance assignment on the basis of ^{13}C- and ^{15}N-resolved NOESY data [29]; however, it has not been tested on datasets originating from large, exclusively methyl-labelled proteins.

Here, we tested a fully automatic, user-unbiased protocol for methyl resonance assignment using CYANA. We used CYPICK to automatically detect NOE signals in a 3D $^{13}C/^{13}C$-separated NOESY spectrum of a 36 kDa homodimer of regulatory subunits of aspartate transcarbamoylase (ATCase-r_2). The CYPICK-generated NOESY peak lists were directly fed to the structure-based FLYA assignment protocol. We show that, on this dataset, FLYA achieves similar levels of assignment coverage (~60%) and accuracy (~95%) as with manually prepared NOESY peak lists, and that these results compare favorably to alternative methyl resonance assignment approaches.

2. Methods

Fig. 1 outlines the CYPICK-FLYA coupled automatic methyl resonance assignment strategy. The two protocols (CYPICK and FLYA) used in the study were not modified from their original implementations [27,28]. An alteration was introduced in the consolidation step of the FLYA assignment result analysis, as detailed in section 2.2.3 below.

2.1. CYPICK calculations

The noise level (L) of the spectrum was estimated automatically by CYPICK and multiplied by a user-defined baseline factor β for defining the intensity of the lowest contour line, B:

$$B = I_0 = \beta L.$$

The intensity I_i of contour line i is calculated from

$$I_i = B\gamma^i.$$

In our study, we used β values of 2, 3, and 5 while keeping γ fixed at 1.3, since moderate changes of γ did not show any significant influence on the resulting peak lists. The scaling factors for the spectral dimensions [27] were set to $\sigma_1 = 0.18$ ppm for the $^{13}C_1$ dimension, $\sigma_2 = 0.16$ ppm for $^{13}C_2$, and $\sigma_3 = 0.036$ ppm for $^1H_1$. The $^{13}C/^{13}C$-separated NOESY spectrum was picked using the recently implemented restricted peak picking mode of CYPICK [27]. A manually prepared 2D [^1H,^{13}C]-HMQC peak list was used as a frequency filter in CYPICK, restricting the peak picking of the $^{13}C/^{13}C$-separated NOESY spectrum to specific spectral regions. Only local maxima in the $^{13}C/^{13}C$-separated NOESY that were found within a given tolerance range of a manually identified 2D HMQC peak were further analyzed in the contour line analysis routine. This tolerance was set to 0.1 ppm for ^{13}C and 0.01 ppm for ^1H. Local maxima within the tolerance range that fulfilled the circularity and convexity criteria, as described in Würz *et al.* [27], were considered as peaks and stored in a peak list.

The peak picking performance was evaluated based on *find*, *artefact*, and *overall* scores with respect to a reference peak list. The definition of these scores is described in [27]. As a reference, we used a manually prepared $^{13}C/^{13}C$-separated NOESY peak list [25] with a tolerance of 0.04 ppm for ^1H and 0.4 ppm for heavy atoms. *Overall* scores were computed using an artefact weight of 0.2.

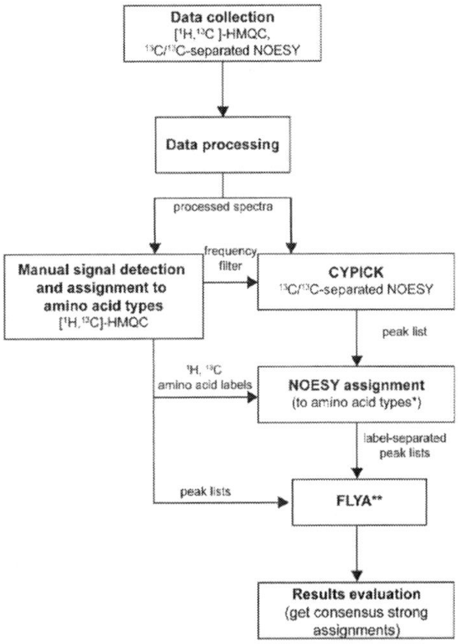

Figure 1. Outline of the CYPICK-FLYA protocol for fully automated methyl resonance assignment. *NOE peaks are assigned to types according to their amino acid assignment: Ile-Ile, Ile-Leu, Ile-Val, Leu-Ile, Leu-Leu, Leu-Val, Val-Ile, Val-Leu, Val-Val. **Automatically analysed and amino acid type-assigned NOE peak lists can be fed to FLYA, or other automatic methyl resonance assignment programs that support peak lists as input.

2.2. *FLYA calculations*

2.2.1. *Preparation of peak lists for FLYA*

FLYA supports input data from a variety of through-bond or through-space NMR experiments, which are fed to the algorithm as appropriately formatted peak lists [28,29]. The data presented in this work consisted of a 3D $^{13}C/^{13}C$-separated NOESY (CCNOESY) spectrum and a 2D [^{1}H,^{13}C]-HMQC spectrum, both of which were collected on a uniformly deuterated, Ile-[$\delta 1^{13}CH_3$], Leu-[$\delta^{13}CH_3$,$^{12}CD_3$], Val-[$\gamma^{13}CH_3$,$^{12}CD_3$] labelled protein sample of the r-chains of aspartate transcarbamoylase (ATCase-r_2) [18].

To reduce the high dimensionality of the combinatorial search for methyl resonance assignments, following the strategy outlined by MAGMA [25], the 2D [^{1}H,^{13}C]-HMQC peak list was first split according to the amino acid types of the methyl groups (in this case Ile, Leu, Val); i.e. prior knowledge of this information was assumed.

The amino acid type assignment of every ^{1}H-^{13}C cross-peak from the 2D [^{1}H,^{13}C]-HMQC peak list was used to automatically assign NOE interactions in the three-dimensional NOESY peak list to amino acid types. For each entry in the NOESY peak list, the difference between $^{13}C_1$, $^{1}H_1$ shifts and the reference ^{1}H, ^{13}C frequencies in the HMQC peak list is computed:

$$\Delta_1 = \sqrt{a(C_1^{NOE} - C_{ref}^{HMQC})^2 + (H_1^{NOE} - H_{ref}^{HMQC})^2}$$

as well as the absolute difference between $^{13}C_2$ NOESY shifts and the ^{13}C shifts from the reference HMQC peak list:

$$\Delta_2 = \left| C_2^{NOE} - C_{ref}^{HMQC} \right|$$

Each NOE is then attributed the C_1-H_1 and C_2 amino acid types of the reference HMQC peak for which Δ_1 and Δ_2 are minimal, respectively. This results in the NOESY peak list type assignments as shown in Fig. 2b. The CCNOESY peak list is then split according to the types of NOE interactions into separate Ile-Ile, Ile-Leu, Ile-Val, Leu-Leu, Leu-Ile, Leu-Val, Val-Val, Val-Ile, Val-Leu NOESY peak lists. The expected peaks computed by FLYA based on the input crystal structure are also classified according to the above types. This restricts matching of measured NOESY peaks to expected NOESY peaks of the same NOE interaction type during the FLYA automatic assignment calculation (Fig. 1).

2.2.2. *FLYA assignment calculations*

All FLYA calculations were carried out using the same parameters. To generate expected methyl-methyl NOE peaks, ^1H-^1H distances were computed from input crystal structures. For ATCase-r_2, a representative structure of the T-state of the enzyme (PDB ID: 1TUG [30]) was used, as previously proposed by Velyvis and Kay [18]. In addition, FLYA was run with a representative structure of the R-state of the enzyme (PDB ID: 1D09 [31]).

In FLYA, the ^1H-^1H distance for a pair of methyl groups is computed as the r^{-6} sum over the nine individual ^1H-^1H distances:

$$d_{\text{eff}} = \left(\sum_{i=1}^{3} \sum_{j=1}^{3} d_{ij}^{-6}\right)^{-1/6}$$

where d_{eff} stands for the effective (FLYA-computed) distance, the sums run over the ^1H atoms of methyl groups 1 and 2, respectively, and d_{ij} is the Euclidean distance between the individual methyl protons i and j in the input structure. For instance, if we assume all d_{ij} distances to be approximately equal, then $d_{\text{eff}} \approx 9^{-1/6} d_{ij} = 0.693\, d_{ij}$. Considering this approximate scaling factor, the distance cut-off $d_{\text{eff}} \leq d_{\text{cut}}$ for generating expected methyl-methyl NOE peaks was set to 5 Å, corresponding to an average maximal observable ^1H-^1H distance of $5.0/0.693\,\text{Å} \approx 7.2\,\text{Å}$. The optimal distance cut-off was established by running calculations with distance cut-offs of 4–10 Å in steps of 0.5 Å. The optimal value was chosen as the distance at which the highest percentage of expected peaks were assigned (Fig. A2).

In FLYA, the probability for expected methyl-methyl NOE peaks was set to 0.2 for all distances [28,29]. The optimality of this value was established in a more systematic effort on a larger benchmark of methyl-methyl NOESY data (data not shown). The chemical shift tolerances for the chemical shift assignment calculations were 0.4 ppm for ^{13}C and 0.04 ppm for ^1H chemical shifts. These values were reduced in the evaluation step (see next section) to 0.2 ppm and 0.02 ppm. The size of the population for the evolutionary algorithm of FLYA was set to 200, which was previously reported as optimal for NOESY-only FLYA calculations [29]. To increase the reliability of assignment solutions, 20 independent FLYA calculations were performed in parallel with the same input data but different random number generator seeds [28,29].

2.2.3. FLYA assignment consolidation. For methyl resonance assignments, a change in the FLYA shift consolidation function was made to allow for simultaneous 'consolidation' of both ^1H and ^{13}C chemical shifts for each assigned methyl group. Normally, FLYA computes a consensus chemical shift for every assigned atom based on the values from (typically 20) separate, independent runs of the algorithm (see above). To allow for simultaneous consolidation of ^1H -^{13}C chemical shift pairs, we performed k-means clustering [32,33] in two dimensions, with the number of centroids set to three. FLYA defines an atom assignment as strong if at least 80% of the chemical shift values, obtained for that atom from the independent algorithm runs, deviate less than the allowed matching tolerance (see above) from the consensus value. Consequently, we here define a methyl assignment as strong if the largest k-means cluster has at least 16 members (80% of 20) or if multiple smaller clusters, whose centroids are separated by less than the tolerance value in ^{1}H and ^{13}C dimensions, amount to at least 16 members. Shift assignments that do not fulfil these criteria are considered weak and therefore unreliable. The k-means clustering option of assignment results for FLYA chemical shift consolidation will be available with the next release of CYANA.

2.2.4. FLYA calculation replicates. We performed an additional 20 replicates of the entire FLYA calculation (which in turn comprises 20 independent, parallel runs of the optimization algorithm) with the optimal parameter values to check for the consistency between different sets of the 20 parallel calculations.

2.3. MAP-XS and FLAMEnGO calculations

MAP-XSII and FLAMEnGO2.0 protocols were run and their results analysed according to instructions outlined in the original publications [21–24]. For both programs, a range of distance thresholds for computing theoretical inter-methyl NOE restraints based on a high-resolution protein structure were considered (Figs. A3, A4). Calculations were run with both the structure of the T-state (PDB ID: 1TUG [30]) and the R-state (PDB ID: 1D09 [31]) of the ATCase-r$_2$ dimer. The optimal distance thresholds for MAP-XSII and FLAMEnGO2.0 calculations were determined as 9 Å and 15 Å, respectively (Figs. A3, A4). For FLAMEnGO2.0, 10000 Monte Carlo (MC) steps were run for a range of distances to obtain the graph in Fig. A4. Subsequently, 100 parallel calculations, each with 100000 MC steps were run at the optimal distance threshold of 15 Å. With MAP-XSII, 20 parallel calculations were run with the weighting between chemical shifts and NOEs set to the default value of 0.2.

3. Results and Discussion

We tested different values of the CYPICK baseline factor parameter β on a 3D ^{13}C/^{13}C-separated NOESY spectrum of ATCase-r$_2$. As established previously [27], an increase in the baseline factor reduces the extent of artefact picking (Table 1), which in turn leads to a better outcome of the automatic resonance assignment with FLYA (Fig. A1). In contrast, enhancing the baseline factor reduces the number of real peaks that are identified. The effect of changes in the contour line factor γ on the peak picking scores was negligible, and this parameter was therefore kept fixed throughout this study (see Methods). It was previously shown that CYPICK performs well in combination with FLYA when applied to the assignment of backbone resonances of uniformly ^{15}N/^{13}C-labelled proteins with up to ~20 kDa molecular mass. Using exclusively ^{13}C-edited and ^{15}N-edited NOESY data analyzed with CYPICK, FLYA resonance assignments were accurate enough to calculate structure bundles in a fully automated fashion with an RMSD bias below 2 Å [27]. Similarly, we note that, in this study, FLYA achieved the best methyl assignments for ATCase-r$_2$ using the CYPICK peak list with the lowest artefact score (12.9%, Table 1, Fig. A1).

Table 1. CYPICK results for the ATCase-r$_2$ 3D CCH-NOESY spectrum

Baseline factor (β)	Find score (%)	Artefact score (%)	Overall score (%)
2	86.5	40.5	74.7
3	84.7	25.3	79.0
5	76.6	12.9	74.4

The high dimensionality of the search space and the sparsity of restraints render methyl resonance assignment a difficult combinatorial problem. For optimal performance of automatic algorithms, it is essential to include the maximal amount of experimentally attainable information in the search. The FLYA search space can be considerably reduced if methyl resonances are separated into distinct groups according to their amino acid types (e.g. Ile, Leu, Val). Ile methyl resonances are readily recognized in

[^{1}H,^{13}C]-HMQC spectra due to their upfield (i.e. lower chemical shift) ^{13}C frequencies (Fig. 2). Discrimination between the overlapping regions of Leu and Val resonances can be achieved by preparing separate Leu-only or Val-only methyl-labelled protein samples [34,35]. Segregation of the search space according to amino acid type can also be applied to the interpretation of methyl-methyl NOESY data (see Methods). With this information available, FLYA then maps inter-methyl contacts (expected peaks) extracted from a protein structure only to methyl-methyl NOEs of the same type (Fig. 2b). In the protocol of Fig. 1, the classification of methyl-methyl NOEs is done automatically, based on the amino acid labelling of frequencies from the [^{1}H,^{13}C]-HMQC peak list (see Methods). As such, the protocol effectively automates both NOE peak picking and amino acid type classification (Fig 2). Although presented here on an example featuring selective Ile-δ1, Leu-δ, and Val-γ labelling, the protocol can easily be extended to any of the other available methyl labelling schemes including, for instance, labelling of Ile-γ2, Ala-β, Met-ε, or Thr-γ methyls.

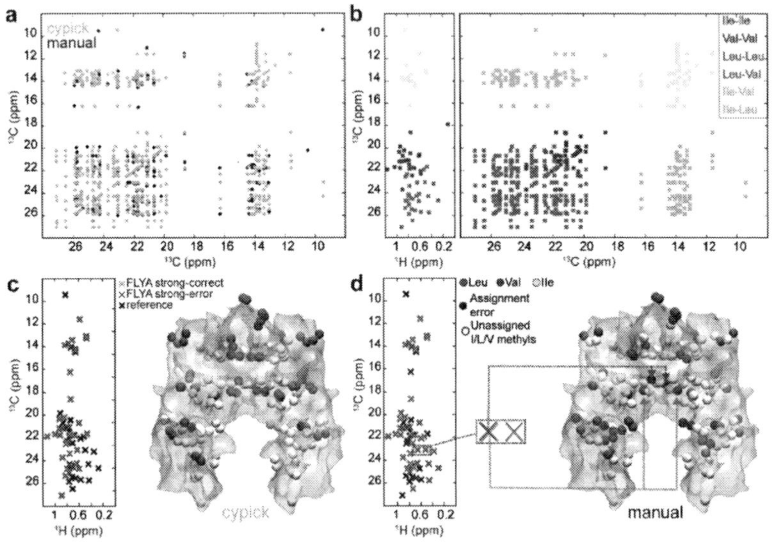

Figure 2. FLYA automatic methyl resonance assignment results using manually prepared and automatically analysed ^{13}C/^{13}C separated NOESY data. (**a**) Overlay of ^{13}C/^{13}C projections of CYPICK (cyan) and manually generated 3D NOESY peak lists (black). (**b**) Separation of methyl-methyl NOEs according to amino acid types. Positions of methyl ^{1}H-^{13}C correlations for Ile (yellow), Val (blue), and Leu (red) in the two-dimensional [^{1}H,^{13}C]-HMQC spectrum (left) and the ^{13}C/^{13}C projection of the 3D NOESY (right). (**c, d**) Left, overlays of the FLYA 'strong' assigned ^{1}H, ^{13}C chemical shifts with chemical shifts of ATCase-r_2 measured in the [^{1}H,^{13}C]-HMQC spectrum (black) using CYPICK (c) or manually analysed NOE data (d). Correct and erroneous assignments are shown in green and red, respectively. Right, representation of the FLYA assignment coverage on the T-state structure of ATCase-r_2. Methyl carbons are shown as balls; yellow, blue, and red balls correspond to correct strong FLYA assignments of Ile, Val, and Leu methyls, respectively. Unassigned methyls (i.e. weak FLYA assignments) are shown in white, and FLYA assignment errors in black.

The optimal distance cut-off for the computation of expected methyl-methyl NOE cross-peaks was found to be ~5 Å (Fig. A2). This threshold may appear low, considering the larger range of inter-methyl distances shown to be measurable for selectively methyl-protonated, uniformly deuterated proteins [25,26]. However, this FLYA derived distance reflects r^{-6} summation over all ^{1}H-^{1}H distances of two methyl groups, and thus corresponds to an actual distance of ~7.2 Å between individual protons (see Methods). For comparison, for the same dataset, a 10 Å carbon-carbon distance was found to be optimal for the computation of expected NOEs by MAGMA [25]. Adding the length of the C-H bond (~1.1 Å) [36] on both ends, in the simplest manner with no geometry considerations, the optimal FLYA C-C distance cut-off is only slightly shorter than that of MAGMA (7.2 Å + 2 × 1.1 Å = 9.4 Å), showing a good agreement between the two methods.

In Fig. 2, FLYA results for CYPICK (c) and manually prepared (d) methyl-NOESY peak lists are compared. The FLYA-calculated strong chemical shift assignments (green/red) overlay well with the reference [^{1}H,^{13}C]-HMQC frequencies of ATCase-r_2 (black). Using a CYPICK-generated NOESY peak list, FLYA assigned 58% of methyl resonances as 'strong' (i.e. confident) with complete accuracy (Fig. 2c). The manually acquired NOESY peak list resulted in 68% of confidently assigned resonances, albeit with one assignment error (Figs. 2d, 3). Closer inspection of the result shows that the erroneously assigned methyl group (Fig. 2d, red) shares the ^{1}H-^{13}C frequency position with another methyl (Fig. 2d, green), which is the correct assignment for that position. Moreover, in this case, the two methyls are spatially proximal in the protein structure (Fig. 2d). Such 'assignment overlaps' may occur occasionally in FLYA assignment results, as FLYA allows multiple expected peaks to be mapped to a measured peak, and should be given special attention in the inspection of FLYA results.

It should be noted that FLYA uses an approximate approach (genetic algorithm) to calculate resonance assignments. Therefore, despite consolidating assignment results across a set of multiple (typically 20), independent runs of the optimization algorithm, different sets could in principle feature some assignment differences. We thus repeated a full set of 20 parallel runs multiple times to evaluate the consistency in coverage and accuracy of methyl assignments determined by FLYA (see Methods). On ATCase-r_2 data, on average, the fully automated CYPICK-FLYA protocol yields strong (confident) assignments for 68 ± 4% of the methyls with 95 ± 2% accuracy. Using manually prepared lists, 66 ± 6% strong methyl resonance assignments are obtained with 94 ± 1% accuracy.

For the FLYA calculations presented here (Figs. 2, 3), we used a structure of the r-chain dimer (r_2) extracted from a structure of the T-state of the enzyme [30]. The T-state is predominantly populated in the apo form of the enzyme, which is consistent with the work of Velyvis *et al.* [18] from which the reference methyl assignments for the ATCase-r_2 dimer were obtained. However, when extracted from the holoenzyme context, the isolated ATCase-r_2 dimer likely features some structural and dynamical differences, as suggested by a comparison of the [^{1}H,^{13}C]-HMQC spectrum of the selectively ILV-labelled r-chain in the holoenzyme with that of the isolated r-chain dimer [18]. Given that the exact solution structure of the isolated ATCase-r_2 is not known and, as previously concluded in the MAGMA study, it may not be sufficient to consider a single conformation in a structure-based assignment strategy [25], we also checked the performance of FLYA on ATCase-r_2 extracted from an R-state structure of the enzyme and analyzed the consistency of the methyl assignments across both structural forms (Fig. A5). FLYA generates more 'strong' (i.e. confident) assignments for the R-state ATCase-r_2, albeit with considerably lower accuracy (87% compared to 100% for the T-state; Fig. A5). If strong assignments that are consistent across both structures are considered, the coverage of strong assignments falls to ~40%; however, all of these assignments are accurate. The FLYA 'preference' for the T-state form of ATCase-r_2 differs from that reported for MAGMA, where completely accurate, confident assignment for 31% of methyl residues was achieved using a structure extracted from the R-state, albeit with manually prepared input NOE restraints [25]. This discrepancy is likely a combination of differences in the input restraints (automatically vs. manually picked NOEs), and algorithmic approaches (genetic vs. exact graph matching). This difference in conformational preference between MAGMA and FLYA for this dataset will therefore be subject to further investigation.

Finally, we compared the performance of FLYA on CYPICK-generated NOESY peak lists to the automatic methyl assignment protocols MAP-XSII and FLAMEnGO2.0 (Figs. 3, A5). All three protocols show comparable (FLYA), or slightly better (MAP-XSII, FLAMEnGO2.0) performance when using CYPICK- versus manually-prepared peak lists as input (Fig. 3). Using the dimer structure extracted from the T-state of the enzyme, FLYA and MAP-XSII produced comparable levels of confident methyl assignments, 58% and 60%, respectively, with FLYA showing higher assignment accuracy (Fig. 3). With the same input data, FLAMEnGO2.0 generated considerably less confident methyl assignments (31%, 10% of which are erroneous). All three protocols show a significant drop in the fraction of confidently assigned methyls when assignment consistency between T- and R-state dimer structure is considered (Fig. A5). Under such circumstances, 39% of methyls could be confidently assigned by FLYA, followed by 34% using MAP-XSII, and 13% using FLAMEnGO2.0. Despite the drop in assignment coverage, the consideration of methyl assignment consistency over different structural forms is strongly advised, as it reflects favorably on the assignment accuracy (Fig. A5). It will be important to verify these findings in the context of methyl resonance assignment using MAGMA, which is currently the most accurate automatic methyl assignment protocol available [25]. At the present time, MAGMA does not support unassigned NOESY peak lists, but requires manually prepared, arbitrarily assigned, and appropriately filtered NOE contact lists as input [25]. The ambiguous nature of the automatically analyzed 3D NOESY data, reflected in the presence of numerous frequencies that fall within a small range in the ^{13}C (NOE) dimension, presently makes the CYPICK-derived 3D NOESY peak lists incompatible with the MAGMA input requirements. Future developments, including automatic assignment and filtering of CYPICK-derived NOESY peak lists could allow for the combined usage of CYPICK and MAGMA.

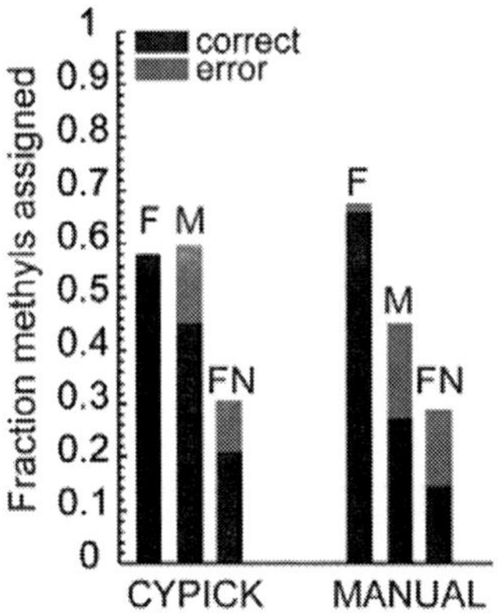

Figure 3. Comparison of automatic methyl resonance assignment protocols using CYPICK- or manually-generated NOESY peak lists. Abbreviations: F, FLYA; M, MAP-XSII; FN, FLAMEnGO2.0.
In the comparison, the input structure was the T-state form of ATCase-r_2 (PDB ID: 1TUG [30]).

4. Conclusions

We outlined a protocol for fully automated methyl resonance assignment based on sparse NOESY restraints collected on a uniformly deuterated, selectively methyl-labelled, high molecular weight protein. We combined the recently published contour-based CYPICK algorithm for signal identification with FLYA for automatic methyl resonance assignment. We showed that using CYPICK, high quality peak lists can be obtained for a 3D $^{13}C/^{13}C$-separated NOESY spectrum of a 36 kDa homodimer. Methyl resonance assignments obtained subsequently by FLYA are of comparable quality to those obtained using manually prepared peak lists (Figs. 2, 3). Comparison to dedicated automatic methyl assignment approaches that support unassigned NOESY peak lists as input (MAP-XSII and FLAMEnGO2.0) shows that, on this dataset, FLYA could achieve at least as many assignments as the alternatives, with considerably higher accuracy (Figs. 3, A5).

In the future, the outlined protocol must be tested on a larger benchmark of experimental datasets to evaluate its robustness and general applicability. Any automatic methyl assignment protocol that relies on information from NOESY peak lists could benefit from CYPICK (Fig. 1). As such, an important future goal is to combine CYPICK with MAGMA.

5. Acknowledgements

We thank Prof. AJ Baldwin for the ATCase-r_2 data, and TR Alderson for critical reading and feedback on the manuscript.

6. Appendix

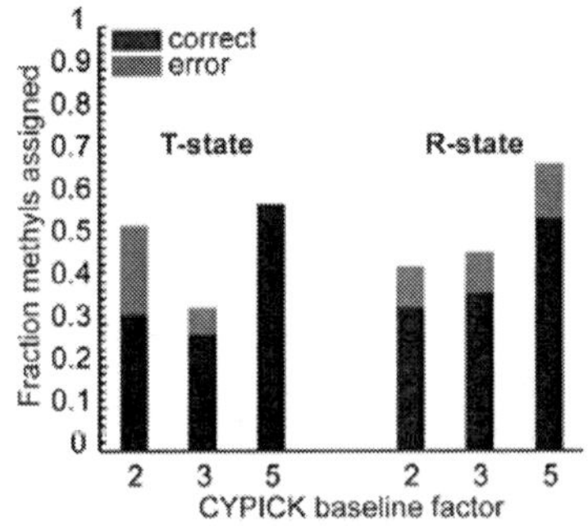

Figure A1. Performance of FLYA on CYPICK peak lists generated with different values of the baseline factor. For the highest baseline factor (5), the least artefacts were picked and the most 'strong' assignments were achieved with FLYA.

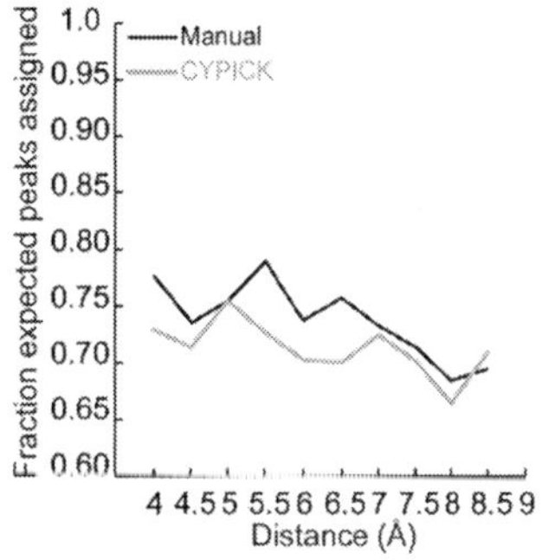

Figure A2. Determination of the distance threshold for the FLYA calculation on ATCase-r_2. The percentage of expected peaks assigned by FLYA at different distance thresholds was used to compute the expected methyl-methyl NOEs. The highest percentage is reached at a 5 Å distance with CYPICK data, or at 5.5 Å with the manually prepared NOESY peak list.

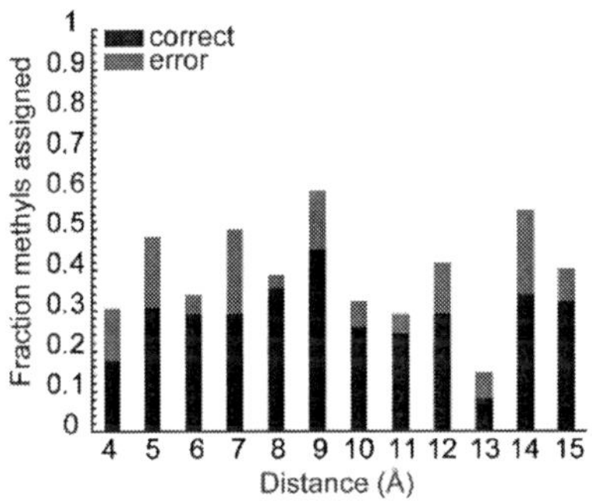

Figure A3. Determination of the optimal distance cut-off for MAP-XSII calculations on ATCase-r_2, as proposed by Xu *et al.* [21,22].

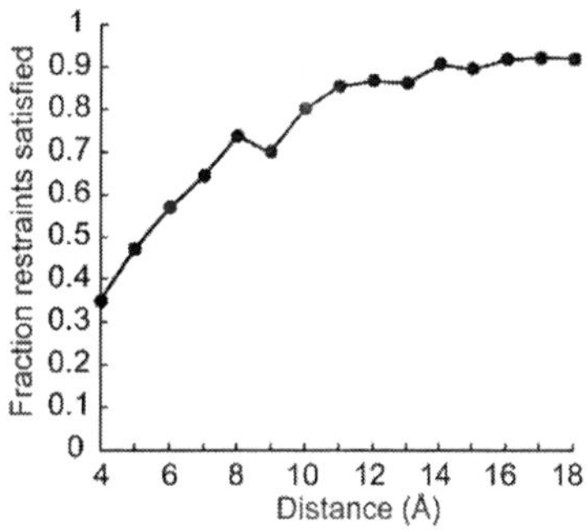

Figure A4. Determination of the optimal distance cut-off for FLAMEnGO2.0 calculations on ATCase-r_2, as proposed by Chao *et al.* [23,24]

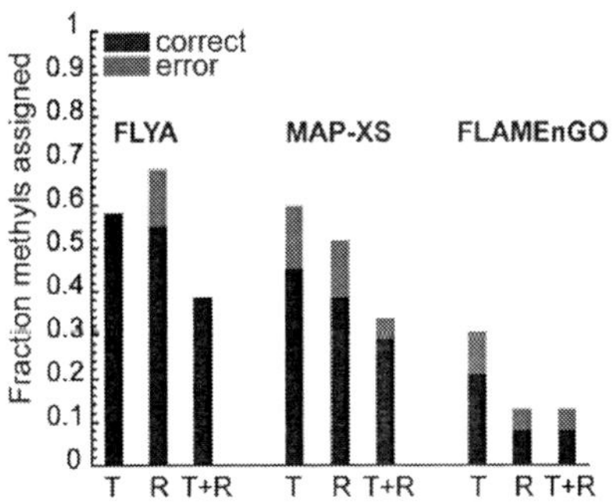

Figure A5. Comparison of automatic methyl assignment protocols on 3D $^{13}C/^{13}C$ separated NOESY data from ATCase-r_2, automatically analysed by CYPICK. The performance was compared using the T-state (PDB ID: 1TUG [30]) and R-state (PDB ID: 1D09 [31]) structures. Assignments consistent with both R- and T-states are evaluated in the T+R column. Correct assignments are indicated in black, erroneous ones in red.

7. References

[1] Wüthrich K 2003 *Angew. Chemie - Int. Ed* **42** 3340–63
[2] Kainosho M and Güntert P 2009 *Q. Rev. Biophys.* **42** 247–300
[3] Bax A 1994 *Curr. Opin. Struct. Biol.* **4** 738–44
[4] Sattler M, Schleucher J and Griesinger C 1999 *Prog. Nucl. Magn. Reson. Spectrosc.* **34** 93–158
[5] Pervushin K, Riek R, Wider G and Wüthrich K 1997 *Proc. Natl. Acad. Sci.* **94** 12366–71
[6] Gardner K H and Kay L E 1998 *Annu. Rev. Biophys. Biomol. Struct.* **27** 357–406
[7] Wiesner S and Sprangers R 2015 *Curr. Opin. Struct. Biol.* **35** 60–7
[8] Proudfoot A, Frank A O, Ruggiu F, Mamo M and Lingel A 2016 *J. Biomol. NMR* **65** 15–27
[9] Clark L, Zahm J A, Ali R, Kukula M, Bian L, Patrie S M, Gardner K H, Rosen M K and Rosenbaum D M 2015 *J. Biomol. NMR* **62** 239–45
[10] Janin J, Miller S and Chothia C 1988 *J. Mol. Biol.* **204** 155–64
[11] Ollerenshaw J E, Tugarinov V and Kay L E 2003 *Magn. Reson. Chem* **41** 843–52
[12] Tugarinov V, Hwang P M, Ollerenshaw J E and Kay L E 2003 *J. Am. Chem. Soc.* **125** 10420–8
[13] Mainz A, Religa T L, Sprangers R, Linser R, Kay L E and Reif B 2013 *Angew. Chemie - Int. Ed.* **52** 8746–51
[14] Sprangers R and Kay L E 2007 *Nature* **445** 618–22
[15] Sprangers R, Velyvis A and Kay L E 2007 *Nat. Methods* **4** 697–703
[16] Baldwin A J, Religa T L, Hansen D F, Bouvignies G and Kay L E 2010 *J. Am. Chem. Soc.* **132** 10992–5
[17] Kato H, van Ingen H, Zhou B-R, Feng H, Bustin M, Kay L E and Bai Y 2011 *Proc. Natl. Acad. Sci. U. S. A.* **108** 12283–8
[18] Velyvis A, Schachman H K and Kay L E 2009 *J. Am. Chem. Soc.* **131** 16534–43
[19] Rosenzweig R and Kay L E 2014 *Annu. Rev. Biochem.* **83** 291–315
[20] Venditti V, Fawzi N L and Clore G M 2011 *J. Biomol. NMR* **51** 319–28
[21] Xu Y, Liu M, Simpson P J, Isaacson R, Cota E, Marchant J, Yang D, Zhang X, Freemont P and Matthews S 2009 *J. Am. Chem. Soc.* **131** 9480–1
[22] Xu Y and Matthews S 2013 *J. Biomol. NMR* **55** 179–87
[23] Chao F-A, Shi L, Masterson L R and Veglia G 2012 *J. Magn. Reson.* **214** 103–10
[24] Chao F A, Kim J, Xia Y, Milligan M, Rowe N and Veglia G 2014 *J. Magn. Reson.* **245** 17–23
[25] Pritišanac I, Degiacomi M T, Alderson T R, Carneiro M G, Ab E, Siegal G and Baldwin A J 2017 *J. Am. Chem. Soc.* **139**
[26] Sounier R, Blanchard L, Wu Z and Boisbouvier J 2007 *J. Am. Chem. Soc.* **129** 472–3
[27] Würz J M and Güntert P 2017 *J. Biomol. NMR* **67** 63–76
[28] Schmidt E and Güntert P 2012 *J. Am. Chem. Soc.* **134** 12817–29
[29] Schmidt E and Güntert P 2013 *J. Biomol. NMR* **57** 193–204
[30] Stieglitz K, Stec B, Baker D P and Kantrowitz E R 2004 *J. Mol. Biol.* **341** 853–68
[31] Jin L, Stec B, Lipscomb W N and Kantrowitz E R 1999 *Proteins Struct. Funct. Genet.* **37** 729–42
[32] Arthur D and Vassilvitskii S 2007 *Proceedings of the eighteenth annual ACM-SIAM symposium on Discrete algorithms* pp 1027–1025
[33] Pedregosa F and Varoquaux G 2011 *J Mach Learn Res* **12** 2825-2830
[34] Lichtenecker R J, Weinhäupl K, Reuther L, Schörghuber J, Schmid W and Konrat R 2013 *J. Biomol. NMR* **57** 205–9
[35] Lichtenecker R J, Coudevylle N, Konrat R and Schmid W 2013 *ChemBioChem* **14** 818–21
[36] Maksić Z B and Randić M 1970 *J. Am. Chem. Soc.* **92** 424–5

IOP Conf. Series: Journal of Physics: Conf. Series **1036** (2018) 012009 doi:10.1088/1742-6596/1036/1/012009

High-Performance Data Analysis on the Big Trajectory Data of Cellular Scale All-atom Molecular Dynamics Simulations

Isseki Yu[1,2*], Michael Feig[5], Yuji Sugita[1,2,3,4]

1) iTHES Research Group, RIKEN, Saitama, Japan
2) Theoretical Molecular Science Laboratory, RIKEN, Saitama, Japan
3) Laboratory for Biomolecular Function Simulation, RIKEN Quantitative Biology Center, Kobe, Japan
4) Computational Biophysics Research Team, RIKEN Advanced Institute for Computational Science, Kobe, Japan
5) Department of Biochemistry and Molecular Biology, Michigan State University, East Lansing, United States

E-mail: isseki@riken.jp

The inside of a cell is highly crowded with a large number of macromolecules together with solvents and metabolites. To know the molecular-level behaviour of biomolecules in such dense crowding environment, we constructed full atomistic model of the cytoplasm of bacteria, and performed massive all-atom molecular dynamics (MD) simulations. On the other hand, to analyse such big MD data, we need significant computational power and efficient calculation methodology. Here, we introduce what and how we analyse the biomolecule properties from the big trajectory data produced by cellular scale all-atom MD simulations.

1. Introduction

Molecular dynamics (MD) simulations are widely used to investigate the microscopic behaviour of biomolecules. Recently, the scale of the MD simulation has rapidly expanded both spatially and temporally. One of the largest targets is the cellular environments in which various kinds of proteins, RNAs, metabolites are interacting under significantly crowded conditions (in fact, 20～40% of the volume is occupied by biomolecules in the cell[1-4]). How variable interactions within dense cellular environments may affect the structure and dynamics of biomolecules, and ultimately their function, is one of the most exciting questions in life science[5-10]. Recently, we constructed a full atomistic model of the cytoplasm of a minimal bacterium[11]. Using the model, we performed massive all-atom molecular dynamics (MD) simulation, and succeeded in reproducing the molecular-level behaviour of biomolecules in the cell[12]. On the other hand, the extraction of dynamic features and insight into the interactions of biomolecules from extremely big and complex data was another challenging issue. Conventional analysis tools for MD trajectories cannot easily handle a trajectory of such a big system[13-15]. In this paper, we introduce the kind of physicochemical properties of biomolecules that we typically analyse from the big MD data of cellular crowding systems and describe how to calculate them using high-performance computer based on spatial decomposition techniques.

-
* To whom any correspondence should be addressed.

IOP Conf. Series: Journal of Physics: Conf. Series **1036** (2018) 012009 doi:10.1088/1742-6596/1036/1/012009

2. Models of crowded systems

By integrating data from a variety of experimental sources, we constructed a full atomistic model of the cytoplasm of a bacterium (*Mycoplasma genitalium*) including all of the molecular components, i.e., proteins, RNA, metabolites, ions, and solvent, that are mapped on the complete biochemical pathways[11]. The size of the system is 100 nm x 100 nm x 100 nm, which greatly exceeds the size of typical molecular dynamics (MD) simulations, covering about 10% of the volume of an entire cell (**MGh** in Figure 1 A). Model cytoplasms at middle (**MGm**) and small (**MGs**) sizes were additionally constructed. These models were subjected to MD simulation using the highly parallelized MD program GENESIS[16] on the supercomputer K[12]. The resulting data sizes of the MD simulations generated for each model are in the 5-20 TB range as shown in Table 1.

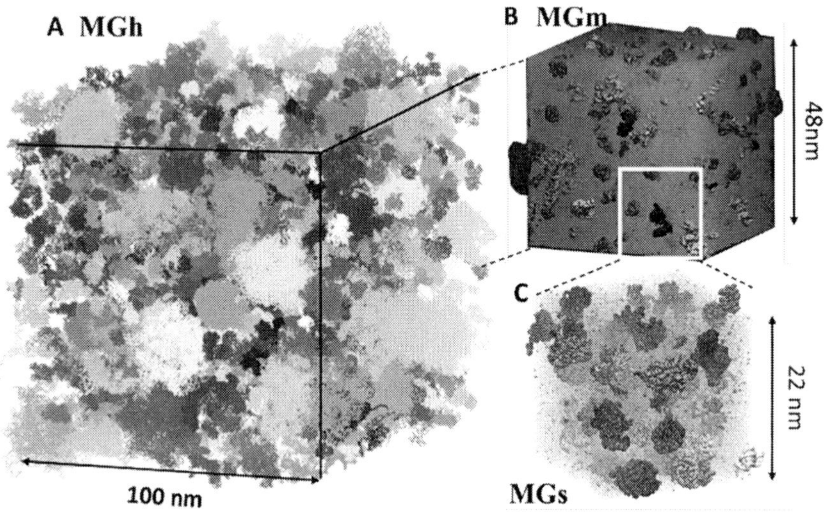

Figure 1. All atom models of the cytoplasm of *Mycoplasma genitalium*. **A**: The largest model contains about 100 million atoms (**MGh**). **B**: A middle size model containing about 10 million atoms (**MGm**). **C**: A small size model containing about 1 million atoms (**MGs**). Each macromolecules is shown with randomly assigned colours. The interspace of macromolecules are filled with water (blue) and metabolites and ions (for **MGh**, only the macromolecules are shown).

Table 1. Data sizes of MD simulations for different bacterial cytoplasm models. For **MGh** and **MGm**, atomic coordinates were stored every 1 ps. For **MGs**, atomic coordinates were stored every 10 ps.

System	Total number of atoms	Total length of MD sim. (ns).	Data size of total frame (TB)
MGh	103,708,785	20	22
MGm	11,737,298	140	17
MGs	1,082,358	1,000 x 4	5.2

IOP Conf. Series: Journal of Physics: Conf. Series **1036** (2018) 012009 doi:10.1088/1742-6596/1036/1/012009

3. Results and Discussions

3.1 Analysis of the kinetic properties of macromolecules

How fast do macromolecules (proteins, RNAs, and huge complex, such as ribosomes or GroEL) move through the crowded environment in a cell? This is one of the most fundamental questions in life science. Here, we focus on the translational and rotational diffusive motion of macromolecules. The influence of crowding on these kinetic properties is also discussed.

3.1.1 Translational diffusion coefficient of macromolecules

The translational diffusion coefficient D_{tr} is one of the most fundamental kinetic properties, quantifying the mobility of macromolecules. D_{tr} is usually calculated from the square displacement (SD) of target molecules. The time evolution of the SD of a macromolecule α is obtained by tracking the center of mass of α. Multiple profiles of SD for α can be obtained by sliding windows with certain intervals. These profiles are then averaged to obtain mean square displacements (MSD). To obtain translational diffusion coefficient D_{tr}, a linear function is fitted to the MSD curve and D_{tr} is subsequently computed from the slope of the fitted line according to the Einstein relation,

$$D_{tr}(\alpha) = \frac{\langle r^2(\alpha,\tau)\rangle_i}{6\tau} \tag{1}$$

where $r^2(\alpha,\tau)$ denotes the SD of the macromolecule α at time τ from the beginning of one of the windows i. Further details of this analysis are explained elsewhere[12].

As an example, D_{tr} of each macromolecule in **MGm** was calculated and is compared with experimentally measured diffusion coefficients for green fluorescence proteins (GFPs)[17](**Figure 2 A**). The resulting values are correlated with the size of the different proteins (i.e., their Stokes radii *Rs*). From this analysis, the agreement with the experimental data or the dependency of D_{tr} on the molecular size under crowded conditions can be evaluated.

3.1.2 Rotation of macromolecules

The rotation of macromolecules is strongly influenced by protein-protein interactions (PPI) with the surrounding molecules. In addition, the rotational dynamics (such as the rotational relaxation time, the rotational diffusion coefficient, and the axis of rotation) can be directly compared with NMR data. Thus, the properties of rotation can be a useful reference for the elucidation of the PPI or to tune the interaction parameters in MD simulation[18]. To analyze the overall tumbling motion of a macromolecule α, the rotation matrix $\boldsymbol{R}$ that defines the rotation of α at $t = t_i$ to the target orientation at $t = t_i + \tau$ is used. Then, the rotational correlation function (RCF) in a given time window i as a function of τ ($c(\alpha,i,\tau)$) is obtained by applying the rotation matrix $\boldsymbol{R}$ on the *principal axis of inertia* or the *NH vector of protein backbone* or *randomly distributed unit vectors attached to the protein structure*[19]. Time-averaged RCF, $\langle c(\alpha,\tau)\rangle_i$ are then obtained using sliding windows as in the calculation of the translational diffusion coefficients. The isotropic rotational relaxation time τ_{rel} was obtained by fitting a single (or multiple) exponential

$$\langle c(\alpha,\tau)\rangle_i \propto \exp(-\tau/\tau_{rel}) \tag{2}$$

Finally, the isotropic rotational diffusion coefficient of α is obtained as

$$D_{rot}(\alpha) = 1/2\tau_{rel} \tag{3}$$

The instantaneous rotation angle θ and the rotation axis $\boldsymbol{v}$ (v_x, v_y, v_z) can be obtained by converting the rotation matrix $\boldsymbol{R}$ to the quaternion $\boldsymbol{q}$. The relation between four elements of $\boldsymbol{q}$ and θ, $\boldsymbol{v}$ is as follows,

$$q = (q_w, q_x, q_y, q_z) = \left(\cos(\theta/2),\ \sin(\theta/2)v\right) \tag{4}$$

Figure 2B shows the time-averaged angular velocity of each macromolecule in **MGm** as a function of their size, *Rs*. The rotation of macromolecules also displayed a strong molecular size dependency as for translational motion.

3.1.3 Influence of local crowding on the translation and rotation of macromolecules

Because different macromolecules are exposed to different local crowding environments, their dynamics is influenced differently even though they have the same size and structure. For example, there are 25 copies of tRNA in **MGm**. Each tRNA has different values of D_{tr} and ω (see red squares in Figure 2). To measure the local degree of crowding around a given target molecule α, we used the number of backbone C_α and P atoms in other macromolecules within the cutoff distance R_{cut} = 50 Å from the closest C_α and P atoms of α at a given time t as the *instantaneous coordination number* of crowder atoms, $N_c(\alpha,t)$. Time averages of $N_c(\alpha,t)$ were then calculated over 10 ns windows. The obtained values of N_c are correlated with D_{tr} or ω in the corresponding 10 ns windows, and histogram-averaged values of D_{tr} and ω are shown in 100 interval of N_c (see small figures inserted in **Figures 2 A** and **2B**). These analysis show how the degree of local crowding retards the dynamics of macromolecules.

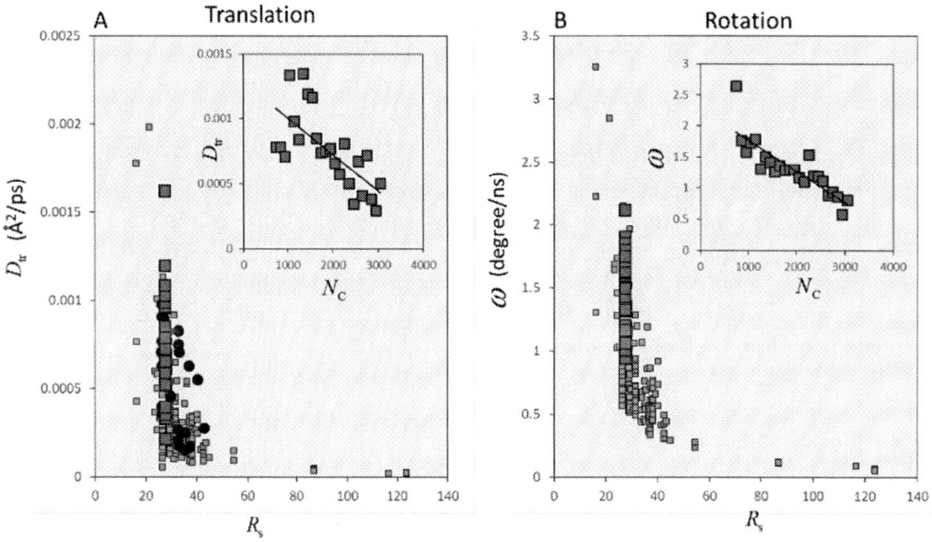

Figure 2. **A**: Translational diffusion coefficients D_{tr} of macromolecules in a model bacterial cytoplasm (**MGm**) as a function of their Stokes radii R_s (grey squares). The MSD was obtained using multiple 10 ns windows. D_{tr} for the multiple (25) copies of tRNA are highlighted with red squares. The correlation between D_{tr} and the local extent of crowding N_c for tRNAs is inserted at the top right. The experimental values of D_{tr} for GFPs (Green Fluorescence Protein) are shown as black filled circles. **B**: Angular velocity of macromolecules in **MGm** as a function of R_s. The averaged rotation angle θ was calculated using multiple 10 ns windows. ω was obtained as θ/10.0ns. ω for the multiple (25) copies of tRNA are highlighted as red squares. The correlation between ω and the local extent of crowding N_c for tRNAs is inserted at the top right.

IOP Conf. Series: Journal of Physics: Conf. Series **1036** (2018) 012009 doi:10.1088/1742-6596/1036/1/012009

3.2 Analysis of the spatial distribution of solvent and metabolites

In section 3.1, the analysis of kinetic properties (translational and rotational diffusion) of macromolecules is discussed. As the data size is greatly reduced (e.g., instead of all-atom coordinates only the centres of mass are considered), these analyses do not require very large computational resources. On the other hand, properties related to inter-molecular distances, or spatial distributions can involve significant computational costs. One typical application that presents significant challenges is the calculation of the density distribution of solvent molecules around the macromolecules (see Fig. 3). To analyse the number density of solvent as a function of the distance from the closest macromolecule $\rho(r)$, one has to calculate *i)*, the volume of the hypothetical layer at a distance r (with a certain thickness Δr) from the macromolecule (we refer to this volume as the available volume $V(r)$; see red layers in **Figure 3**), and *ii*), the number of water molecule that are present in a given layer at distance r, $N(r)$.

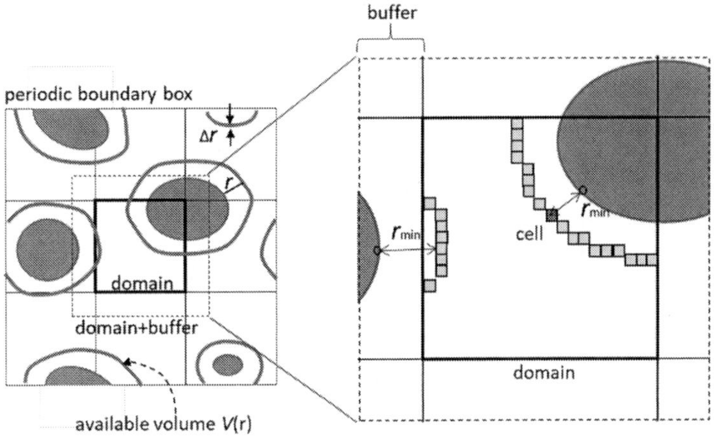

Figure 3. Schematic representation of the analysis of solvent density around macromolecules (blue objects) based on spatial decomposition techniques. The available volume $V(r)$ at a distance r from the closest macromolecule is represented as red layers. The system is equally divided into domains having enough buffer region. Each domain is further divided into smaller cells. The profile of $V(r)$ is approximated by taking the histogram of the number of cells with the minimum distance to the macromolecules r_{min}.

Because one needs to calculate the distance between vast numbers of sites and macromolecular atoms, the calculation of $V(r)$ needs significant CPU power and large amounts of memory. To overcome these difficulties, we developed a hybrid (MPI/OpenMP) parallelization scheme based on the spatial decomposition technique. The whole system (usually corresponding to a box under periodic boundary conditions) is decomposed into smaller domains. Each domain has a buffer region with enough thickness to obtain the profile of $\rho(r)$ up to a given target distance. A domain is further decomposed into smaller cells. Each MPI process then assigns atoms inside the *domain + buffer* region to cells. The calculation for each domain is done by each MPI process, and the calculation for cells is decomposed into OpenMP threads. For each time step (t), the minimum distance (r_{min}) from a given cell in a gvien *domain* to any atoms of macromolecules in the *domain + buffer* region is determined. Such a calculation is repeated for all cells in a given domain. Then, the histogram of the number of cells as a function of the r_{min} at time t ($N_{cell}(r, t)$) is obtained with a certain bin size Δr by accumulating the results from different domains. The total number of target solvent atoms in the cells assigned to each bin ($N_{atom}(r, t)$) is also counted. Finally, $\rho(r)$ is calculated as follows,

$$\rho(r)=\langle\rho(r,t)\rangle_t=\left\langle\frac{N_{\text{atom}}(r,t)}{V(r,t)}\right\rangle_t=\left\langle\frac{\sum_i^{N_{\text{cell}}}N_{\text{atom}}(i,r,t)}{N_{\text{cell}}(r,t)\cdot V_{\text{cell}}}\right\rangle_t \tag{5}$$

where V_{cell} is the volume of a cell.

In **Figure 4**. the example of the (normalized) density distribution $\rho(r)$ obtained by this scheme. $\overline{\rho}(r)$ is shown for several small molecules (such as water, phosphates and amino acids) around macromolecules in **MGm**. From these profiles, it is possible to understand how strongly these molecules associate with macromolecular surfaces.

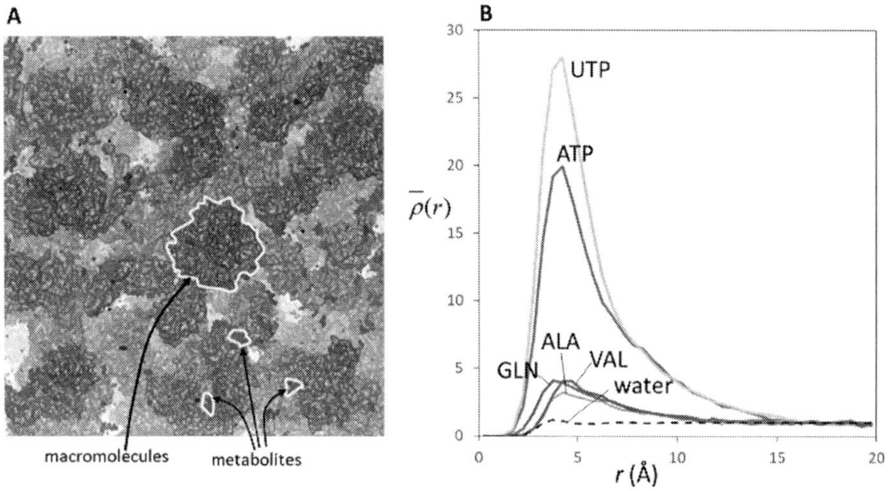

Figure 4. A: Snapshot of metabolites around macromolecules in **MGm**. red: phosphates (such as ATP: adenosine triphosphate, UTP: uracil triphosphate), green: amino acids (such as ALA: alanine, VAL: valine, and GLN: glutamine), and blue: other types of metabolites (such as NAD: nicotinamide adenine dinucleotide). **B**: heavy atom number density of five types of metabolites as a function of the distance to the closest macromolecule heavy atom $\rho(r)$. Each profile is normalized (denoted as $\overline{\rho}(r)$) by the density at the most distant region (r=20.0 Å). For the calculation of $V(r)$, the resolution (i.e., a cell size) is set to the 1.0 Å^3. $\overline{\rho}(r)$ of water oxygen atoms is also shown for comparison with a dashed line.

Figure 5 shows benchmark timing results for the calculation of $\overline{\rho}(r)$ for water oxygen (dashed line in **Figure 4B**). The performance numbers of the calculation were obtained on RIKEN's supercomputer system HOKUSAI GreatWave (CPU: SPARC64, performance: 1Pflops). As Figure 5 shows, the calculation is linearly accelerated with an increasing number of CPU cores.

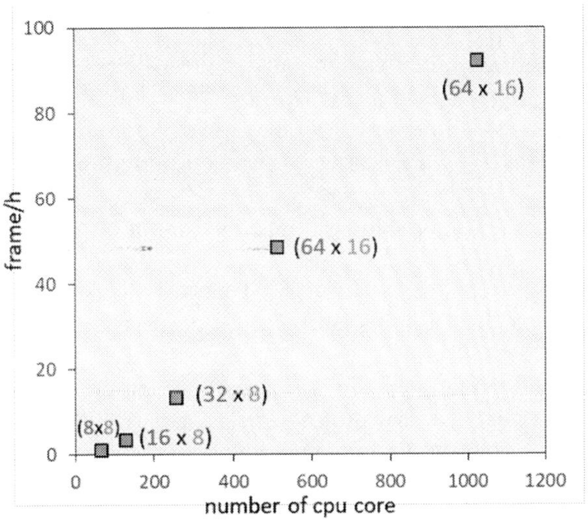

Figure 5. Benchmark calculation of $\overline{\rho}(r)$ for water (dashed line in **Figure 4B**). The number of frames (snapshots) processed per hour is shown vs. the number of CPU cores that were used. The number of MPI processes (red) and OpenMP threads (green) are shown beside each marker (blue).

4. Conclusions

We have presented analysis techniques for large all-atom MD trajectories of cellular crowding systems. In addition to the calculation of kinetic properties of macromolecules, we discuss the analysis of the spatial density distribution of solvents and metabolites which requires significant computer power. To accelerate the calculation of such a time-consuming analysis, we developed a hybrid (MPI/OpenMP) parallelization framework based on the spatial decomposition technique. This method exhibits good scalability to more than 1,000 CPU cores on a suitable supercomputer system.

The developed framework can be applicable not only the calculation of solvent density analysis, but also to the analysis of many physicochemical property related to local quantities of a given target molecule or local spatial properties in the system. For example we applied the same framework for the calculation of solvent accessible surface areas (SASA) of macromolecules, protein-protein interactions, and the extraction of hydrogen bonds in the large crowded systems. The analysis methods described here are implemented in one of the analysis modules (SPANA: SPatial decomposition ANAlysis) of the MD software GENESIS[16].

References

[1] Minton, A. P. The influence of macromolecular crowding and macromolecular confinement on biochemical reactions in physiological media. *J. Biol. Chem.* (2001), ***276***, 10577-10580.
[2] Ellis, R. J.; Minton, A. P. Cell biology: join the crowd. *Nature* (2003), ***425***, 27-28.
[3] Wang, Y.; Li, C.; Pielak, G. J. Effects of proteins on protein diffusion. *J. Am. Chem. Soc.* (2010), ***132***, 9392-9397.
[4] Monteith, W. B.; Pielak, G. J. Residue level quantification of protein stability in living cells. *Proc. Natl Acad. Sci. USA* (2014), ***111***, 11335-11340.
[5] Inomata, K.; Ohno, A.; Tochio, H.; Isogai, S.; Tenno, T.; Nakase, I.; Takeuchi, T.; Futaki, S.; Ito, Y.; Hiroaki, H.; Shirakawa, M. High-resolution multi-dimensional NMR spectroscopy of proteins in human cells. *Nature* (2009), ***458***, 106-109.
[6] Feig, M.; Sugita, Y. Variable interactions between protein crowders and biomolecular solutes are important in understanding cellular crowding. *J. Phys. Chem. B* (2012), ***116***, 599-605.
[7] Harada, R.; Sugita, Y.; Feig, M. Protein crowding affects hydration structure and dynamics. *J. Am. Chem. Soc.* (2012), ***134***, 4842-4849.
[8] Feig, M.; Sugita, Y. Reaching new levels of realism in modeling biological macromolecules in cellular environments. *J. Mol. Graph. Model.* (2013), ***45***, 144-156.
[9] Harada, R.; Tochio, N.; Kigawa, T.; Sugita, Y.; Feig, M. Reduced native state stability in crowded cellular environment due to protein-protein interactions. *J. Am. Chem. Soc.* (2013), ***135***, 3696-3701.
[10] Monteith, W. B.; Cohen, R. D.; Smith, A. E.; Guzman-Cisneros, E.; Pielak, G. J. Quinary structure modulates protein stability in cells. *Proc. Natl Acad. Sci. USA* (2015), ***112***, 1739-1742.
[11] Feig, M.; Harada, R.; Mori, T.; Yu, I.; Takahashi, K.; Sugita, Y. Complete atomistic model of a bacterial cytoplasm for integrating physics, biochemistry, and systems biology. *J. Mol. Graph. Model.* (2015), ***58***, 1-9.
[12] Yu, I.; Mori, T.; Ando, T.; Harada, R.; Jung, J.; Sugita, Y.; Feig, M. Biomolecular interactions modulate macromolecular structure and dynamics in atomistic model of a bacterial cytoplasm. *eLife* (2016), ***5:e19274***.
[13] Feig, M.; Karanicolas, J.; Brooks, C. L. MMTSB Tool Set: enhanced sampling and multiscale modeling methods for applications in structural biology. *J. Mol. Graph. Model.* (2004), ***22***, 377-395.
[14] Humphrey, W.; Dalke, A.; Schulten, K. VMD: Visual molecular dynamics. *Journal of Molecular Graphics* (1996), ***14***, 33-38.
[15] Roe, D. R.; Cheatham, T. E. PTRAJ and CPPTRAJ: Software for Processing and Analysis of Molecular Dynamics Trajectory Data. *J. Chem. Theory Comput.* (2013), ***9***, 3084-3095.
[16] Jung, J.; Mori, T.; Kobayashi, C.; Matsunaga, Y.; Yoda, T.; Feig, M.; Sugita, Y. GENESIS: a hybrid-parallel and multi-scale molecular dynamics simulator with enhanced sampling algorithms for biomolecular and cellular simulations. *Wiley Interdiscip. Rev. Comput. Mol. Sci.* (2015), ***5***, 310-323.
[17] Nenninger, A.; Mastroianni, G.; Mullineaux, C. W. Size dependence of protein diffusion in the cytoplasm of Escherichia coli. *J. Bacteriol.* (2010), ***192***, 4535-4540.
[18] Michael Feig, G. N., Isseki Yu, Po-hung Wang, Yuji Sugita. Challenges and opportunities in connecting simulations with experiments via molecular dynamics of cellular environments. *Proceedings of the international meeting on "High-Dimensional Data-Driven Science"* (2017).
[19] Wong, V.; Case, D. A. Evaluating rotational diffusion from protein MD simulations. *J. Phys. Chem. B* (2008), ***112***, 6013-6024.

ACKNOWLEDGMENTS

The simulations and analysis were carried out using the RIKEN Integrated Cluster of Clusters (RICC) and RIKEN HOKUSAI supercomputer systems, and HPCI strategic research project (hp140229, hp150233) and HPCI general trial use project (hp150145, hp160120) and FLAGSHIP 2020 project focused area 1 "Innovative drug discovery infrastructure through functional control of biomolecular systems (hp160207)". This work was supported in part by RIKEN QBiC, iTHES and pioneering project "Dynamic structural biology"(to YS), a Grant-in-Aid for Scientific Research on Innovative Area "Novel measurement techniques for visualizing 'live' protein molecules at work" (No. 26119006) (to YS), a grant from JST CREST on "Structural Life Science and Advanced Core Technologies for Innovative Life Science Research" (to YS), a Grant-in-Aid for Scientific Research (C) from MEXT (No. 25410025) and Incentive Research Projects from RIKEN (to IY), and support from the U.S. National Institutes of Health (NIH, GM092949, GM084943) and the U.S. National Science Foundation (NSF, MCB 1330560) (to MF).

IOP Conf. Series: Journal of Physics: Conf. Series **1036** (2018) 012010 doi:10.1088/1742-6596/1036/1/012010

Challenges and opportunities in connecting simulations with experiments via molecular dynamics of cellular environments

Michael Feig[1,2*], Grzegorz Nawrocki[1], Isseki Yu[3,4], Po-hung Wang[3], Yuji Sugita[2,3,4,5]

[1]Department of Biochemistry and Molecular Biology, Michigan State University, East Lansing, MI, 48824 USA, [2]Quantitative Biology Center, RIKEN, Kobe, Japan, [3]Theoretical Molecular Science Laboratory, RIKEN, Wako, Japan, [4]iTHES Research Group, RIKEN, Wako, Japan, [5]Advanced Institute for Computational Science, RIKEN, Kobe, Japan.

E-mail: feig@msu.edu

Abstract. Computer simulations are widely used to study molecular systems, especially in biology. As simulations have greatly increased in scale reaching cellular levels there are now significant challenges in managing, analyzing, and interpreting such data in comparison with experiments that are being discussed. Management challenges revolve around storing and sharing terabyte to petabyte scale data sets whereas the analysis of simulations of highly complex systems will increasingly require automated machine learning and artificial intelligence approaches. The comparison between simulations and experiments is furthermore complicated not just by the complexity of the data but also by difficulties in interpreting experiments for highly heterogeneous systems. As an example, the interpretation of NMR relaxation measurements and comparison with simulations for highly crowded systems is discussed.

1. Introduction

Computer simulations have become a central element in modern science as a bridge between experiments and theory [1]. Simulations are commonly applied to describe the evolution of complex systems in time and space based on theoretical models but in regimes where direct analytical or even numerical solutions are not feasible. The level of realism that is achieved by such simulations depends on the nature of the underlying models and may range from idealized conceptual views to physically highly [1]accurate descriptions of the systems that are being studied. The most sophisticated simulations can, at least in principle, rival experiments and provide complete spatio-temporal information although the availability of computer resources limits the scales that can be accessed.

A key advantage of simulations is that essentially any question can be interrogated irrespective of practical limitations that may hinder experiments, including hypothetical scenarios that could not even be realized experimentally under any circumstances; however, computer simulations remain a fundamentally theoretical approach that is most valuable when new hypotheses or predictions are generated that can subsequently be subjected to experimental validation. A common strategy is thus an iterative approach where simulations and experiments rely on each other in the development of

* To whom any correspondence should be addressed: 603 Wilson Rd., 218 BCH, East Lansing, MI 48824, USA

new mechanistic insights. While computer simulations are widely used today in all fields of science, they are especially valuable in biology where a high degree of system complexity challenges both experiments and theory [2, 3]. In fact, molecular dynamics simulations of biological macromolecules such as proteins and nucleic acids have become a staple in modern biological science having contributed much to our detailed mechanistic understanding of biomolecular processes [4]. Very recently, such simulations have been extended to cellular scales and simulations of entire cells in molecular detail will soon become reality [5].

The main results of computer simulations are three-dimensional coordinate trajectories over time for a given system. Molecular systems are often described at an atomistic level of detail although higher (quantum-mechanical) or lower (coarse-grained) resolutions are possible. In addition to the system of immediate interest, the environment often also needs to be considered. For many molecular systems, especially in biology, this involves aqueous solvent so that water, ions, and co-solvents are part of the system, typically also in atomistic detail; therefore, system sizes between 50,000 and 1 M atoms are common nowadays for simulations of single macromolecules or macromolecular complexes [6], but much larger systems with as many as 100 M atoms have been reported for studies of many interacting molecules for the cytoplasm of a bacterial cell [7] (see figure 1). The time scales covered by such simulations are now routinely reaching 1 μs and in exceptional cases as much as 1 ms [8]. Depending on how often coordinates are saved, this means that a single simulation may generate data on terabyte to petabyte scales. The large amounts of data coupled with the high degree of complexity in many systems presents formidable data management and analysis challenges and it is also becoming increasingly difficult to compare with experiments. A further discussion of these challenges is the topic of this article.

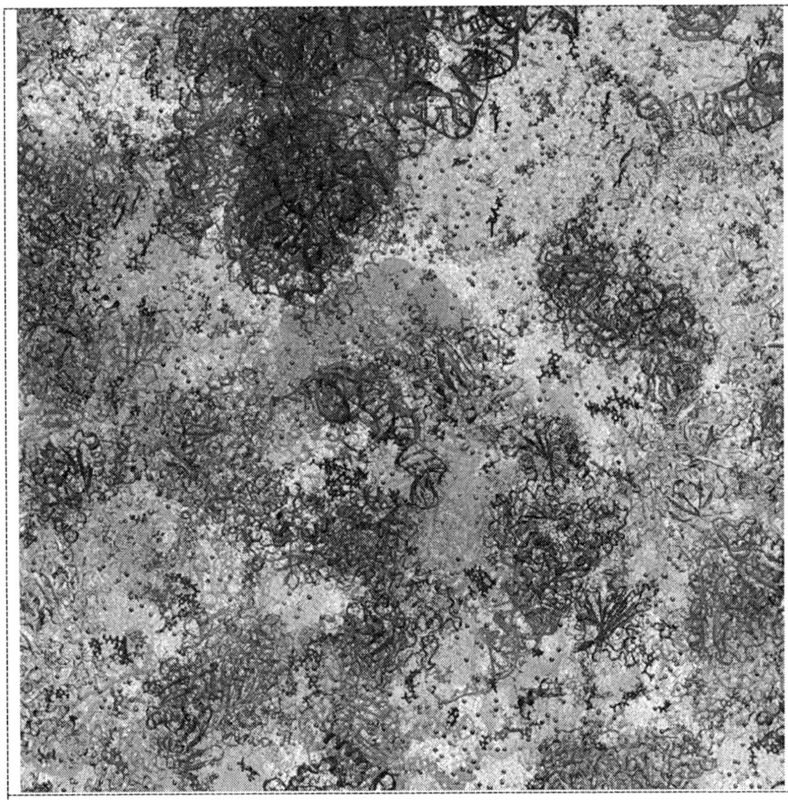

Figure 1. Model of bacterial cytoplasm in atomistic detail.

2. Data size challenges

The large amount of data generated by computer simulations intrinsically presents big data challenges. At the logistical level, the storage, management, and dissemination of terabyte-scale trajectory data is still not trivial even as the performance and capacity of storage resources continues to increase [9]. While it has become common at least in most areas of biological science that primary data is made publicly available when research findings are published, this is generally not the case for simulation studies [10]; furthermore, while there are public databases for essentially every kind of biological data generated by experiments or from computational analysis, there is no widely used resource for molecular dynamics simulations [11] despite many efforts to develop such databases [10, 12-16]. The main reason is that a database where the original trajectories are collected and simply made available for download is not practical because of network bandwidth limitations. This is especially true for the large cellular-scale simulations that are beginning to emerge, as it will likely remain impossible in the foreseeable future to efficiently transfer petabyte-scale data sets over the Internet.

One way to overcome such limitations is to reduce the data. This can be done by using only snapshots at infrequent time points and/or by removing less important parts of the system such as solvent. It is also possible to store the system of interest at a coarse-grained level even though the original simulations were carried out in atomistic detail with the idea that an atomistic level of resolution could be reconstructed on the fly if needed [17]. A different strategy is to maintain the full data sets but develop tools that allow remote analysis so that only the results of such analysis have to be transmitted instead of the actual data [10, 12]; however, this requires significant software and hardware infrastructure and may limit the flexibility in terms of what analysis can be carried out. The same challenges also apply even within a computational laboratory when there is not enough storage space to maintain all of the generated simulations available for direct access so that re-analysis and comparative studies of large sets of simulations become difficult tasks although it would be prudent to do so given the high computational costs to generate the data in the first place.

3. Information wealth challenges

A different challenge is how to fully capitalize on the wealth of information that is provided by simulations where every molecule is represented in atomistic detail. Early molecular dynamics simulations of biological macromolecules, where a single small protein or piece of DNA was studied over pico- to nanosecond time scales, could often be understood qualitatively by inspecting molecular movies generated from the trajectories but nowadays this is rarely a productive approach for analyzing simulation trajectories. In fact, in the largest cellular-scale simulations published recently that contain thousands of proteins [7, 18] it would take days just to look at every molecule for a minute each. Simulations are often carried out with a specific scientific question in mind and one way to navigate the large amount of information is to only focus on that question during the analysis in a strictly hypothesis-driven fashion; however, large-scale simulations of complex biological systems allow scientific discovery beyond the motivating question(s) that led to the simulations to be carried out initially.

One approach to attempt such discovery in a system where one cannot simply 'look' at what is happening is to carry out a battery of standard analyses to characterize structural and dynamic features followed by automated feature analysis to identify, for example, if the secondary structure in a given protein is lost compared to an experimental reference structure. It is more difficult, however, to recognize causal relationships whereby the loss of secondary structure in this example may be related to interactions with other molecules or locally altered solvent properties. Not knowing what to focus on *a priori* and being faced with too many possibilities when considering first correlations and then causal connections between different components in a system with as many as 100 M atoms presents a classical data science challenge. There is an ideal opportunity for the application of machine learning and artificial intelligence techniques to interpret the increasingly rich information that is being generated nowadays via simulation. While machine learning has seen some applications in the analysis of molecular dynamics data [19, 20], the unsupervised analysis of complex biomolecular simulations

IOP Conf. Series: Journal of Physics: Conf. Series **1036** (2018) 012010 doi:10.1088/1742-6596/1036/1/012010

remains a key challenge that needs to be addressed. Molecular dynamics simulations of entire cells with billions of atoms and tens of thousands of macromolecules are on the horizon and the traditional manual analyses will become entirely inadequate for such data sets.

4. Connections between simulation and experiment

As computer simulations are based on theoretical models, the connection with experiments is crucial. At the onset, experiments are needed to define the composition of a given system and resolve molecular structures to provide initial coordinates from which simulations can be started. Experimental findings also usually generate the initial motivation to carry out simulations with the goal of developing a deeper understanding based on details extracted from the simulations that are not accessible experimentally. Once simulations have been carried out, comparisons with experiment are needed for validation and to follow up on predictions to further advance knowledge. As important as the connection between simulations and experiments is, there are many challenges: experimental conditions and scales are often quite different from what is being simulated and experiments cannot typically provide full atomistic resolution and picosecond time resolution at the same time as in the simulations.

Atomistic resolution in experiments is often obtained via extensive time- and ensemble averaging, as in X-ray crystallography, nuclear magnetic resonance spectroscopy, or cryo-electron microscopy. Simulations often rely on the ergodic hypothesis that states that the long-time average of a single system is equivalent to the ensemble average of the same system [21]; however, since the accessible time scales in the simulations are still relatively short, a single simulation of a single system rarely provides fully-converged conformational space averages. The situation can be improved by running multiple replicates of a given system, or by employing enhanced sampling strategies to accelerate the exploration of the conformational energy landscape with given computational resources [22-24]. Simulations of multiple components such as in simulations of cellular environments, on the other hand, do allow ensemble averaging if there are multiple copies of the same molecule present [7]. Because different copies in such systems experience different local environments and sample different regions of phase space, extensive conformational averaging may be achieved with such simulations even if the overall simulation lengths for such large systems are much more limited than for smaller, single-molecule systems.

Different copies of the same molecule experiencing different environments may also increase the complexity of the conformational energy landscape when interactions with the environment modulate biomolecular structure. There is increasing evidence that non-specific protein-protein interactions modulate biomolecular structure [25], but simulations suggest that such effects may be limited to only a small subset of molecules [7, 26]; for example, in a simulation of a bacterial cytoplasm, a few copies of the molecule pyruvate dehydrogenase, subunit A, were seen to unfold due to protein-protein interactions, whereas the majority of copies remained stably folded in their native state [7]. While such rare events can be easily discerned in simulations, experiments that rely on averages often cannot detect events that occur for only a small percentage of molecules making it difficult to compare simulations and experiments in such cases.

The comparison of kinetic and diffusive properties between simulation and experiments requires that both approaches cover the same time scales. Simulations are so far limited to millisecond time scales for single molecules and to the microsecond range for the largest cellular systems. Experiments may describe dynamics from femtoseconds to hours or longer depending on the method that is being used. NMR experiments are especially suitable to cover a wide range of time scales [27, 28]. NMR data is often compared with molecular dynamics simulations because processes occurring on different time scales can be isolated and matched to the simulation time scales [29-32]; however, the standard interpretation of NMR data usually depends on certain assumptions that can become problematic in highly crowded cellular systems. Using dynamic relaxation of protein backbone amide N-H vectors as an example (see figure 2), the resulting dynamics is a result of combining internal dynamics with rotational and translational diffusion. All three processes can be easily separated in the analysis of simulation data. NMR on the other hand measures the overall relaxation rates that combine fluctuations due to internal motions with rotational tumbling. In interpreting the experimental data, the usual assumption is that both the solvent environment and the rotational tumbling are isotropic; furthermore, a separation of time scales between internal and rotational motions is assumed. It is then possible to determine the extent of internal motions on short and long time scales as well as rotational diffusion times via so-called model-free analysis [33]; however, none of these assumptions may be true in highly crowded heterogeneous cellular systems because of anisotropy and coupling of internal and diffusional motions. Although one can formally carry through with the standard analysis of the experimental data, the results are likely going to become problematic in such a scenario. A better strategy would be the direct calculation of relaxation times from the simulation as it avoids the assumptions made in the experimental analysis [34]; however, this is also not without challenges, the simulations would then be required to not just reach sufficiently long time scales for convergence but also they also have to accurately capture the entire dynamic spectrum as a result of different processes, each with their respective time scales estimated correctly. Because of methodological limitations this is often not the case; for example, while the time scale of internal protein motions are perhaps accurate with current force fields, the diffusional motions that depend on the water model used in the simulations would be accelerated if the popular TIP3P model is used [35]. While a separate analysis of diffusion and internal dynamics could correct such artifacts, this is not easily done when calculating spectra for the overall dynamics. While this example focuses on NMR data, similar arguments can be made for comparisons with single-molecule fluorescence data, electron paramagnetic (EPR) spectroscopy, or other types of spectroscopic measurements.

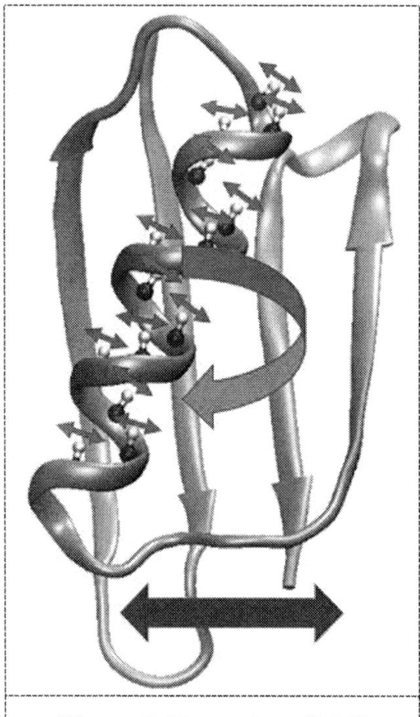

Figure 2. Dynamics of N-H vectors: internal motion (blue), rotational diffusion (green), translational diffusion (purple)

Beyond structural and dynamic information, simulations also allow the estimation of various energetic terms; for example, it is possible to calculate free energies of crowding, the free energy for transferring a given molecule from dilute solvent to a crowded environment, or interaction free energies within cellular environments [26, 36]. There are so far no good examples for experimental measurements of such quantities although this may not be impossible with innovative experimental setups.

Finally, it goes almost without saying that meaningful comparisons between experiments and simulations require that the same systems are being studied *in silico* and *in vitro* or *in vivo*. For traditional studies of a single molecule under dilute conditions that is relatively easy to accomplish because the goal of both experiments and simulations would be to study pure systems with a minimum of contaminants; however, cellular-scale biological systems are highly complex and while the simulated systems are well-defined, corresponding experimental systems are less controlled which makes it difficult to follow up on predictions made in simulation studies for such systems. For example, it would be next to impossible to carry out experiments on the same exact model cytoplasm that is shown in figure 1. Even although it may be possible to match the same exact initial molecular composition, active metabolism, protein and nucleic acid synthesis and degradation, and molecular diffusion and osmotic effects would likely lead to significant fluctuations in concentrations and molecular composition. On the other hand, simulations of complete cells with all of the biological function intact are not going to be feasible in the foreseeable future, therefore, matching simulated and experimentally studied cellular systems is a major challenge that will remain difficult to overcome even as the scale and complexity further increase.

5. Conclusions

The extension of computer simulations to cellular-scale system is exciting but also presents significant challenges for fully taking advantage of the vast data generated in such efforts. Efficient management and mechanisms of public sharing of the resulting large data sets is the first issue that needs to be addressed but the bigger issues are how to analyze and interpret such data sets effectively. As traditional approaches to the analysis of simulations do not scale well to systems with thousands of macromolecules, a greater emphasis on machine learning and artificial intelligence will be required in the future. The comparison with experimental data, a vital component in any simulation study, is furthermore complicated by a variety of factors, even though time and spatial scales are increasingly overlapping between simulations and experiments. A major issue is the complexity of cellular-scale systems where the interpretation of experimental data becomes more difficult and the increasing importance of rare events that can be observed in simulations but are difficult to see experimentally. There are also challenges with exactly matching systems between simulations and experiments, which will make it increasingly difficult to follow-up experimentally on predictions made by simulations of cellular environments. Understanding and addressing these challenges will be essential in maintaining the synergy between simulations and experiments in studying biological systems at increasingly larger scales.

6. References

[1] van Gunsteren W F, Berendsen H J C. 1990 *Angew. Chemie* **29** 992-1023
[2] Karplus M, Petsko G A. 1990 *Nature* **347** 631-9
[3] Hospital A, Goñi J R, Orozco M, Gelpi J L. 2015 *Adv. Appl. Bioinform. Chem.* **8** 37-47
[4] Karplus M, McCammon J A. 2002 *Nat. Struct. Biol.* **9** 646-52
[5] Feig M, Yu I, Wang P-h, Nawrocki G, Sugita Y. 2017 *J. Phys. Chem. B*
[6] Perilla J R, Goh B C, Cassidy C K, Liu B, Bernardi R C, Rudack T, Yu H, Wu Z, Schulten K. 2015 *Curr. Opin. Struct. Biol.* **31** 64-74
[7] Yu I, Mori T, Ando T, Harada R, Jung J, Sugita Y, Feig M. 2016 *eLife* **5** e19274
[8] Shaw D E, Maragakis P, Lindorff-Larsen K, Piana S, Dror R O, Eastwood M P, Bank J A, Jumper J M, Salmon J K, Shan Y B, et al. 2010 *Science* **330** 341-6
[9] Kumar A, Grupcev V, Berrada M, Fogarty J C, Tu Y-C, Zhu X, Pandit S A, Xia Y. 2014 *J. Big Data* **2** 9

[10] Hospital A, Andrio P, Cugnasco C, Codo L, Becerra Y, Dans P D, Battistini F, Torres J, Goñi R, Orozco M, et al. 2016 *Nucleic Acids Res.* **44** D272-D8
[11] Thibault J C, Roe D R, Facelli J C, Cheatham T E. 2014 *J. Cheminf.* **6** 4
[12] Feig M, Abdullah M, Johnsson L, Pettitt B M. 1999 *Future Gener. Comput. Syst.* **16** 101-10
[13] Tai K, Murdock S, Wu B, Ng M H, Johnston S, Fangohr H, Cox S J, Jeffreys P, Essex J W, P. Sansom M S. 2004 *Org. Biomol. Chem.* **2** 3219-21
[14] Meyer T, D'Abramo M, Hospital A, Rueda M, Ferrer-Costa C, Pérez A, Carrillo O, Camps J, Fenollosa C, Repchevsky D, et al. 2010 *Structure* **18** 1399-409
[15] van der Kamp M W, Schaeffer R D, Jonsson A L, Scouras A D, Simms A M, Toofanny R D, Benson N C, Anderson P C, Merkley E D, Rysavy S, et al. 2010 *Structure* **18** 423-35
[16] Thibault J C, Facelli J C, Cheatham T E. 2013 *J. Chem. Inf. Model.* **53** 726-36
[17] Cheng Y-M, Gopal S M, Law S M, Feig M. 2012 *IEEE Trans. Comput. Biol. Bioinf.* **6** 476-86
[18] Feig M, Harada R, Mori T, Yu I, Takahashi K, Sugita Y. 2015 *J. Mol. Graph. Modell.* **58** 1-9
[19] Glazer D S, Radmer R J, Altman R B. 2008 *Pac. Symp. Biocomput.* 332-43
[20] Rydzewski J, Nowak W. 2016 *J. Chem. Theory Comput.* **12** 2110-20
[21] Feig M. Molecular simulation methods. Computational modeling in lignocellulosic biofuel production. Acs symposium series. 1052: American Chemical Society; 2010. p. 155-78.
[22] Paul W, Muller M. 2002 *Comput. Phys. Commun.* **146** 113-7
[23] Okamoto Y. 2004 *J. Mol. Graph. Modell.* **22** 425-39
[24] Abrams C, Bussi G. 2014 *Entropy* **16** 163-99
[25] Charlton L M, Barnes C O, Li C, Orans J, Young G B, Pielak G J. 2008 *J. Am. Chem. Soc.* **130** 6826-30
[26] Harada R, Tochio N, Kigawa T, Sugita Y, Feig M. 2013 *J. Am. Chem. Soc.* **135** 3696-701
[27] Kay L E. 1998 *Biochem. Cell Biol.* **76** 145-52
[28] Kay L E. 2016 *J. Mol. Biol.* **428** 323-31
[29] Virk A S, Stait-Gardner T, Willis S A, Torres A M, Price W S. 2015 *Front. Phys.* **3** 1
[30] Graf J, Nguyen P H, Stock G, Schwalbe H. 2007 *J. Am. Chem. Soc.* **129** 1179-89
[31] Roll C, Ketterle C, Faibis V, Fazakerley G V, Boulard Y. 1998 *Biochemistry* **37** 4059-70
[32] Chen J H, Brooks C L, Wright P E. 2004 *J. Biomol. NMR* **29** 243-57
[33] Lipari G, Szabo A. 1982 *J. Am. Chem. Soc.* **104** 4546-59
[34] Peter C, Daura X, van Gunsteren W F. 2001 *J. Biomol. NMR* **20** 297-310
[35] Yeh I C, Hummer G. 2004 *J. Phys. Chem. B* **108** 15873-9
[36] Feig M, Sugita Y. 2012 *J. Phys. Chem. B* **116** 599-605

Acknowledgments

Funding was provided by the National Institute of Health Grants R01 GM084953 and GM103695 (to MF), from the National Science Foundation Grant MCB 1330560 (to MF), the Fund from the High Performance Computing Infrastructure (HPCI) Strategic Program (hp120309, hp130003, hp140229, hp150233) and HPCI general trial use project (hp150145, hp160120) and FLAGSHIP 2020 project focused area 1 "Innovative drug discovery infrastructure through functional control of biomolecular systems" (hp160207, hp170254) of the Ministry of Education, Culture, Sports, Science and Technology (MEXT) (to YS), a Grant-in-Aid for Scientific Research on Innovative Areas "Novel measurement techniques for visualizing 'live' protein molecules at work" (No. 26119006) (to YS), a grant from JST CREST on "Structural Life Science and Advanced Core Technologies for Innovative Life Science Research" (to YS), RIKEN QBiC, iTHES, and Dynamic Structural Biology (to YS), a Grant-in-Aid for Scientific Research (C) from MEXT (No. 25410025) (to IY).

IOP Conf. Series: Journal of Physics: Conf. Series **1036** (2018) 012011 doi:10.1088/1742-6596/1036/1/012011

TimeTubes: Visual Exploration of Observed Blazar Datasets

Issei Fujishiro, Naoko Sawada, Masanori Nakayama
Center for Information and Computer Science, Keio University,
3-14-1 Hiyoshi, Kohoku-ku, Yokohama 223-8522, Japan

E-mail: {fuji, naoko.sawada, nakayama}@fj.ics.keio.ac.jp

Hsiang-Yun Wu
Center for Information and Computer Science, Keio University,
3-14-1 Hiyoshi, Kohoku-ku, Yokohama 223-8522, Japan
Current address: Institute of Computer Graphics and Algorithms, TU Wien,
Favoritenstr. 9 / 5. Stock (5th floor), A-1040 Vienna, Austria

E-mail: hsiang.yun.wu@acm.org

Kazuho Watanabe
Department of Computer Science and Engineering, Toyohashi University of Technology,
1-1 Hibarigaoka, Tempaku-cho, Toyohashi 441-8580, Japan

E-mail: wkazuho@cs.tut.ac.jp

Shigeo Takahashi
School of Computer Science and Engineering, University of Aizu,
Aizu Wakamatsu, Fukushima 965-8580, Japan

E-mail: takahashis@acm.org

Makoto Uemura
Astrophysical Science Center, Hiroshima University,
1-3-1 Kagamiyama, Higashi-Hiroshima 739-8526, Japan

E-mail: uemuram@hiroshima-u.ac.jp

Abstract. Blazars are attractive objects for astronomers to observe in order to burrow into the magnetic field in the relativistic jet. This paper presents TimeTubes as a novel visualization scheme that allows astronomers to interactively explore characteristic temporal variation patterns in observed blazar datasets. In the TimeTubes spatialization, the two Stokes parameters and their errors with a common timestamp are transformed into an ellipse. A series of such ellipses are aligned in parallel along the timeline to form a 3D volumetric tube. The resulting tube is then colorized by the observed intensities and colors of the blazar, and finally volume-rendered. A designated user interface is provided with visual exploration functions according to Shneiderman's Visual Information Seeking Mantra. In the latest version, an auxiliary mechanism, called visual data fusion, was incorporated to ameliorate data- and mapping-inherent uncertainties for more efficient and effective visual exploration.

1. Introduction

Blazars are active galactic nuclei whose relativistic jets ejected from the central black hole are pointing toward the Earth [1] (see Figure 1). Astronomers have observed blazars for a long period of time to acquire time-dependent multivariate datasets for optical polarization, since they can obtain a clue to the magnetic field structure in the jet from the observed temporal variation features of the position angle of polarization. To visually analyze the datasets, they have commonly used *animated scatterplots* [2], as shown in Figure 2. The conventional visualization scheme suffers from visual clutter and low interactivity, and thus the limitations unfortunately compromise astronomers' ability to carefully scrutinize the original dataset.

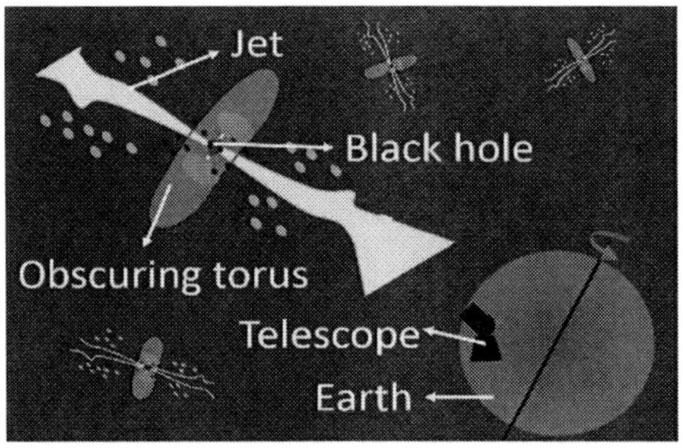

Figure 1. Observation of a blazar. The ellipse and its center represent an obscuring torus and a black hole, respectively. The jets ejected from the central black hole are pointing toward the Earth. The position angle of the optical polarization is perpendicular to the direction of the magnetic field of the emitting region in the optically-thin regime.

In our initial report [3], we attempted to present a novel visualization scheme, termed *TimeTubes*, whose geometrical transformation of the time-dependent multivariate data and designated interactions are provided for astronomers, allowing them to effectively perform their visual exploration. More specifically, TimeTubes spatializes the blazar data into a 3D volumetric tube structure, in which an elliptic shape is used to encode the two Stokes parameters and their errors with a common timestamp and align a series of such ellipses along the timeline to form a tube in the 3D space (see Figure 3). The resulting tube is then colorized by observed intensities and colors, and finally volume-rendered. The tube makes it easy for astronomers to detect universal patterns in the trajectory of the polarization parameters, which may correlate with the intensity and color variation (see Figure 4). Some new behaviors of blazars have been found through the use of TimeTubes and reported in [4, 5].

In previous reports [6, 7], we improved the TimeTubes tool to help astronomers analyze datasets more effectively and obtain more information at a time by incorporating a novel mechanism, *visual data fusion*, into the 3D spatialization. By visually fusing another dataset from a different observatory but for the same blazar into the originally targeted dataset, TimeTubes can ameliorate two types of *uncertainty*: data-inherent uncertainty, which arises from errors and missing periods in the dataset; and mapping-inherent uncertainty, which arises as a result of interpolation when TimeTubes transforms the data into a continuous geometry. The visual data fusion enables astronomers to validate the datasets in a meticulous manner.

Following a brief review of related work in the next section, the spatial design of TimeTubes is described in Section 3, and designated user interface and exploration functions are introduced in Section 4. Section 5 is devoted to the description of the basic mechanism of visual data fusion. Finally, Section 6 concludes the paper with a few remarks on on-going challenges.

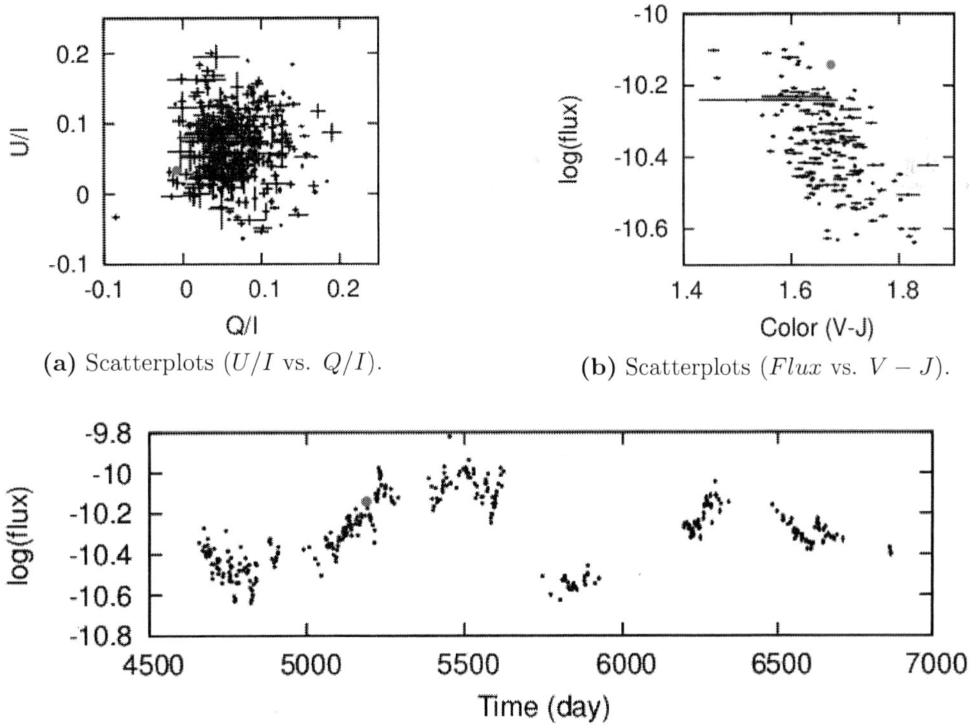

(a) Scatterplots (U/I vs. Q/I).

(b) Scatterplots ($Flux$ vs. $V - J$).

(c) Light curve ($Flux$ vs. JD).

Figure 2. Animated scatterplots as a conventional way to visualize time-dependent multivariate blazar datasets. The descriptions of the variables are tabulated in Table 1. In (a), crucifix glyph is used to represent polarization parameters as well as their error intervals. The three scatterplots are mutually linked with current dots highlighted in red as a cross-reference.

2. Related Work

Various techniques have been developed for visualizing time-dependent data, some of which have been extended to deal with multivariate data. For example, Bach et al. [8] developed *Time Curves*, which is a general approach to visualizing time-dependent data. In this approach, data similarity is naturally expressed by folding a timeline. *ThemeRiver*, which was proposed by Havre et al. [9], depicts thematic variation of documents over time using a smooth stacked-graph layout. *Storyflow*, which was developed by Liu et al. [10], improved the ThemeRiver by enabling it to visualize storylines legibly.

In addition, 3D visualization techniques have been developed for visualizing time-dependent multivariate data. The *Great Wall*, which was proposed by Tominski and Schulz [11], visualizes spatiotemporal data with reference to 2D geographical space and 1D linear time. To express time variation, Gruend et al. [12] expanded *parallel coordinate plots* into 3D space. *Perspective Tunnel* [13] exploits 3D space through mapping information on the ceiling, floor, and walls of a tunnel in a perspective manner. It was developed for overcoming occlusion of overlapping items, ambiguity of symbols in representation, and physical display limitations.

However, these techniques do not provide an optimal way to explore the blazar datasets in an astrophysics context; indeed, blazars fluctuate drastically in a wide time range, which can be from a few minutes to several years, and the observed datasets intrinsically become error-prone. This has motivated us to come up with a new visualization scheme: TimeTubes.

3. Spatial Design

In this section, we describe the target blazar datasets and spatial mapping of TimeTubes [3].

3.1. Target dataset

As tabulated in Table 1, the target dataset consists of six representative variables, which have been chosen from eight variables in the original blazar dataset to facilitate effective blazar classification. All the variables are measured in Julian day (JD).

Optical polarization, which is a significant property of a blazar, is described by the degree (PD) and angle (PA) of polarization. The parameters Q/I and U/I are two linear polarization parameters, where I is the total flux of the blazar and Q and U are Stokes parameters. These quantities are related in the following equations:

$$PD = \sqrt{(\frac{Q}{I})^2 + (\frac{U}{I})^2} \quad (1)$$

$$PA = \frac{1}{2}arctan(\frac{U}{Q}) \quad (2)$$

In the conventional method of visualizing the dataset, three specific kinds of animated scatterplots have been used in combination with one another, as shown in Figure 2. Two scatterplots are used to correlate Q/I and U/I (a) and color and flux (b), respectively. Note that in (a), the errors of the parameters Q/I and U/I are indicated by a crucifix glyph at the point $(Q/I, U/I)$. The last scatterplots, called *light curves* in (C), is used to see the temporal variation of observed flux. When these three scatterplots are animated, current dots highlighted in red are used as a cross-reference. Since the points become too crowded and mutually overlap, especially when the observation period gets longer, these representations suffer from visual clutter artifacts and low interactivity. These limitations unfortunately compromise astronomers' ability to carefully scrutinize the dataset.

Table 1. Representative time-varying variables for observed blazars.

Variable	Description
$Flux(V)$	Observed intensity of blazar
Q/I	Linear polarization component (0 or $\pi/2$ degrees)
E_Q/I	Error on Q/I
U/I	Linear polarization component ($\pi/4$ or $3\pi/4$ degrees)
E_U/I	Error on U/I
$V - J$	Observed color of blazar

3.2. Spatialization

We have respected the concepts of the Great Wall and Perspective Tunnel described in Section 2 to design how TimeTubes spatializes the six variables in Table 1.

TimeTubes relies on 3D geometry to visually encode the Stokes parameters with each timestamp as an ellipse, whose central coordinate is given as $(x, y, z) = (Q/I, U/I, JD)$ and whose radius values on major and minor axes are given as E_Q/I and E_U/I, respectively. As shown in Figure 3(a), the confidence interval of each data sample can be illustrated intuitively with the ellipse width. Then, assuming a sort of temporal coherence of the observed parameters, the neighboring ellipses are connected to each other by five centripetal Catmull-Rom splines (one for connecting the ellipse centers and the remaining four for connecting four right perimeter points of the ellipses), to approximate a 3D volumetric tube, as shown in Figure 3(b). Note that, unlike other variants, a centripetal Catmull-Rom spline can avoid self-intersection and cusp [14].

For substantializing the tube in 3D space, *pseudo* volume rendering was employed to add a sort of uncertainty visualization flavor. The tube is actually volume-rendered with multiple (empirically eight) concentric tubes with semi-translucent colors, as illustrated in Figure 3(b). Moreover, to enable users to observe the tubes from any angle in 3D space, these concentric tubes are built as the slices in *shell* texture, which has often been used in fur rendering [15]. To colorize the tube with the color $V - J$ and the intensity $Flux(V)$, a 2D color transfer function was introduced. To reflect the reliability of observations, an associated opacity transfer function was introduced as well. The translucency of the outer tube is generally defined to be larger than that of the inner. Small errors result in a sharply rendered tube, whereas a tube associated with a large range of errors has a vague appearance.

The time-varying nature of the six variables in Table 1 can be clarified by evaluating how fast and how much the tube twists, and the colored trajectory can be observed intuitively in 3D space, even from a still image.

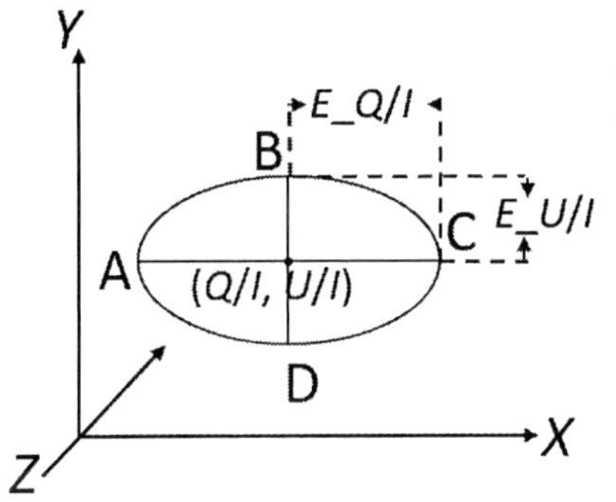

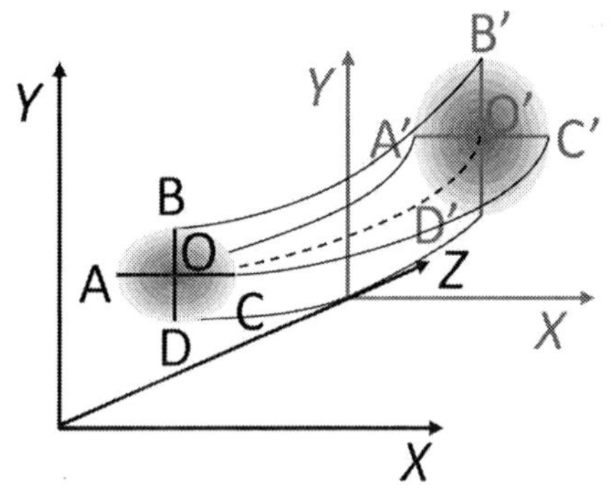

Figure 3. Spatial mapping in TimeTubes: (a) Stokes parameter samples with each timestamp are shown as an ellipse. (b) The neighboring ellipses are connected by centripetal Catmull-Rom splines.

4. Exploration Functionality

A prototype of TimeTubes has been implemented using Delphi on a laptop with an Intel Core i7-4510U CPU, an NVIDIA GeForce GT750M GPU, and 16GB RAM. Throughout the draft, we chose the dataset for a specific blazar named **3C 454.3** to show all visualization results.

Figure 4 gives a screenshot of the up-to-date user interface of TimeTubes. Figure 4(a) shows the visualization result of the dataset, where users are allowed to interactively change several viewing parameters of the tube. For example, they might want to observe the tube from different viewpoints or get detailed information about a specific part of the tube on demand. If the orthographic projection is chosen, the resultant view uses an aggregate of elliptical glyphs to imitate the Stokes QU plane scatterplots in Figure 2(a). On the other hand, Figure 4(b) shows federated (linked) scatterplots, whose axes can be selected freely by users. A light curve as in Figure 2(c) is actually plotted here. The menu shown in Figure 4(c) allows for camera control, switching the display of the cruciform axes, and placing a glyph that expresses the contour of the tube at each data sample. The operation panel shown in Figure 4(d) enables users to perform their visual analysis through overview, filtering, zooming, details on demand, and relating with the scatterplots. In the upper part of the panel, the selected 2D transfer function is shown,

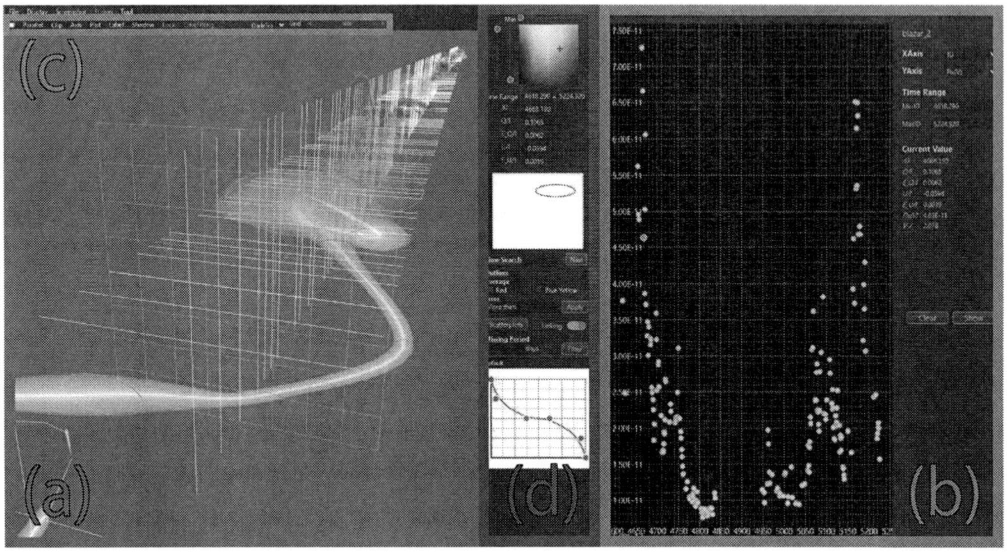

Figure 4. User interface of TimeTubes: (a) Visualization window, (b) federated scatterplots window, (c) menu, and (d) operation panel.

in which we reflected the change in colors commonly used in astrophysics like the scatterplots ($Flux$ vs. $V - J$) in Figure 2(b), while users are allowed to modify the 2D color transfer function for their own illustrative purposes. In the lower part of the panel, the editable opacity transfer function is also delineated. A built-in transfer function editor allows users to change the opacity distribution by dragging points on the definition curve. Users may choose from three standard opacity distributions: linear (default), flat, and valley-shaped.

Astronomers have evaluated several versions of TimeTubes, and we have obtained positive feedback from them, as reported in [3].

In the following sections, we will describe the effects of the basic exploration functionality of TimeTubes according to Shneiderman's *Visual Information Seeking Mantra* [16].

4.1. Overview

Figure 5 gives an overview of the dataset. The projection adjustment allows users to switch between parallel and perspective projections. The default is set to parallel projection in Figure 5(a), according to the preference of measurability. We can take advantage of the perspective projection in Figure 5(b) in a focus+context manner. To confirm the current snapshot for analysis, the grid plane is shown. The grid opacity can be adjusted so as not to interrupt the analysis itself.

4.2. Zooming & Filtering

Figure 6(a) is a zoomed view of Figure 5(b). When zooming in the concentric tubes, color and/or opacity distribution can be adjusted. In Figure 6(a), brightness has been accentuated to highlight flare patterns (see Section 6).

On the other hand, Figure 6(b) is a filtered view of Figure 5(b). To help users inquire about the outliers of datasets, TimeTubes offers specific filtering styles. We can locate the data samples whose Q/I and U/I are both larger than the averages over the whole time domain with red axes. Furthermore, users can view the data samples where only Q/I or only U/I is larger than its average, and the data samples will be shown with blue and yellow axes, respectively.

IOP Conf. Series: Journal of Physics: Conf. Series **1036** (2018) 012011 doi:10.1088/1742-6596/1036/1/012011

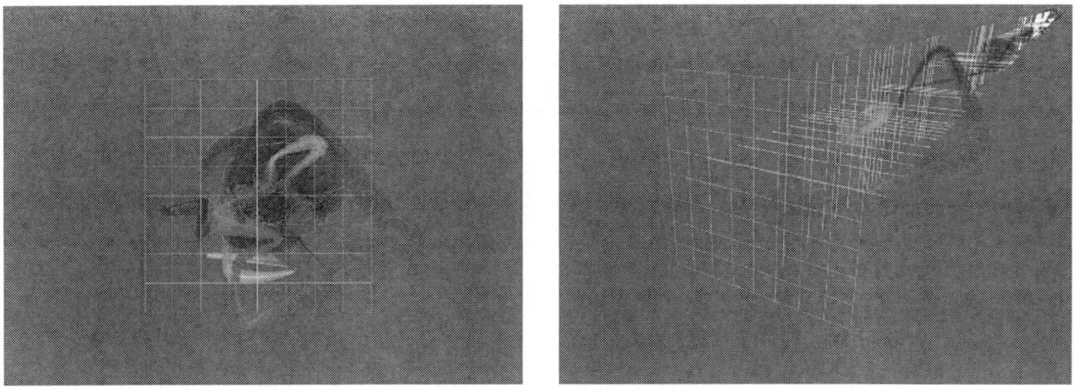

(a) Orthographic view (b) Perspective view

Figure 5. Overview.

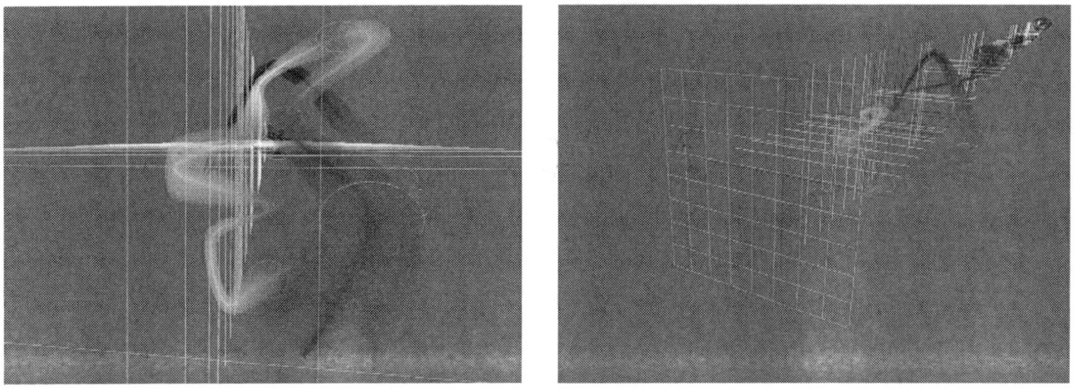

(a) Brightness-adjusted semantic zooming (b) Filtering by Stokes parameter thresholding

Figure 6. Zooming and filtering.

4.3. Details on demand

Figure 7 shows a details-on-demand view of the dataset. Users may view the tube along the timeline by scrolling a mouse, whereas time navigation allows users to randomly search for the data with a specific timestamp. As seen in Figure 7, the data probe panel returns detailed data and the sectional shape of the current ellipse. The information will change as the green grid moves along the timeline.

5. Visual Data Fusion

As described above, the observed blazar datasets exhibit uncertainty due to their errors and missing periods, whereas periods interpolated by TimeTubes result in a different type of uncertainty. In the previous reports [6, 7], such data- and mapping-inherent uncertainties were ameliorated by *visual fusion* of datasets for the same blazar from two different observatories: one from Hiroshima University (HU) [2] and the other from the University of Arizona (AU, http://james.as.arizona.edu/p̃smith/Fermi/).

Some differences can be found between HU and AU datasets. The astronomical telescope at HU allows us to simultaneously observe all six time-dependent variables in Table 1, whereas the observation timestamps of polarization and intensity in the AU dataset do not match. Therefore, we need to interpolate each of these variables respectively to treat the AU datasets the same way we treat the HU datasets.

IOP Conf. Series: Journal of Physics: Conf. Series **1036** (2018) 012011 doi:10.1088/1742-6596/1036/1/012011

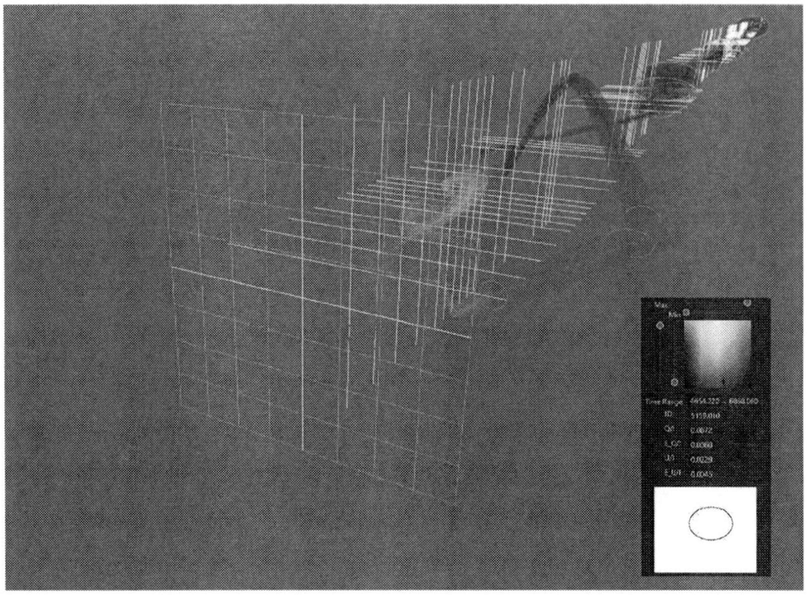

Figure 7. Details-on-demand display.

To visually identify the existence of data samples, and if the shape of the tube is formed by observed or interpolated values, the latest version of TimeTubes places white crosses at all observation timestamps, and elliptical glyphs at timestamps where polarization was observed. Moreover, to clarify whether the currently-focused-on JD in TimeTubes has observed data or not, the colors of the grid, cross, and elliptical glyph will change when the currently-focused-on JD coincides with an observation time in the dataset. Note that the color of the tube generated from the AU dataset is determined in an ad hoc manner; this is because the color data is not available from the data source.

Two modes of visual data fusion are introduced. Note that the *Merge* mode corresponds to visual data fusion at the filtering stage of the conventional visualization pipeline: fusing the datasets *numerically* to ameliorate both data- and mapping-inherent uncertainties by increasing the amount of observed data used to form a single tube. In contrast, the *Juxtaposition* mode corresponds to the mapping stage: fusing the datasets *optically* to ameliorate data-inherent uncertainty by comparing the datasets from two observatories.

5.1. Juxtaposition mode

The Juxtaposition mode offers the most straightforward method of visual data fusion, and it would be best to begin with this essential mode. The observation values are not completely reliable due to their errors like $E\_Q/I, E\_U/I$, but we can increase their reliability by using datasets from two different observatories in a comparative manner. Moreover, users can identify differences between multiple datasets with the Juxtaposition mode, which may lead to the estimation of the true observed value.

With the *Side-by-side* option, users can compare two observed datasets in separate windows, as shown in Figure 8(a). This option can be used for datasets not only for the same blazar, but also for different blazars. With the *Union* option in Figure 8(b), users can overlap multiple tubes in the same window. The color of the tubes can be defined separately. This option is suitable for comparing datasets for the same blazar, where the arbitrarily focused time of the datasets in the visualization space fits automatically.

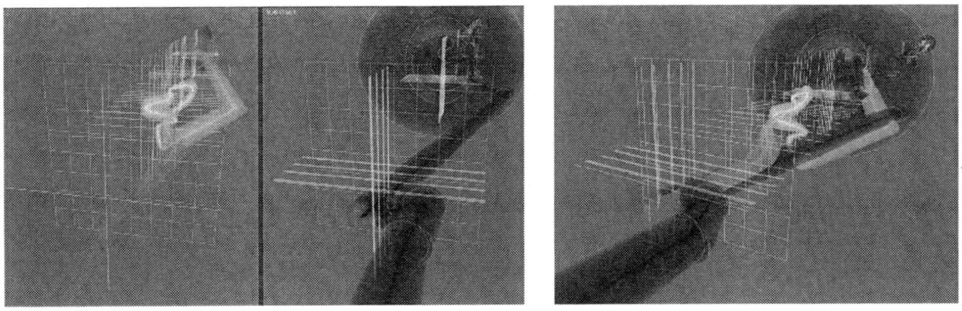

(a) Side-by-side (b) Union

Figure 8. Visual data fusion by the Juxtaposition mode.

If the values of the observed data from different observatories are very close, we can consider these values to be more reliable than those from a single observatory. If they are dissimilar, we can estimate a true value by allowing for errors and observation conditions.

Because of the differences between the observation conditions, the global centers of observed datasets may not coincide with each other. We can get a hint for proper calibration by comparing multiple observed datasets from different observatories with the Juxtaposition mode [7].

5.2. Merge mode

The number of data samples will increase when the datasets from two observatories are joined using the *Merge* mode. As a result, the missing period of observation can be shortened virtually, ameliorating the uncertainties of the missing periods. In the Merge mode, data samples from the AU and HU datasets have been merged into a single, chronological dataset before the tube is rendered. If multiple data samples contain the same observation timestamp, TimeTubes chooses the one with smaller errors. The color hue of the glyph expressing the tube contour distinguishes data sources: Green indicates HU and pink indicates AU. For the analysis of blazars, denser observation is required because blazars fluctuate very rapidly. However, there are many intrinsically unobservable periods in a dataset, due to daylight, bad weather, and so on, which may cause astronomers to miss important events of the blazar. Merging the two observed datasets, we can densify information rather than visualizing them separately. Using this mode, new features of the blazars are expected to be found.

With the *All* option, which is selected by default, all the data samples from the two observatories are merged into a single tube. This option leads to the densest information. The *Selective* option allows users to select arbitrary portions of the datasets with scatterplots. Outliers and/or unexpected values may be included in the dataset for various reasons. Visualizing the dataset without those values, users can analyze more correct visualization results. Moreover, when both HU and AU have data samples at an equal JD, users can choose to visualize either of them with this option, allowing them to visually validate the observed data.

We merged the observed datasets from HU and AU for the same blazar (that is, **3C 454.3** around $JD4,856$) using the all option. Figures 9(a) and (b) visualize the two datasets separately. The HU dataset does not have data samples for the period indicated by the red arrow in Figure 9(a), whereas the AU dataset does have the data samples for the same period, as indicated by the red circle in Figure 9(b). We relied on merge to fill in the missing period in the HU dataset with the data samples from the AU dataset. Contrasting the result shown in Figure 9(c) with the original shown in Figure 9(a) indicates that the uncertainty caused by the missing period is ameliorated and astronomers can get more information from one view than the result of visualizing a single dataset.

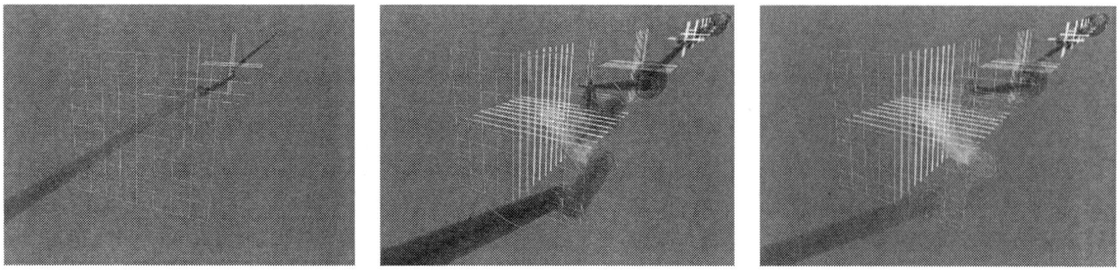

(a) HU (b) AU (c) Merge HU with AU

Figure 9. Visual data fusion by the Merge mode: Filling in the missing period.

6. Concluding Notes

The use of TimeTubes has led to the recognition of noteworthy patterns in blazar variations [4, 5]. For example, Figure 10 (a) and (b) visualize the blazar dataset around *JD* 5,156, where a counterclockwise polarization *rotation* feature is clearly illustrated, which appears around the lower-right region of the *QU* plane for one month. Owing to TimeTubes, we can quickly find that the *flare* (a light burst where the intensity of a blazar suddenly becomes stronger) and color variation are closely associated with the rotation. Such a discovery could hardly be achieved solely by the conventional animated scatterplots in Figure 2, whereas the visual exploration with TimeTubes encourages us to investigate the phenomena more carefully by returning to the conventional scatterplots as shown in Figure 10(c).

After browsing the tube in the direction of time, a similar event was manually found also around *JD* 5,504, as shown in Figure 11. If we could derive a specific metric from the time-varying multi-dimensional variables to quantize the relationship between the polarization rotation and flare, a new function of visual query for similarity search would be realized in a future TimeTubes framework.

(a) Front view (b) Perspective view (c) Associated light curve

Figure 10. Polarization rotation and flare around *JD* 5,156.

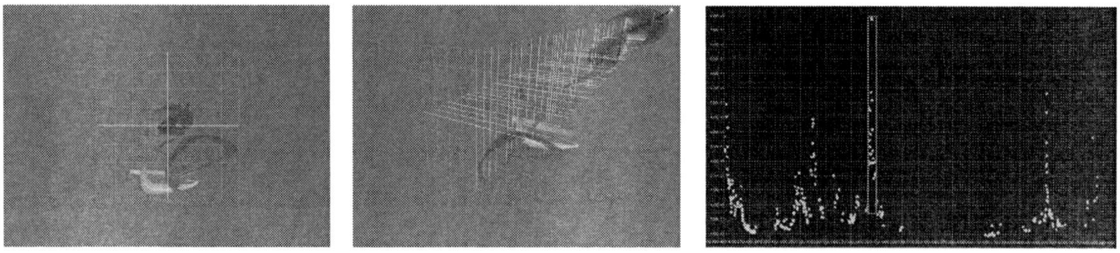

(a) Front view (b) Perspective view (c) Associated light curve

Figure 11. Polarization rotation and flare around *JD* 5,504.

From a visual design perspective [17], the TimeTubes spatialization could be thought of as a *2.5D* design. This is because the elliptical plane respects observable light patterns, whereas we treat depth very differently to chase temporal alteration of the variables. Thanks to the flexibility in projection and the effect of view frustum culling, each of the ellipses can be observed with minimum occlusion.

In order to support more effective analytics tasks, we have recently attempted to emphasize the depth cues of the tube images in an immersive analytics environment for individuals. A novel method, relying primarily on the psychological factors of depth perception—mainly through *anamorphosis* and *motion parallax*—allows the viewer to easily perceive a 3D popup effect with the naked eye using only two standard display monitors which are orthogonally arranged, and a single Web camera [18], as shown in Figure 12. In anamorphosis, the distorted projection requires a viewer to occupy a specific vantage point so that he/she reconstitutes the intended image. However, such a constraint can be uncurbed by keeping track of the viewer's line of sight in real time. By combining the relaxed anamorphosis with repeated off-axis projections of a 3D geometry content with spotlight rendering, we succeeded in inducing a sort of motion parallax. *Good continuation* of the projected content guarantees *illusion of immersion* for encouraging continuous and precise observation of TimeTubes.

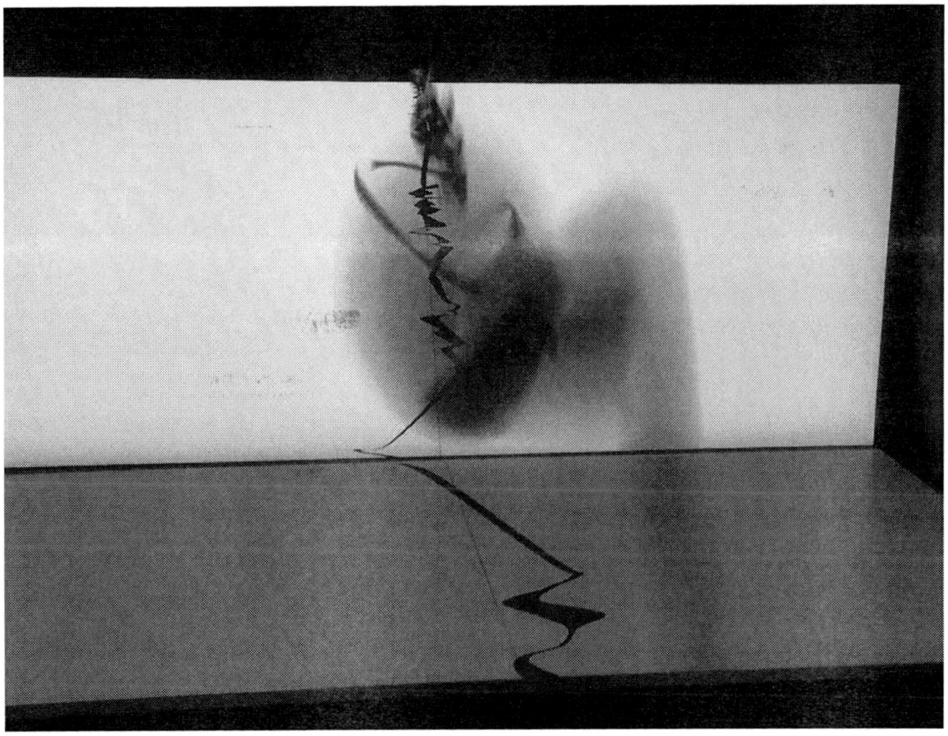

Figure 12. TimeTubes in an immersive display environment: A preliminary result.

Acknowledgments
The present work has been financially supported in part by MEXT KAKENHI Grants-in-Aid for Scientific Research on Innovative Areas No. 25120014 and No. 25120007. Akihiro Imada from Keio University assisted in porting our TimeTubes code to the immersive display environment in Figure 12.

References

[1] Antonucci R 1993 *Annual Review of Astronomy and Astrophysics* **431** 473
[2] Ikejiri Y *et al* 2011 *Publications of the Astronomical Society of Japan* **63** 3 639
[3] Xu L *et al* 2016 *Proc. NICOGRAPH International 2016* pp 15–21
[4] Uemura M *et al* 2016 *Galaxies* **4** 3 23
[5] Uemura M *et al* 2017 *Publ. Astron. Soc. Japan* (Preprint: https://arxiv.org/abs/1709.02524)
[6] Sawada N *et al* 2017 *Proc. IEEE PacificVis 2017* pp 336–337·
[7] Sawada N *et al* 2017 *CGI'17 Proc. Computer Graphics International Conference* 14
[8] Bach B *et al* 2016 *IEEE Transactions on Visualization and Computer Graphics* **22** 1 559
[9] Havre S *et al* 2000 *Proc. IEEE Symposium on Information Visualization* pp 115–123
[10] Liu S *et al* 2013 *IEEE Transactions on Visualization and Computer Graphics* **19** 12 2436
[11] Tominski C and Schulz H-J 2012 *Proc. Workshop on Vision, Modeling, and Visualization* pp 199–206
[12] Gruendl H *et al* 2016 *Computer Graphics Forum* **35** 3 321
[13] Mitchell K and Kennedy J 1997 *Proc. 8th Eurographics Workshop on ViSC* pp 31–39
[14] Barry P J and Goldman R N 1988 *ACM Computer Graphics* **22** 4 199
[15] Lengyel J E 2000 *Proc. Eurographics Workshop on Rendering 2000* pp 243–256
[16] Shneiderman B 1996 *Proc. IEEE Symposium on Visual Languages 1996* pp 336–343
[17] Ware C 2008 *Visual Thinking for Design* (Burlington MA: Morgan Kaufmann)
[18] Isaka T and Fujishiro I 2016 *ITE Journal* **70** 6 J143

Zonal Reduced-Order Modelling Toward Prediction of Transitional Flow Fields

T Misaka[1] and S Obayashi[2]
[1]Frontier Research Institute for Interdisciplinary Sciences, Tohoku University, Sendai 980-8577, Japan
[2]Institute of Fluid Science, Tohoku University, Sendai 980-8577, Japan

E-mail: misaka@edge.ifs.tohoku.ac.jp

Abstract. The utilization of measurement data is becoming attractive in various fields due to the massive growth of sensing and networking technologies. It is expected to utilize such a data-rich environment to improve engineering simulations in computer-aided engineering (CAE). Data assimilation is one of methodologies to statistically integrate a numerical model and measurement data, and it is expected to be a key technology to take advantage of measured data in CAE. However, the additional cost of data assimilation is not always affordable in CAE simulations. In this study, we consider the cost reduction of numerical flow simulation by the help of a reduced-order model. Since the prediction accuracy of existing ROMs are limited in transitional flow problems such as two- to three-dimensional flow transition, we investigate here a zonal hybrid approach of a full-order model and a reduced-order model.

1. Introduction

Computer-aided engineering (CAE) represented by numerical simulation play an indispensable role in modern research and development (R&D) processes. CAE realizes the reduction of design and development lead-time and enables design optimization for realizing better performance of products. Increase of computing power further improves the accuracy and ability of CAE, however, it also exposes the importance of the treatment of uncertainties contained in a CAE model, computational conditions and measurement data. On the other hand, downsizing and commoditization of sensors with the help of micro-electro-mechanical system (MEMS) technology are now making easy to gather measured data to a computing server. The movement of the internet of things (IoT) would make the collection of measurement data easier. One of the approaches to utilize measurement data for the improvement CAE simulation is the use of a data assimilation method. Modern data assimilation methods such as ensemble Kalman filter, particle filter and four-dimensional variational method demand several to several tens more computational resources than the original CAE simulation, therefore, there exists a wall that prevents the implementation of statistical methods such as data assimilation in the real-world R&D process.

To alleviate the above-mentioned difficulty, two major approaches can be considered, i.e., the development of efficient data assimilation techniques or the cost reduction of the original simulation model. We consider here the latter approach, which is so-called reduced-order model (ROM) approach [1,2]. The reduced-order modelling would be considered as the interpolation of a state vector in time or in a parameter space, therefore, it is not possible to predict/extrapolate a field which is not included in a learning data set such as spatial bases. In fluid problems, proper orthogonal decomposition

(POD)-based ROM is often used, where a flow field is reconstructed by using the POD spatial bases. Fast and accurate prediction is possible especially in the case of periodic flows if the pre-constructed spatial bases are appropriate. However, it is difficult to represent a transitional field without having appropriate bases before/after the transition. To overcome this, a hybrid approach, where a full-order model (FOM) and a ROM are switched to track the transition of a state, was proposed in a relatively simple problem [3]. In their research, FOM and ROM are switched in time for a whole domain. In the case of a complex flow field, the modelling fidelity can be switch in space using the framework of multi-block computational infrastructure [4].

In this study, we consider a simple flow around a square cylinder, where the flow field can be two- or three-dimensional depending on a Reynolds number. More specifically, the flow field changes from two-dimensional vortex shedding to three-dimensional complex vortical structure by increasing Reynolds number from approximately 100 to 1000. There also exists a major transition phenomenon in fluid dynamics called laminar to turbulent transition, which goes well beyond the scope of this research. First, a zonal ROM is considered. The accuracy of a ROM is then assessed by evaluating the error of full-order Navier-Stokes and the ROM predictions in the wake of the square cylinder.

2. Numerical methods

In this section, we briefly explain computational fluid dynamics simulation of a flow around a square cylinder. It is also discussed about the POD and a ROM using POD bases.

2.1. Computation flow simulation

For flow simulation, we employ the incompressible Navier-Stokes equations:

$$\frac{\partial u_j}{\partial t} + u_k \frac{\partial u_j}{\partial x_k} = -\frac{1}{\rho}\frac{\partial p'}{\partial x_j} + 2\nu \frac{\partial S_{jk}}{\partial x_k}, \tag{1}$$

$$\frac{\partial u_j}{\partial x_j} = 0, \tag{2}$$

where u_j and p' represent the velocity components in three spatial directions (j, k = 1, 2, or 3), the pressure deviation from the reference state $p = p_0 + p'$, respectively. $S_{jk} = \left(\partial u_j/\partial x_k + \partial u_k/\partial x_j\right)/2$ denotes a strain rate tensor. The summation convention is used for the velocity components u_j. ρ and ν denote density and kinematic viscosity, respectively. Equation (1) can be normalized by introducing the Reynolds number: Re $= UL/\nu$, which represents the ratio of inertial and viscous effects in a flow field and governs the flow structure including the onset of flow transition.

A computational domain is composed of a parallelepiped domain with a square cylinder of unit sides inside as in figure 1. The longitudinal length of the square cylinder is set to two and periodic boundary condition is employed in the longitudinal direction. A uniform flow is specified in the inlet boundary (the left boundary in figure 1) and the convection boundary condition is used on the outlet boundary (the right boundary). The Neumann boundary condition is applied to the top and bottom boundaries in figure 1. The wall boundary of the square cylinder is represented by specifying the opposite velocity vector inside the wall, which realizes the non-slip boundary condition on the wall with the linear velocity distribution near the wall.

Figure 2 shows flow fields with the Reynolds number of 100 and 1000, where the flow field has two-dimensional structure in the former case and the flow field shows three-dimensional structure with small disturbances in crossflow direction in the latter. This is one of flow transition phenomena occurs in a small Reynolds number regime. There is also an important flow transition phenomenon which is taken place in relatively high Reynolds numbers: Re = $2.5\text{x}10^5$ for a flow around a sphere. This type of flow transition is often due to the laminar-turbulent transition and it plays important roles in many industrial applications. It is because skin friction and heat transfer on the wall becomes several times larger in a turbulent boundary layer than those of a laminar boundary layer.

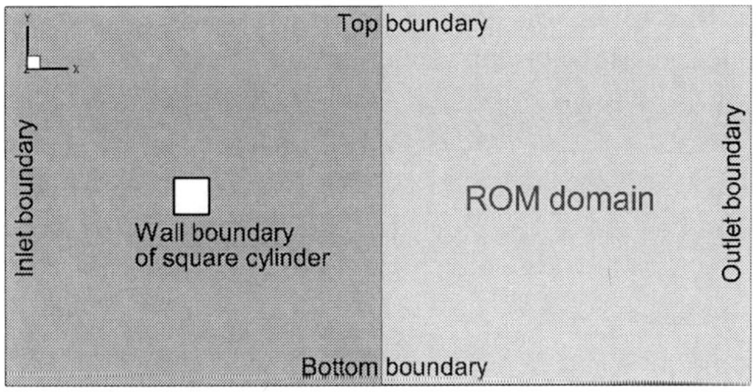

Figure 1. Computational domain and the region of interest for reduced-order modelling.

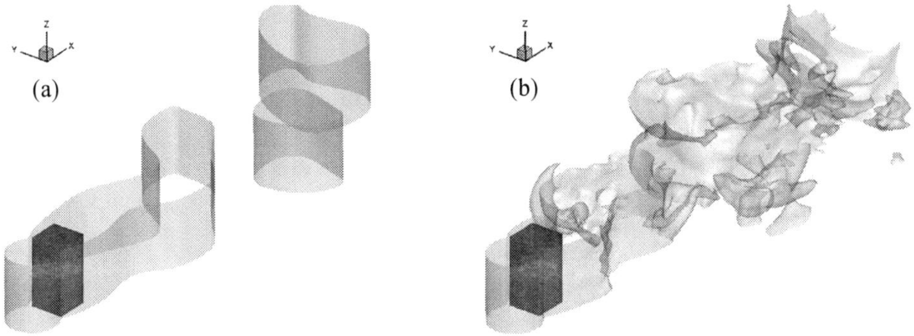

Figure 2. Flow fields obtained by Navier-Stokes equations for Reynolds numbers of (a) Re = 100 and (b) Re = 1000 visualized by a streamwise velocity.

2.2. Proper orthogonal decomposition (POD)

The proper orthogonal decomposition (POD) is a methodology similar to the principal component analysis [5]. The POD decomposes a flow field based on the variance of deviation, that is,

$$\arg\min_{\boldsymbol{\varphi}} \|X - \boldsymbol{\varphi}\boldsymbol{\varphi}^T X\|, \tag{3}$$

where X contains a certain data set and $\boldsymbol{\varphi}$ is a POD basis vector. In the case of a flow simulation data set, the matrix X tends to be very large, therefore, a snapshot-based approach is employed. During an unsteady flow simulation, a flow field is extracted every several time steps as a snapshot. The matrix X is composed as,

$$X = \begin{bmatrix} u_1{}^1 & \cdots & u_1{}^m \\ \vdots & & \vdots \\ u_n{}^1 & \cdots & u_n{}^m \end{bmatrix}, \tag{4}$$

where a superscript m indicates the number of snapshots and n is the number of variables (number of grid points multiplied by the number of variables) in the case of flow simulation. The minimization in equation (3) results in an eigenvalue problem in equation (5).

$$X^T X \boldsymbol{\varphi}_i = \lambda_i \boldsymbol{\varphi}_i, \tag{5}$$

Using m snapshots of a flow field, we have m eigenvalues λ_i and eigenvectors $\boldsymbol{\varphi}_i$. The POD basis vector which has the same size as the original field variable is defined as,

$$\boldsymbol{\Psi}_i = X\boldsymbol{\varphi}_i/\sqrt{\lambda_i}. \tag{6}$$

This POD basis vector can be used to construct the reduced-order model as described in the following section. Using the basis vectors, the original velocity field can be reproduced by,

$$\boldsymbol{u}(x,t) = \overline{\boldsymbol{u}}(x) + \sum_{i=1}^{r} a_i(t)\boldsymbol{\Psi}_i(x). \tag{7}$$

where $\overline{\boldsymbol{u}}(x)$ is the average field subtracted from the matrix X before composing the matrix $X^T X$ and r denotes the number of POD modes used for reconstructing the flow field. The temporal coefficients $a_i(t)$ can be obtained from the original snapshots by the inner-product of the velocity vector and the POD basis vector as,

$$a_i(t) = (\boldsymbol{u}(x,t) - \overline{\boldsymbol{u}}(x)) \cdot \boldsymbol{\Psi}_i(x). \tag{8}$$

The temporal coefficients obtained from the original snapshots are used to construct surrogate-based reduced-order models as described in the following section.

2.3. *Reduced-order model (ROM)*

A well-known approach to construct a reduced-order model from an unsteady flow field is the Galerkin projection method. The inner product of the POD bases and governing equations is calculated to generate a set of ordinary equations, which can be in the form of,

$$\frac{da_i}{dt} = f(a_1, a_2, \cdots, a_r). \tag{9}$$

Due to the orthogonality of POD bases, the derived ordinary equations are easy to solve and the number of equations corresponds to that of POD bases considered. Using the obtained time-varying coefficients a_i, a velocity field is reproduced by equation (7). The Galerkin projection approach is attractive since the mathematical background is well established, however, there exist several drawbacks such as a computational cost for calculating inner products, a stability problem for high Reynolds number flows. One alternative would be the use of surrogate models for the time development of the time coefficients. A radial basis function (RBF) is one of such surrogate models to develop the temporal coefficients in time [6], i.e., the temporal coefficients of POD mode i can be represented as a linear combination of r radial basis functions ϕ,

$$a_i^n = f_i(a_i^{n-1}) = \sum_{j=1}^{r} w_{i,j} \cdot \phi(r_j). \tag{10}$$

where $w_{i,j}$ are weight coefficients and r_j is a distance $r_j = \|a_i^{n-1} - \hat{a}_i^j\|$ from a set of data points $\hat{a}_i^j$ calculated from the snapshots by equation (8). The temporal coefficient of mode i at time level $n-1$, a_i^{n-1} is advanced to that at time n with equation (10). In this study, we adopted a multi-quadratic

IOP Conf. Series: Journal of Physics: Conf. Series **1036** (2018) 012012 doi:10.1088/1742-6596/1036/1/012012

function is employed: $\phi(r_j) = \sqrt{r_j{}^2 + r_0{}^2}$, where r_0 is a parameter which can be determined, for example, by the leave-one-out cross-validation. Ensuring that the interpolated value matches the given $r+1$ data points from the snapshots, equation (10) can be rewritten in the matrix form as,

$$A_i \boldsymbol{w}_i = \boldsymbol{y}_i, \tag{11}$$

where

$$A_i = \left[\begin{pmatrix} \phi(\|a_i^1 - \hat{a}_i^1\|) & \cdots & \phi(\|a_i^1 - \hat{a}_i^r\|) \\ \vdots & \ddots & \vdots \\ \phi(\|a_i^r - \hat{a}_i^1\|) & \cdots & \phi(\|a_i^r - \hat{a}_i^r\|) \end{pmatrix}\right] \tag{12}$$

$$\boldsymbol{w}_i = \left[w_{i,1}, w_{i,2}, \cdots, w_{i,r}\right]^T, \tag{13}$$

$$\boldsymbol{y}_i = [a_i{}^1, a_i{}^2, \cdots, a_i{}^r]^T, \tag{14}$$

Equation (11) is solved by a matrix solver using the singular value decomposition (SVD). The resulting weights $w_{i,j}$ determine the equations for temporal coefficients.

3. Zonal mode analyses

In this section, we compare POD bases by changing a domain of the decomposition. Figure 3 shows the first POD mode for a flow field with Re = 100, where a Karman vortex street is formed behind a square cylinder. Using the same flow field, the POD analysis is conducted in the whole domain as shown in figure 3(a), a top-half domain in (b), a downstream half domain in (c), and the one-fourth domain in (d). Since each POD basis is normalized so that the magnitude of a POD basis is equal to one, a time coefficient is multiplied to a corresponding POD basis for comparison in those figures. The result shows that the distributions appear similarly, and the magnitude is slightly different in figures 3(c) and (d). The difference may come from the accuracy of the matrix computation related equation (5). The phase and period of the vortices are very similar for all the cases. This tendency was confirmed for other POD bases, however, the difference related to the accuracy of the matrix computation becomes more noticeable in high frequency (low energy) POD modes.

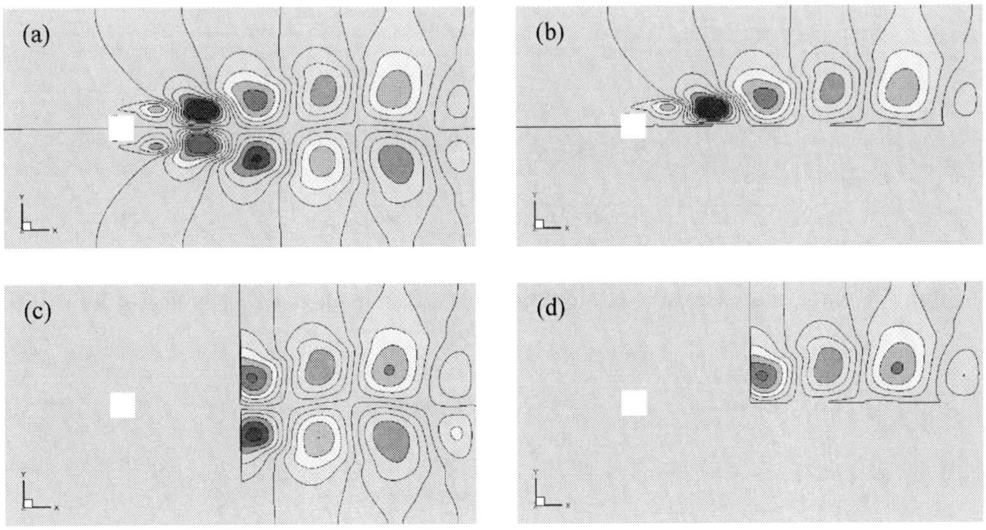

Figure 3. A POD mode obtained by limited domains, (a) whole domain, (b) top-half domain, (c) downstream half domain, and (d) one-fourth domain.

4. Zonal hybrid flow prediction

In this section, we conduct a zonal hybrid flow prediction, where a downstream part of the domain is predicted by the POD-RBF-ROM and the upstream part is simulated by the full-order Navier-Stokes equations (FOM, hereafter). The error of the hybrid flow prediction is evaluated by root mean square error (RMSE) and by visualized flow fields. To simplify the evaluation of the hybrid approach, we update the flow field in an offline manner, i.e., we prepare a set of snapshots by the FOM simulation. The evolution of the flow field is then represented simply by switching the snapshots. The POD-RBF-ROM prediction replaces the downstream half domain of the snapshot and RMSE is evaluated.

4.1. Two-dimensional flow field for Re = 100

We here evaluate the RMSE of the velocity field between full-order Navier-Stokes prediction and the POD-RBF-ROM prediction in the downstream half of the domain. Figure 4 shows the RMSE for cases with different values of POD energies, E_{ROM} = 99.99, 99.0 and 95.0%. In addition, the RMSE evaluated on the boundary between an upstream FOM domain and a downstream ROM domain. The error on the domain boundaries can be utilized in actual applications where the FOM results is not available in the ROM domain. The results show that the RMSE decreases as the POD energy increases, which corresponds to the more number of ROM equations. The RMSE on the boundary has similar magnitude as the RMSE of the downstream half domain, therefore, it can be used to evaluate the error of the ROM prediction by the comparison of FOM and ROM results on the boundary.

Figure 5 shows the streamwise velocity distribution for three different POD energy thresholds, (a) E_{ROM} = 99.99%, (b) E_{ROM} = 99.0%, and (c) E_{ROM} = 95.0%. The distribution is smoothly connected in the case of E_{ROM} = 99.99%, while discontinuity of the distribution is seen in E_{ROM} = 95.0%. Small distortion of the contour lines can be observed in the case of E_{ROM} = 99.0%. The number of POD bases (ROM equation) are 11, 6 and 3 for the cases of E_{ROM} = 99.99, 99.0 and 95%, respectively. For the POD, 40 snapshots are used as input.

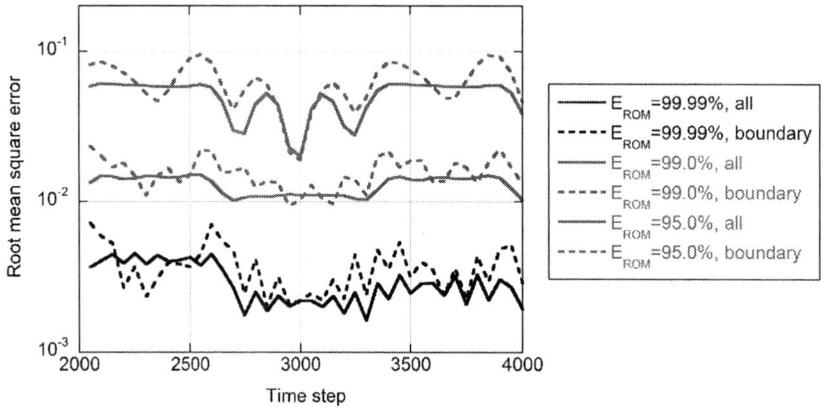

Figure 4. Error of a reconstructed flow field for different energy thresholds for Re = 100.

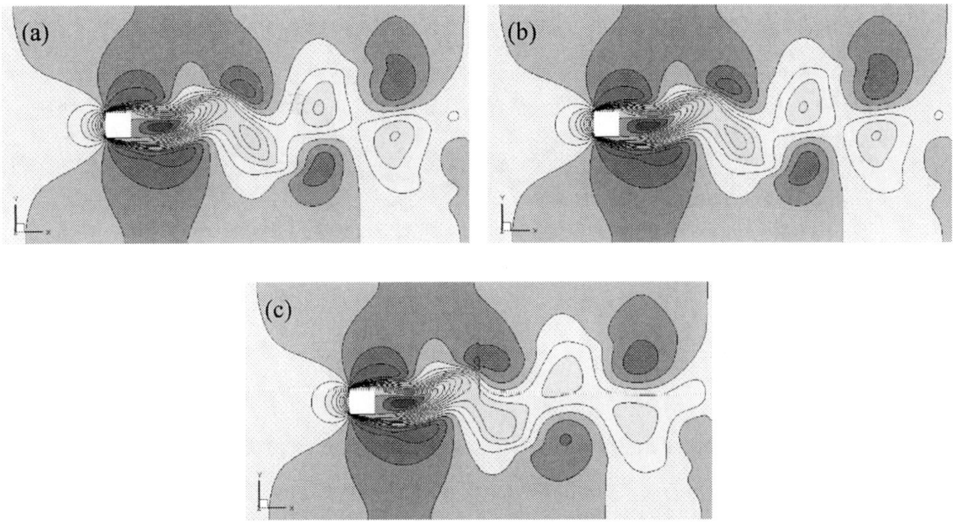

Figure 5. A reconstructed flow field for different energy thresholds, (a) E_{ROM}= 99.99%, (b) E_{ROM}= 99.0%, and (c) E_{ROM}=95.0%.

4.2. Three-dimensional flow field for Re = 1000

Figure 6 shows the same plot as figure 4, but for the cases with Re = 1000. Again, we consider cases of E_{ROM} = 99.99, 99.0 and 95.0%. The RMSE exhibits larger values compared to the cases of Re = 100, which indicates that the accuracy of the ROM prediction is degraded regardless of the same energy thresholds. The number of POD bases used here are 78, 34 and 18 modes for E_{ROM} = 99.99, 99.0 and 95.0%, respectively, which is much larger than the cases of Re = 100. Even so, the accuracy of the ROM prediction is not satisfactory to zonally replace the FOM prediction by the ROM. For the POD, 100 snapshots are considered as input.

Figure 7 shows the streamwise velocity distributions for (a) the FOM prediction, and (b) the ROM prediction with E_{ROM} = 99.99%. The ROM reproduced flow field is significantly different from the FOM results. This is because the time history of temporal coefficients predicted by equation (10) becomes slightly complex for Re = 1000 as shown in figure 8. In the case of Re = 100, the temporal coefficients varies sinusoidally and it is accurately predicted by the POD-RBF-ROM. On the other hand, the sinusoidal variation of the temporal coefficients is distorted in the case of Re = 1000, and this causes the difficulty in predicting using the POD-RBF-ROM. It is known that the projection-based ROM is often unstable [1]. The possibility to overcome this difficulty is to improve ROM or to use the other spatial basis which realizes the sinusoidal variation of temporal coefficients. The dynamic mode decomposition (DMD) would be used to improve the ROM prediction [7].

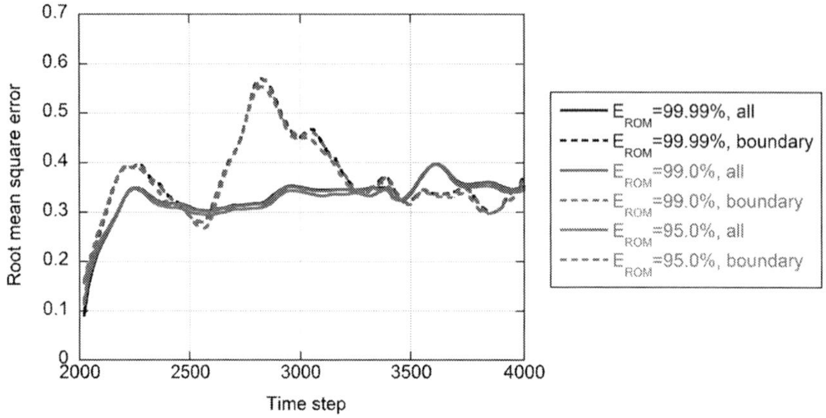

Figure 6. Error of a reconstructed flow field for different energy thresholds for Re = 1000.

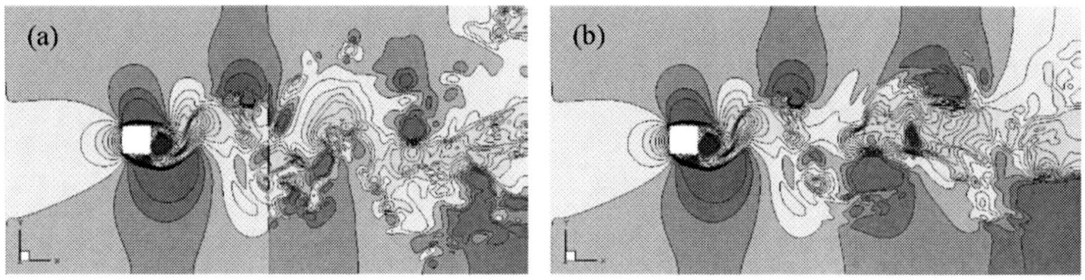

Figure 7. (a) A reconstructed flow field for the energy threshold of E_{ROM}= 99.99%, and (b) a flow field obtained by a full-order Navier-Stokes equations.

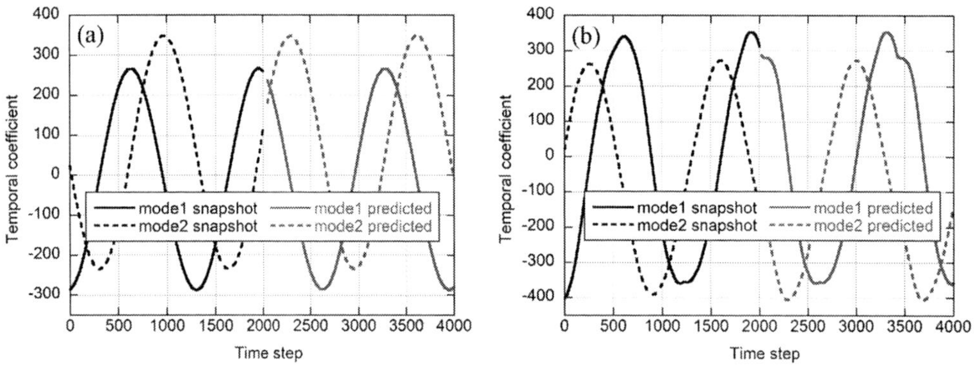

Figure 8. Temporal coefficients of first and second modes for (a) Re =100, and (b) Re =1000

5. Conclusions

In this study, we considered a simple flow field around a square cylinder, where the flow field can be two- or three-dimensional depending Reynolds numbers. First, we showed that the POD analysis can be applied to an arbitrary sub-domain of the flow field. The accuracy of the POD-RBF-ROM was then assessed by evaluating the RMSE of full-order Navier-Stokes and the POD-RBF-ROM predictions in the wake of the square cylinder. The accuracy of the POD-RBF-ROM for Re = 1000 was much degraded compared to the case with Re = 100, because the prediction of temporal coefficients became difficult by increasing Reynolds number. In the future work, we investigate the possibility of zonal model switching to reduce the cost of numerical simulation while retaining the accuracy.

Acknowledgements
This work was supported by a Grant-in-Aid for Scientific Research on Innovative Areas (No. 16H015290) from the Ministry of Education, Culture, Sports, Science and Technology of Japan.

References

[1] Couplet M, Basdevant C and Sagaut P 2005 *J. Comput. Phys* **207** 192,
[2] Kikuchi R, Misaka T and Obayashi S 2016 *Int. J. Comput. Fluid Dyn.* **30** 285
[3] Williams M, Schmid P and Kutz J 2013 *SIAM Multiscale Model. Simul.* **11** 522
[4] Nakahashi K 2005 *Proc. 17th AIAA Computational Fluid Dynamics Conference,* Paper 4876
[5] Sirovich L 1987 *Quart. Appl. Math.* **45** 561
[6] Xiao D, Fang F, Pain C and Hu G 2015 *Int. J. Numer. Methods Fluids* **79** 580
[7] Schmid P 2010 *J. Fluid Mech.* **656** 5

IOP Conf. Series: Journal of Physics: Conf. Series **1036** (2018) 012013 doi:10.1088/1742-6596/1036/1/012013

Bringing Computer and Physics Closer

Yonghao Yue

Graduate School of Frontier Sciences, The University of Tokyo, Kiban building 701, 5-1-5 Kashiwa-no-ha, Kashiwa-shi, Chiba 277-8561, Japan

E-mail: yonghao@edu.k.u-tokyo.ac.jp

Abstract. Physically based simulations are important for predicting the future states of the objects in consideration, and have been one of the major research topics in computer graphics. Applications include optical simulations for realistic image synthesis given a 3D scene, and rigid/fluid dynamics for computer animations. The need for the ultimate realism may sometimes ask for more efficient and accurate methods than those available in other fields. In addition, physically based optimizations are useful for designing and fabricating man-made objects given some desirable functionalities governed by physical laws. The ability to incorporate predictive simulations also allows us to build smarter robots that can optimize for their next movement based on the predictions. We introduce some of our work performed from this perspective.

1. Introduction

Our world is governed by physical laws: the way how light travels from the light source to our eye can be described by the radiative transport equation; the way how the water in a cup behaves as we interact with the cup is described by the Navier-Stokes equations. The ability of predicting such physical behaviors would allow us to estimate how objects would deform and how they would look like, and is hence useful for computer animations.

Physical laws also come into play when we want to design and fabricate man-made objects. For example, understanding how the light refracts is essential for designing lenses, and understanding how the contact and frictional forces act against the gravity is important for designing a stable sculpture that is assembled from a number of small bricks. Optimizations with physically based constraints in consideration are thus fundamental for designing and fabricating functional man-made objects.

With the ability to incorporate predictive simulations in planning, we can also build smarter robots that use vision techniques to acquire the external information, perform predictive simulations to estimate the future states of the objects, and determine the appropriate control based on the predicted states. In this paper, we will review and introduce a series of our past work done from the above perspective.

2. Predictive Simulations

As for predictive simulations, we introduce our work on optics- and mechanics- based methods. For optics based method, we present an efficient and unbiased free path sampling technique for radiative transport [1, 2]. For mechanics based method, we present how to simulate non-Newtonian fluids including shear thinning (cream and foam) and shear thickening (oobleck) fluids [3].

Figure 1. Examples simulated with our method (reprinted from [1]). We observed a 380 times speed up over the delta tracking for the sky scene shown in the right.

2.1. Efficient and Unbiased Free Path Sampling

Modern rendering methods utilize Monte Carlo techniques (e.g., Monte Carlo path tracing [4, 5] or Markov chain Monte Carlo techniques [5, 6]) to draw light paths according to some probability density function related to the contribution of each light path: in methods based on Monte Carlo integration, we integrate for each pixel on the image plane the contribution (radiance) of the light paths that pass through the camera lens and hit the pixel, while in MCMC based methods, the intensity distribution on the image plane is viewed as the marginalized distribution of the radiance of the light paths (up to a normalizing constant).

A typical way for generating light paths in a participating medium is to iteratively sample the scattering direction and the distance (free path) $t \geq 0$ between successive scattering events. Given the current location $\boldsymbol{x}_0$ and the direction $\boldsymbol{d}$ (in a unit vector), the probability density function $p_{\text{fp}}(\boldsymbol{x})$ to generate the next scattering event (or equivalently the free path) is given by $p_{\text{fp}}(\boldsymbol{x}(t)|\boldsymbol{x}_0, \boldsymbol{d}) = \exp(-\tau(\boldsymbol{x}_0, \boldsymbol{x}(t)))\sigma(\boldsymbol{x}(t))$, where $\boldsymbol{x}(t) = \boldsymbol{x}_0 + t\boldsymbol{d}$, σ is the extinction coefficient, $\tau(\boldsymbol{x}_0, \boldsymbol{x}(t)) = \int_0^t \sigma(\boldsymbol{x}_0 + t'\boldsymbol{d})dt'$ is the optical depth, and $\exp(-\tau(\boldsymbol{x}_0, \boldsymbol{x}(t)))$ is the transmittance. $p_{\text{fp}}(\cdot)$ represents the probability that the light travels freely without any interaction in the interval $(\boldsymbol{x}_0, \boldsymbol{x}(t))$ and then gets scattered or absorbed at $\boldsymbol{x}$. The corresponding cumulative distribution function is given by $C_{\text{fp}}(\boldsymbol{x}(t)|\boldsymbol{x}_0, \boldsymbol{d}) = 1 - \exp(-\tau(\boldsymbol{x}_0, \boldsymbol{x}(t)))$.

A typical way to sample the free path t is to use the inversion method: draw a random number $\xi \in [0, 1)$, and solve for t such that $C_{\text{fp}}(\boldsymbol{x}(t)|\boldsymbol{x}_0, \boldsymbol{d}) = \xi$, or equivalently $\tau(\boldsymbol{x}_0, \boldsymbol{x}(t)) = -\ln(1-\xi)$. One way to find t is to use numerical integration to incrementally compute the optical depth (the resulting algorithm is called ray-marching [7] in the graphics community). However, insufficient number of quadrature points may result in an inaccurate optical depth. On the other hand, using many quadrature points may affect the performance.

In the nuclear science community, there is a technique called delta tracking (also known as Woodcook tracking or pseudo scattering) [8] for unbiased sampling of the free path. Delta tracking first computes an upper bound (can be any upper bound) of the extinction coefficient, called the majorant extinction coefficient σ_M. It then fills the entire region with particles that incur null collisions (for convenience we call them 'null particles') of an amount of $\sigma_M - \sigma(\boldsymbol{x})$ [9, 10], so the extinction coefficient is now σ_M everywhere. Those null particles are transparent: a light hitting a null particle will continue its journey along the same direction. Thus, the addition of the null particles will preserve the appearance of the medium. Since the medium is now uniform with the extinction coefficient σ_M, the free path sampling can be performed analytically: the optical depth is now given by $t\sigma_M$, and t can be solved via $t = -\ln(1-\xi)/\sigma_M$. Now, we stop only when the light hits a true particle (in the probability $\sigma(\boldsymbol{x})/\sigma_M$) and otherwise continue the journey in the same direction.

While delta tracking gives unbiased free path, it becomes less efficient for a more inhomogeneous medium, because we need to fill more null particles to have an homogeneous medium, resulting in a lower probability that the light hits a true particle. The idea of our method is to partition the entire space into a set of disjoint subregions, each with its own majorant extinction coefficient, which can be usually set much tighter than that for the entire domain.

To perform unbiased sampling in the partitioned space, our approach is to 'rewind' the sampling once it crosses a partition. Suppose that we are currently in a subregion i, and there is another subregion j adjacent to the subregion i along the ray direction. $\sigma_M^{(i)}$ and $\sigma_M^{(j)}$ are the majorant extinction coefficients for the subregions i and j, respectively. We start delta tracking in the subregion i with $\sigma_M^{(i)}$. If we find a true collision in i, we accept it. Otherwise, the ray will leave i and enter j. If this happens, we reset (rewind) the free path back to the crossing point between i and j along the ray, and perform delta tracking using $\sigma_M^{(j)}$ from there. This approach is proven to give unbiased sampling for the free path [1], and was also used in [11] to treat adjacent medium boundary.

The next question is then how we should partition the space. On one hand, we want to partition the space as much as possible, as long as the partition would bring enough reduction in $\sigma_M^{(i)}$ so we may enjoy performance improvement. But on the other hand, if we partitioned a homogeneous region, the rewinding process for crossing the partition would be a waste compared to the case with no partition. Taking this observation one step further, we have developed a mathematical framework [1, 2] to estimate the performance gain or loss for a given partition, which can then be used to determine the optimal partitioning location. We briefly explain our idea for a one dimensional case.

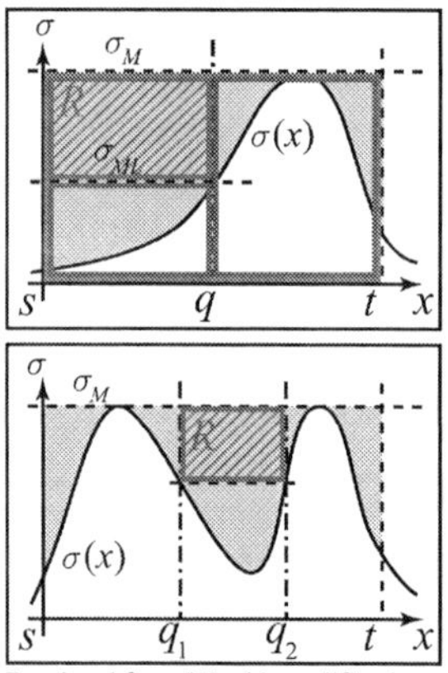

Reprinted from [1] with modifications.

Suppose that σ_M is the maximum value of $\sigma(x)$ in an interval $(s, t]$. In a single iteration of the delta tracking, the free path will proceed $1/\sigma_M$ on average. Thus, it takes $N = (t - s)\sigma_M$ iterations, on average, to go across this interval. This value is exactly the area of the bounding box of $\sigma(x)$ (blue box in the right inset) in the interval $(s, t]$. If we partition this interval at a location q, the average number of iterations for passing through each subregion is the bounding area of each subregion (red boxes). On top of their sum, we need an additional iteration for the rewinding process. Then, the reduction N_r in the expected number of iterations when we partition the interval is given by the reduction of the total box areas (the brown box area R) minus one: $N_r = R - 1$. Thus, the partition that maximizes R is the candidate for the partitioning location. In addition, we should stop partitioning if $R \leq 1$ because it offers no reduction in the cost. This strategy extends to the case when $\sigma(x)$ has multiple peaks. Formally, the problem to find the partitioning location in the one dimensional case reduces to a 2D largest empty rectangle problem [12] for the region bounded by σ_M and $\sigma(x)$ (the gray region). We can also derive an analytical cost function for a 3D volume [2], and make use of various space partitioning data structures.

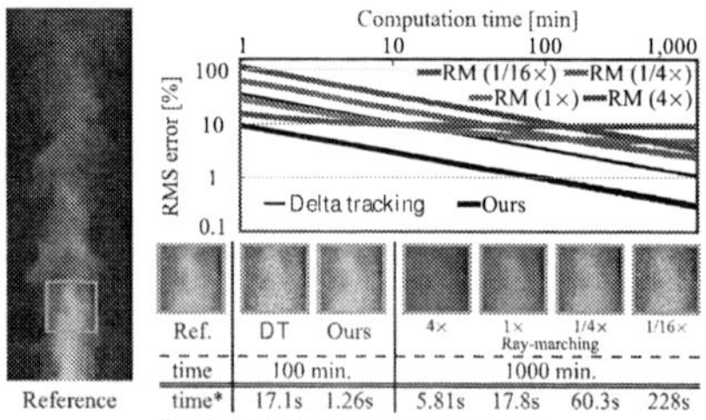

Reprinted from [1] with modifications.

We now briefly summarize our results. With the ray-marching method with fixed sampling intervals, the results may have bias and may not converge to the exact solution (the exact solution was computed using the delta tracking with 16 cores in parallel in 5 days), as in the right inset. 4×, 1×, 1/4× and 1/16× in ray-marching mean the sampling intervals, represented as the multiples of $1/\sigma_M$ of the medium, respectively. 'time*' at the bottom of the right inset indicates the computation time to generate 10^6 rays. Our method is unbiased and is approximately 10 times faster than delta tracking.

With our method, we may obtain an arbitrary speed up, depending on the inhomogeneity of the participating medium; as in the inset in the next page, the performance gain increases as

IOP Conf. Series: Journal of Physics: Conf. Series **1036** (2018) 012013 doi:10.1088/1742-6596/1036/1/012013

Figure 2. Left: a pie throw example with the characteristic clumping and tearing behavior. Right: side by side comparisons between viscoplastic materials (left object in each pair) and shear thickening (oobleck) materials (right object in each pair). While the viscoplastic materials flow immediately after impact with the ground plane, the shear thickening materials rebound elastically after the fast initial impact, and only flow after the second, lower-velocity impact.

the medium is more inhomogeneous. Using our method, we can simulate the light transport for various scenes with multiple scattering in consideration, as in Figure 1. The sky scene contains highly inhomogeneous participating medium: in the real world, the extinction coefficients of the clouds are about 10^2 to 10^3 times as large as those of the atmosphere. In our case, we observed a 380 times speed up over the delta tracking technique.

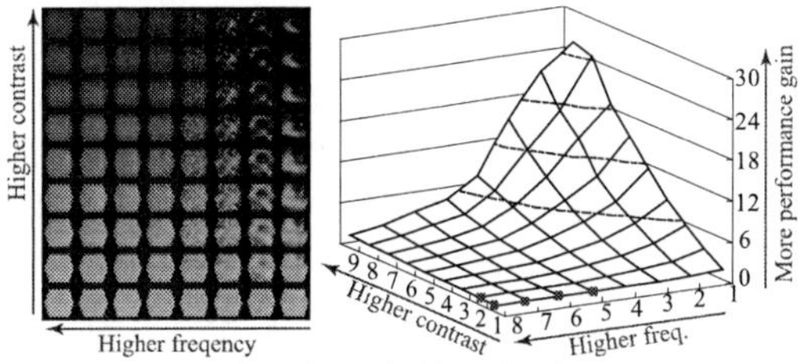

Reprinted from [1] with modifications.

2.2. A Material Point Method for Shear Dependent Flows

Next, we review our work on how to simulate the motion of complex materials. We have focused on foams, because they arise in many problem domains [13], including firefighting, oil recovery, chemical filtration, industrial textile processing, and computer animation. Dense foams consist of a number of small individual bubbles, each in the order of 40μm in diameter [14]. The observed macroscopic behavior of a dense foam is a seamless continuum exhibiting a non-Newtonian force response. Foams behave elastically when the applied stress is small. When the stress is large enough, the microscopic stochastic rearrangement of bubbles results in a macroscopically viscoplastic behavior, and the continuum flows more easily under larger stresses, known as the shear thinning effect. Because of these behaviors, foams may experience extreme shape changes.

We adopt the Material Point Method (MPM) (a hybrid particle/grid method) [15] because of its relative ease of handling large deformations and topological changes [16, 17]. The dynamics of the medium over time t is described by the Euler-Lagrange partial differential equation $\rho\frac{D\boldsymbol{v}}{Dt} = \nabla\cdot\boldsymbol{\sigma} + \rho\boldsymbol{g}$, subject to the mass conservation condition $\frac{D\rho}{Dt} + \rho\nabla\cdot\boldsymbol{v} = 0$, where $\frac{D\cdot}{Dt}$ denotes the material derivative, ρ, $\boldsymbol{v}$, and $\boldsymbol{\sigma}$ are the scalar density, vectorial velocity, tensorial Cauchy stress fields, respectively. $\boldsymbol{g}$ is the gravitational acceleration.

For the treatment of elasticity, we use the elastic part of the left Cauchy-Green tensor $\mathbf{b}^e$. The application of a slight strain on the foam results in the Kirchhoff stress ([18]-§9.2) $\boldsymbol{\tau} = J\boldsymbol{\sigma} = \frac{\kappa}{2}(J^2-1)\mathbf{I} + \mu\,\mathrm{dev}[\bar{\mathbf{b}}^e]$, where κ and μ are respectively the bulk and shear moduli, $J = \sqrt{\det(\mathbf{b}^e)}$, $\mathrm{dev}[\mathbf{x}] = \mathbf{x} - \frac{\mathrm{tr}[\mathbf{x}]}{3}\mathbf{I}$ is the deviatoric operator, and $\bar{\mathbf{b}}^e = J^{-2/3}\mathbf{b}^e$ is the volume preserving left-Cauchy Green tensor. We adopt the von Mises yield condition to determine the limit of the elastic regime: the material behaves elastically as long as $\Phi(s) = s - \sqrt{2/3}\,\sigma_Y \leq 0$ is satisfied, where σ_Y is the yield stress and s is the magnitude of the shear stress $\mathbf{s} = \mathrm{dev}[\boldsymbol{\tau}]$.

We adopt the Herschel-Bulkley model [19] as the plastic constitutive relation for foam [13]. First, the time derivative of $\mathbf{b}^e$ is given by the identity $\dot{\mathbf{b}}^e = (\nabla\boldsymbol{v})\mathbf{b}^e + \mathbf{b}^e(\nabla\boldsymbol{v})^T + \mathcal{L}_{\boldsymbol{v}}\mathbf{b}^e$, where $\mathcal{L}_{\boldsymbol{v}}\mathbf{b}^e$ is the Lie derivative of $\mathbf{b}^e$. While $(\nabla\boldsymbol{v})\mathbf{b}^e + \mathbf{b}^e(\nabla\boldsymbol{v})^T$ captures the change in the strain due to the velocity field, $\mathcal{L}_{\boldsymbol{v}}\mathbf{b}^e$ captures the change relative to the velocity (i.e., due to the plastic flow). In rheology, one adopts a flow rule to describe the plastic deformation, where excess elastic strain flows into plastic strain over time. We use the flow rule of Simo and Hughes (see [18]-§9,

[20]-§1.4) $\mathcal{L}_v \mathbf{b}^e = -\frac{2}{3}\mathrm{tr}[\mathbf{b}^e]\gamma\hat{\mathbf{s}}$, where γ is the flow rate, and $\hat{\mathbf{s}} = \mathbf{s}/s$ is the flow direction. With the Herschel-Bulkley model, the flow rate is given by $\gamma(s) = \max\left(0, \left(\frac{\Phi(s)}{\eta}\right)\right)^{1/h}$, where η is the viscosity. For $\Phi(s) \geq 0$, we can rewrite $\gamma(s) = \frac{\Phi}{\eta^{1/h}\Phi^{1-1/h}}$, where the denominator $\eta^{1/h}\Phi^{1-1/h}$ can be viewed as the effective viscosity coefficient from the vantage point of the standard viscoplastic model. This reveals the advantage of the Herschel-Bulkley model: by changing the parameter h, we can capture a wide range of viscoplastic behaviors. Shear thinning and thickening materials are characterized by $h < 1$ and $h > 1$, respectively. Setting $h = 1$ recovers the viscoplastic (Bingham) model; taking the limit $\eta \to 0$ recovers perfect plasticity.

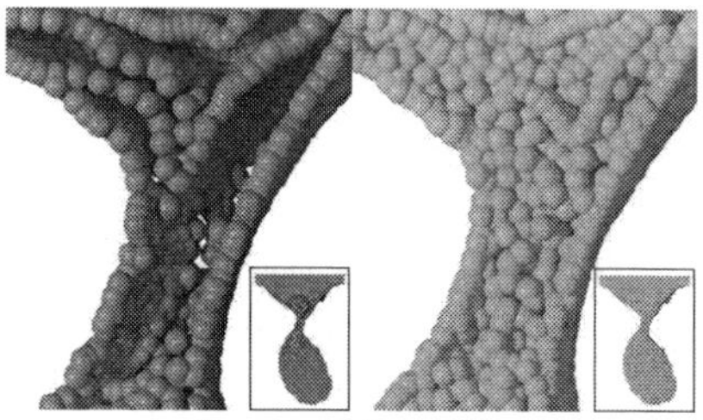

Reprinted from [3].

When we numerically advect material points, their distribution can become highly nonuniform. This happens, for example, when a sample of foam is compressed along one axis and stretched along a perpendicular axis. In this case, the uniformity of the material point distribution can rapidly degrade, producing a sparse sampling along the stretched axis, resulting in numerically induced, nonphysical voids (the left side of the right inset). We developed a resampling approach for the material point method to maintain a uniform point distribution (the right side of the right inset). We employ Poisson disk sampling to insert new points, and detect and merge points that are too close to one another.

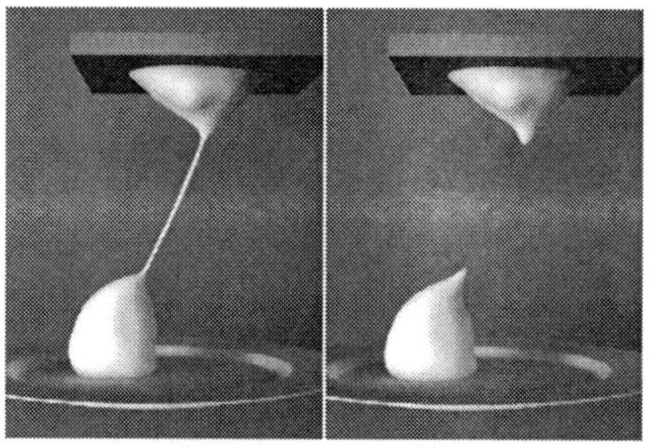

Reprinted from [3].

Finally, we observe that an explicit tearing model is required to allow the foam to tear apart, because the grid maintains artificial topological connections at the grid resolution through the stencil of the shape functions. Without a tearing model, regions of foam can become progressively thinner, yielding artificially-slender threads (the left side of the right inset). We therefore developed an extension to directly detect and tear weak regions based on their accumulated plasticity (the right side of the right inset). We first identify regions that are in a 'weakened' or torn state. In these regions we directly modify the computed forces and the accumulation of strain so that the material no longer resists separation, nor does the geometric stretching accumulate as additional physical strain. After tearing has occurred, we gradually adjust the accumulated plasticity to model the recovery of bubble neighbor connectivity in these regions, to allow the foam to once again resists stretching.

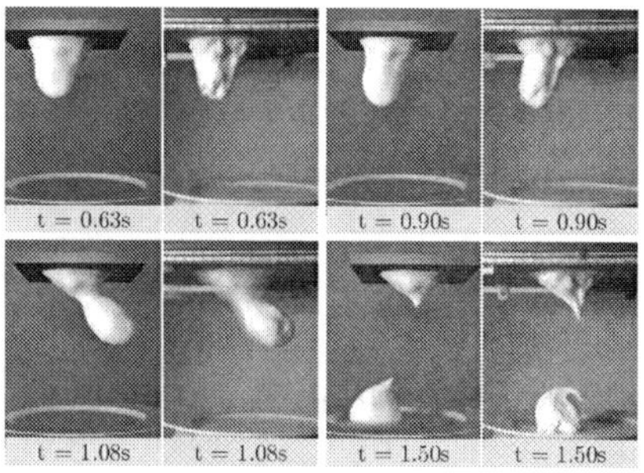

Reprinted from [3].

We demonstrate the method's efficacy with a comparison to recordings of real foam (the right inset), and with several animated examples (Figure 2). In the comparison with the recorded footage, we performed a real experiment by attaching a sample of real shaving foam to the base of a platform, acquiring an approximated shape of the foam using a 3D scanner (Kinect), and recording the motion of the foam subject to an oscillatory motion. Then, we simulated the acquired geometry, and compared the motion of our simulations to the recorded video. Initially, the foam oscillates and slowly begins to stretch downwards due to accumulated plasticity. As the oscillations continue, the foam exhibits typical nonlinear shear thinning behavior, and rapidly tears apart. Our simulations successfully capture the same characteristic behaviors of real foam. As in Figure 2 left, our method can be used to reproduce a pie throw example with a whipped cream, and as in Figure 2 right, our method can handle a (Bingham) viscoplastic material as well as a shear-thickening material (like oobleck).

3. Designing Functional Man-Made Objects

Next, we show how the ability to incorporate physical laws can be used for designing functional man-made objects. We introduce our work on optics- and mechanics- based designs. For optics based design, we present our work on goal based lens design [21], with which we can shape the incident light and project an arbitrarily specified caustic image on a screen. For mechanics based design, we present our work for computing a layout of LEGO bricks given a 3D object model [22], that can be assembled in the real world, accounting for the stability of the sculpture.

3.1. Continuous Surface Generation for Goal Based Caustics

Given a desired image of caustics that we want to project onto a planar screen, our method computes the lens geometry that shapes the incident light via refraction and generates the desired caustics on the screen. In particular, our lens can produce high-resolution and complex patterns, such as natural images, with a continuous dynamic range (high quality even in regions of low intensity). In addition, our lens has a wide focus range: the caustic pattern is stable and perceivable even when the object is not perfectly in focus. This is important for the ease of the setting of the lens, light source and the screen.

Reprinted from [21].

Our method consists of two steps. In the first step, we compute a map between the light reaching the incident surface and that reaching the screen. This map describes how the light should be shaped (through the refraction) to produce the desired pattern. In the second step, we reconstruct the lens surface from this map.

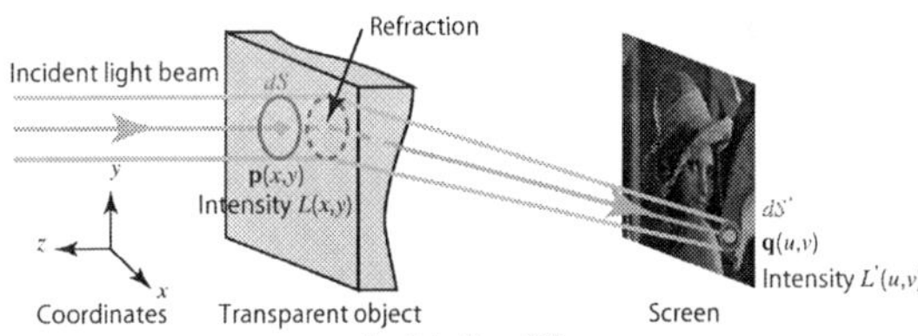

Reprinted from [21].

We consider a collimated setting: the (parallel) light enters the incident surface perpendicularly without refraction (as in the right inset). The light refracts at the other side of the lens, called the refractive surface, and then reaches the screen. Each light path can be described by the relationship between a point $\boldsymbol{p}(x, y)$ on the incident surface and a point $\boldsymbol{q}(u, v)$ on the screen. In our method, we consider a one-to-one relationship between $\boldsymbol{p}(x, y)$ and $\boldsymbol{q}(u, v)$. We want to find from which regions of the incident surface we need to gather light in order to form the intensity distribution of the desired caustic pattern. This problem can be formulated as the problem of finding a mapping from $\boldsymbol{q}(u, v)$ to $\boldsymbol{p}(x, y)$ satisfying the following two conditions. The first condition is that the mapping is continuous (almost everywhere) over the entire domain. With a continuous mapping, the refractive surface also becomes continuous, enabling a stable projection of the caustics image. The second condition is that this mapping preserves the light energy: suppose that the light incident on an infinitesimal region dS around the point $\boldsymbol{p}(x, y)$ on the incident surface has intensity $L(x, y)$, and that the intensity of this light on an infinitesimal region dS' around the point $\boldsymbol{q}(u, v)$ on the screen is $L'(u, v)$, then we have $LdS = L'dS'$.

We represent the mapping from $\boldsymbol{q}(u, v)$ to $\boldsymbol{p}(x, y)$ as a parameterization of x and y using u and v. Under this parameterization, the ratio dS/dS' is equivalent to the determinant J of the Jacobian $\frac{\partial \boldsymbol{p}}{\partial \boldsymbol{q}}$, hence, $LJ = L'$. Directly computing this parameterization is difficult, we use a geometric flow approach instead. We start from an initial parameterization $x = u$ and $y = v$, and then maintain the continuity of the parameterization while gradually modifying it for a better satisfaction of the conservation of light energy. If the parameterization satisfies $LJ = C$ for every light path, where C is the intensity of the caustics, then that mapping is the solution to the problem. Otherwise, there is a difference $D = LJ - C$. We reduce this difference by continuously modifying the point $\boldsymbol{p}(x, y)$ on the incident surface, while fixing the corresponding point $\boldsymbol{q}(u, v)$ on the screen.

First, we observe that there are regions with positive and negative differences D. We regard the positive and negative regions as source and sink, respectively, analogous to those in computational fluid dynamics, where such sources and sinks can be relaxed via pressure gradient: we first solve Poisson's equation $\nabla^2\phi = -D$ to find the pressure field ϕ, and then update the points $\boldsymbol{p}$ along the gradient of ϕ, $\frac{\partial \boldsymbol{p}}{\partial t} = \nabla\phi$ (see (a), (b), and (c) in the right inset for an example of the fields D, ϕ, and $\nabla\phi$, respectively). In our method, we repeatedly update the parameterization by computing the difference $D = LJ - C$, solving the Poisson's equation $\nabla^2\phi = -D$ and updating $\boldsymbol{p}$ through $\frac{\partial \boldsymbol{p}}{\partial t} = \nabla\phi$ (as in Figure 3).

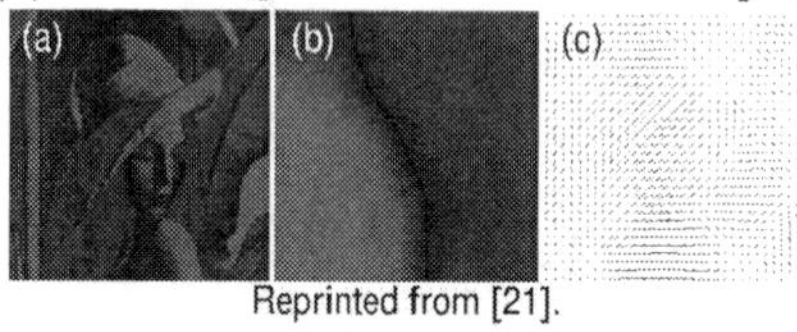

Reprinted from [21].

We can compose our mapping with any divergence free mapping to produce the same desired caustic pattern, as a divergence free mapping preserves the area and hence the light energy. Since the parameterization is initially curl free, and our flow $\nabla\phi$ is curl free, the parameterization obtained using our method contains no divergence-free components. In this sense, our mapping can be regarded as the 'minimum' mapping. We also note that as long as the difference D is non-zero, there will always be a non-zero flow, hence $D = 0$ is the only fixed point of the flow constructed above.

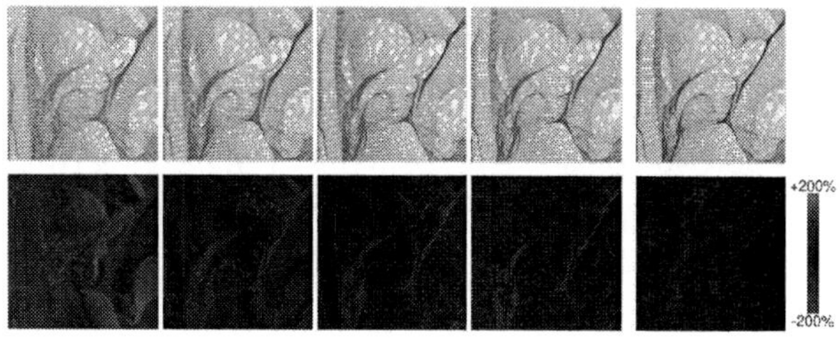

Figure 3. Reprinted from [21]. From left to right: parameterizations after 1, 3, 5, 7 and 25 iterations, respectively. The top row shows the iso-parametric lines of the parameterization, and the bottom row shows the difference D (+100% corresponds to the average intensity of the desired caustics).

To reconstruct the lens surface (right inset), we find the surface in a least square sense, assuming that the refractive surface is a single-valued function, with its z coordinate represented as $z = h(x, y)$. A vector $\boldsymbol{N}$ normal to the refractive surface can be represented as $\boldsymbol{N} = \left(\frac{\partial h}{\partial x}, \frac{\partial h}{\partial y}, -1\right)$. Considering only the x and y components, we have $\boldsymbol{N}_{xy} = \nabla h$. Taking the divergence of both sides, we obtain another Poisson's equation (similar to [23]) $\nabla^2 h = \nabla \cdot \boldsymbol{N}_{xy}$. Using the Snell's law, the normal vector is related to the surface height h via $\boldsymbol{N}_{xy} = \frac{1}{k}(\boldsymbol{q} - \boldsymbol{p})$, where $k = \eta\sqrt{\|\boldsymbol{q} - \boldsymbol{p}\|^2 + (H - h)^2} - (H - h)$, η is the refractive index (1.49 for an acrylate basin), and H is the distance between the screen and the lens. We use an iterative approach to compute h: we first initialize the refractive surface as a plane, and then iteratively compute the normal and update h. This process converges usually only after a few iterations.

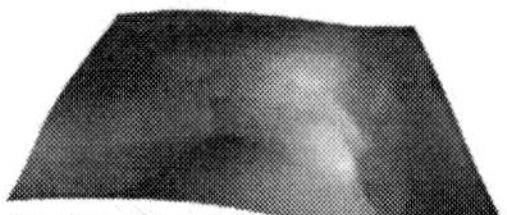

Reprinted from [21].

In the inset at the beginning of this subsection, we show an example of the fabricated lens and its caustics. In the right inset, we show examples with sharp features and characters. In each row, the subfigures from left to right show the photographs of the fabricated objects and their caustics, the simulated caustic patterns, and the caustic patterns of the fabricated objects.

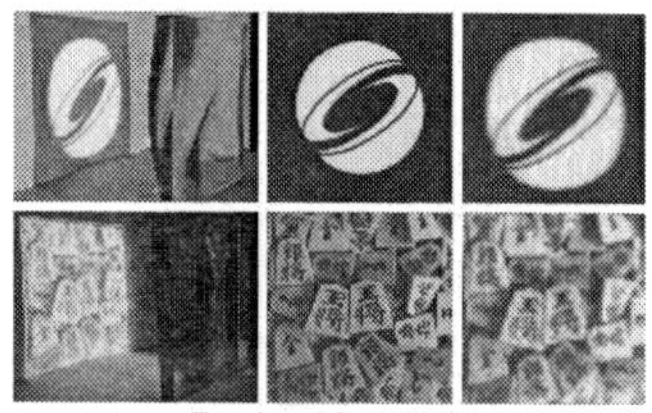

Reprinted from [21].

3.2. Stability Analysis and Layout Optimization for LEGO Sculptures

Next, we present our work on optimizing designs with mechanical constraints in consideration. We have focused on computing brick layouts for LEGO sculptures. This LEGO construction problem is interesting and challenging [24], first because LEGO bricks are a versatile building tool that is useful for many applications including rapid prototyping for personal design [25], yet finding a brick layout that can be used for the assembly in the real world requires us to predict the stability of the sculpture and to place bricks only at discretely possible locations, hence the optimization problem naturally has a combinatoric nature.

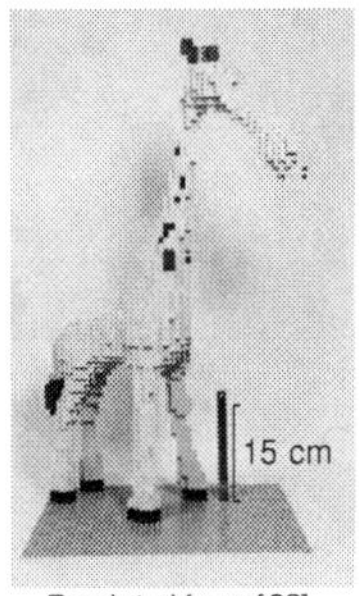

Reprinted from [22].

At the heart of our method is a force-based metric for stability estimation, which allows us to 1) compare the physical stability between different structures, and 2) determine if the resulting structure is stable enough for assembly. In addition, our force-based stability analysis tells us the weakest portion of the structure. Building atop our stability analysis, we iteratively update the layout by locally and randomly reconfiguring the region near the weakest portion. We only accept reconfigurations that give us better stability. This way, we obtain a monotonic increase of the stability. Previous methods [24, 26, 27, 28] lack this force-based metric (they build atop heuristic based metrics like the number of connected bricks), and may fail to predict the stability of the assembled sculpture. Their metrics also do not map monotonically to the physical stability.

Our stability analysis accounts for friction and normal forces: the bricks are held together by friction forces between the knobs (round dots) and cavities (in the back side), and are supported by neighboring bricks through the normal forces. We found it practical to model the maximum friction load using a constant T, rather than explicitly model it as a function of the normal force between the knob and cavity (see Appendix A of [22] for the details). Among the normal forces, there is a special set of forces that work horizontally between knobs and cavities. For the ease of computation, we treat the sum of such horizontal forces between each pair of connected bricks as support forces, and exclude them from the normal forces. This way, normal forces become (only) those have the non-negativity constraints, and support forces can take any value, assuming bricks will never fracture. Suppose that the bricks are placed according to the layout, the gravity will then induce forces and torques on the bricks. If the static friction, normal and support forces are able to perfectly counteract the gravity forces, the sculpture will remain static. Our stability analysis computes the force and torque balances in this static picture.

The force and torque balances give rise to a set of equality constraints, stating that the translational and rotational accelerations for every brick should be zero. Namely, for each brick b_j, we have $ct_T(b_j) : \sum_{\boldsymbol{F}_i \in \mathcal{F}_{b_j}} \boldsymbol{F}_i + m_{b_j}\boldsymbol{g} = \boldsymbol{0}$ as the force balance, and $ct_R(b_j) : \sum_{\boldsymbol{F}_i \in \mathcal{F}_{b_j}} \boldsymbol{L}_i \times \boldsymbol{F}_i = \boldsymbol{0}$ as the torque balance, where $\mathcal{F}_{b_j}$ is the set of forces working on the brick b_j , m_{b_j} is the mass of the brick b_j, $\boldsymbol{g}$ is the gravitational acceleration, and $\boldsymbol{L}_i$ is the arm vector, pointing from the center of the brick to the position where the force is assigned.

The friction and normal forces should satisfy non-negativity (inequality) constraints: $ct_{\mathcal{F}_f}(i) : 0 \leq F_i \in \mathcal{F}_f$ and $ct_{\mathcal{F}_n}(i) : 0 \leq F_i \in \mathcal{F}_n$, where $\mathcal{F}_f$ and $\mathcal{F}_n$ are the sets of all friction and normal forces, respectively. If all the friction forces are within the maximum friction load T, we can conclude the layout is stable. However, this style of stability assessment only tells us a binary result (whether it stays or collapses), and cannot be used to compare the stability between two different layouts. Instead of posing the maximum friction load as an inequality constraint, we consider the capacity $C_i = T - F_i$ of a friction force $F_i \in \mathcal{F}_f$. If $C_i > 0$, the corresponding point can still accept additional forces. We define $C_m = \min_i C_i$ to indicate the smallest (weakest) capacity.

Usually, the answer of the force distribution for a static object is not unique (e.g., consider the case of a table with four legs). As long as the forces can be redistributed to make $C_m > 0$, the LEGO sculpture remains stable. We take this concept of force redistribution one step further to estimate what is the highest C_m we can get. Namely, we find a force distribution $\{F_k^M\}$ that maximizes the smallest capacity C_m subject to the linear equality and inequality constraints discussed above (i.e., ct_T, ct_R, $ct_{\mathcal{F}_f}$, and $ct_{\mathcal{F}_n}$), where $F_k^M \in \mathcal{F}$, $\mathcal{F} = \mathcal{F}_f \cup \mathcal{F}_s \cup \mathcal{F}_n$, and $\mathcal{F}_s$ is

Figure 4. Reprinted from [22] with modifications. Real-life sized table sculpture assembled using the layout generated by our method.

the set of support forces. With this force distribution $\{F_k^M\}$, we can compute the maximum capacity as $C_M = \min_{F_i \in \mathcal{F}_f}(T - F_i^M)$. C_M represents how much additional force the structure can accept for the case $C_M \geq 0$ or how much the forces are overflowing for the case $C_M < 0$, enabling us to compare the stability between different layouts with a larger C_M meaning more stable. Another important aspect of posing the relationships between the maximum friction load and the friction forces as the objective function is that we can now find a solution not only for stable cases but also for unstable ones (there will be no solution if they were posed as inequality constraints). This enables us to guide the layout refinement from an unstable structure towards a stable one. In addition, $C_M > 0$ naturally serves as a threshold for the stability, because it means there is a way to redistribute forces to make all capacities positive. Therefore, we can use C_M as the stability metric for a layout. Furthermore, we can compute $F_i^w = \arg\min_{F_i \in \mathcal{F}_f}(T - F_i^M)$, and identify the two bricks that share the contact point corresponding to F_i^w. These two bricks are the weakest portion of the LEGO sculpture.

With the stability analysis at hand, we find a stable configuration via an iterative process. From a randomly initialized layout with the connection graph of all the bricks being single connected, and the fill out of the bricks conforming to the given shape and color, we iteratively perform random reconfiguration around the weakest portion, until C_M becomes positive. At each random reconfiguration, we only accept a new layout if its C_M value is larger.

With our method, we can create layout-sensitive sculptures like the Giraffe model with a slender neck to sustain a heavy head, as in the inset at the beginning of this subsection. Besides computing for a stable layout for a given shape, our method can be used to incorporate external weights as in the right inset. Figure 4 shows our result of a real-life sized table sculpture. We imposed external weights of 6kg in total equally divided to the table top. With this setting, we can put a laptop or a book on it, and use it as a real table.

(a) Input external force (b) without external force (c) with external force
Reprinted from [22].

4. Physics for Smart Robots

We have shown that the ability to incorporate physics is important and useful for computer animation to predict the objects' appearance and motion, as well as for designing functional man-made objects that can shape the light and deal with mechanical constraints. Now, we briefly introduce a series of our work on developing smarter robots with the physics in consideration. We have focused on robotic folding of garments [29, 30, 31].

IOP Conf. Series: Journal of Physics: Conf. Series **1036** (2018) 012013 doi:10.1088/1742-6596/1036/1/012013

4.1. Unfolding Laundry Garments by a Robot

Deformable objects such as garments are highly unstructured, making them difficult to recognize and manipulate. In the first work, we focused on how to unfold a garment from a random pose to a desired configuration [32, 33, 34]. To unfold a garment, our approach is to grasp the garment at two desired points with the robot arms (as in Figure 5). These desired points are chosen such that we can likely succeed in flattening the garments by simply laying down the garments on the table while grasping at the desired points. For a sweater, for example, these points are at the middle of the left and right elbows. To grasp the garment at the desired grasping points, our idea is to let the robot iteratively try grasping the garment (each trial may fail to grasp the desired point), until the two robot arms simultaneously succeed in grasping the desired points. With this iterative approach, it is easy to cope with the inaccuracy during the process.

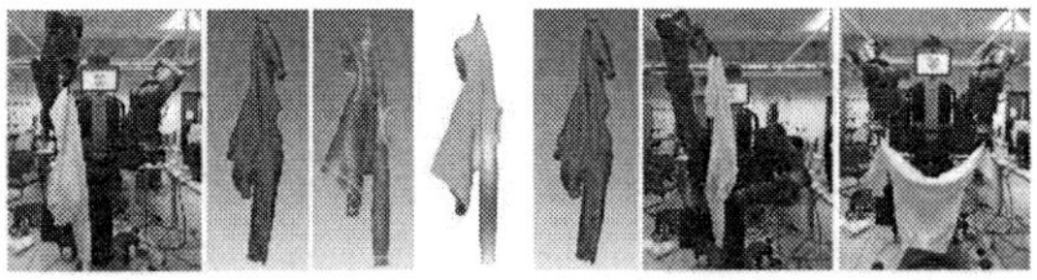

Figure 5. Reprinted from [29]. From left to right: an initial pick up, the 3D reconstructed mesh, the corresponding pose found in the database, the probability distribution of the next desired grasping point, the next desired grasping point registered on the reconstructed mesh, and finally snapshots of regrasping and unfolding.

To perform the grasp and to identify the point the robot is actually grasping, a key challenge is the estimation of the garment pose. For this pose estimation, we prepared a garment database in advance using an offline physically based simulation. The database contains 37 garments with different types, including sweater, pants, shorts, dresses, and scarves, each with manually labeled key grasping points such as sleeve end, elbow, shoulder, chest, and waist. We hang each garment for each grasping point in the simulator and simulated the stationary state of the garment.

During the online manipulation, we use Kinect to scan the garment and reconstruct a 3D mesh. Then, we search for the database for the closest garment shape. Because the database is a sparse sample of (the continuously) possible poses, we further perform a non rigid registration to obtain a map between each point on the garment mesh in the database and the corresponding point in the scanned mesh. This way, we can map a desired point on the mesh in the database to the point in the scanned mesh, and tell the robot which point to try grasping next. After grasping at two desired points, the robot will proceed to place the garment on the table.

4.2. Robotic Ironing

Once the garment is flattened on the table, we perform robotic ironing next. We developed a novel solution to analyze the cloth surface by fusing two surface scan techniques: a curvature scan and a discontinuity scan. The curvature scan is performed via Kinect, which is good at capturing regions with large variation in height (which does not require an ironing), while permanent wrinkles, which correspond to high curvatures or discontinuities in the gradient of the garment shape, can be effectively captured via an illumination based approach.

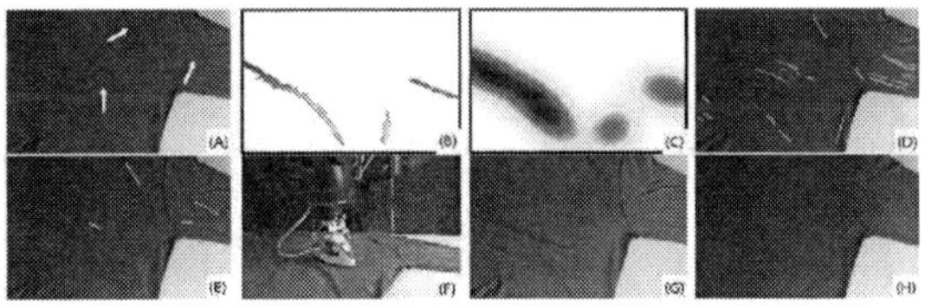

Figure 6. Reprinted from [30]. (A): roughly flattened cloth (white arrows show some obvious wrinkles). (B): Detected height bumps. (C): GMM fit to the bumps. (D): SVM-based discontinuity detection. (E): Permanent wrinkle detection (excluding C from D). (F): Our robot ironing. (G): The ironed result. (H): After manually flattening.

We found that using this combination of a 3D depth scan and an illumination based approach allows us to robustly identify permanent wrinkles from other deformations without a carefully calibrated system. The 3D height map and the 2D illumination image are combined in a probabilistic framework. We then classify and rank the deformations for ironing, and compute

IOP Conf. Series: Journal of Physics: Conf. Series **1036** (2018) 012013 doi:10.1088/1742-6596/1036/1/012013

the trajectory for ironing. The robot aligns the principal axis of the iron with the target wrinkle, following the computed trajectory, and performs ironing. After several iterations, we obtain a desired ironed result with most of the permanent wrinkles removed.

4.3. Folding Garments with Path Optimization

Once the garment is flattened, the next task is to fold the garment. Previous methods [35, 36] have focused on folding the garment according to a predefined folding plan (including the path of the robot arm). However, the layout of the same folding action can vary in terms of the material properties such as cloth hardness and the environment such as friction between the garment and the table. Given the starting and ending folding positions, different folding trajectories will lead to different results. Thus, we have developed a novel method that learns optimal folding trajectory parameters from predicted thin shell simulations of similar garments, which can then be applied to a real garment folding task.

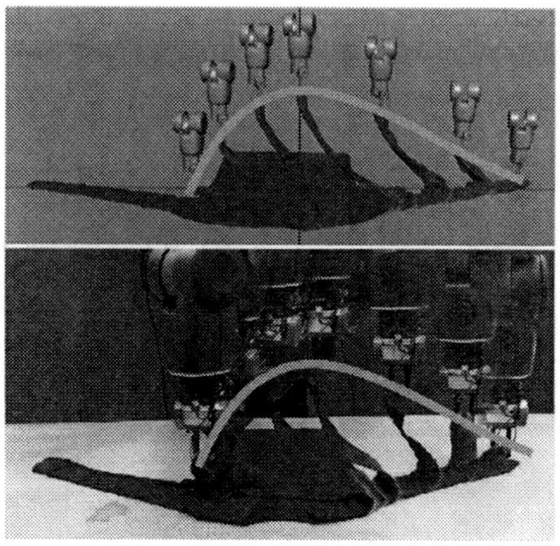

Figure 7. Reprinted from [31]. Our optimized result of robotic manipulation (top) and the real robot implementation (bottom). The green curves show the virtual and real trajectories for folding.

The input to our method is the material parameter of the garment, the friction between the garment and the table, the garment shape flattened on the table, the target folded shape, and the order to fold. To estimate the garment pose (and the label of each feature, like the sleeve) on the table, we capture the image and compute the contour of the garment. Then, we match the contour with predefined 2D templates, using a 2D version of the non rigid registration technique, to identify the labels of the feature points of the garment. To optimize for the trajectory for folding the garment, our prototype approach tries to find the parameters for defining the trajectory curve.

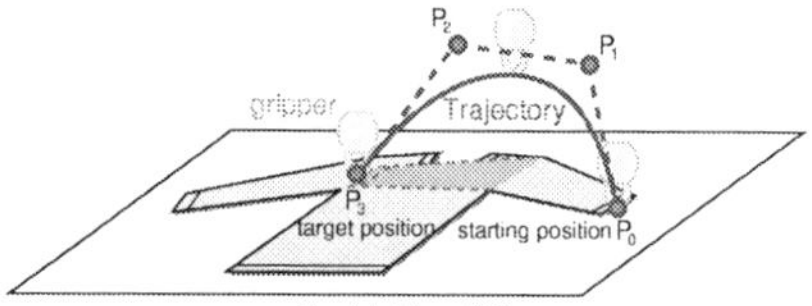

Reprinted from [31].

We use a third order Bézier curve to describe the trajectory, and optimize for the control points of the curve, except for the two end points. Our state variable $\boldsymbol{x}$ thus consists of the positions of the intermediate control points ($\boldsymbol{x} = (\boldsymbol{P}_1^T, \boldsymbol{P}_2^T)^T$). We solve for a minimization problem to find the optimal trajectory $\boldsymbol{x}^*$: $\boldsymbol{x}^* = \arg\min_{\boldsymbol{x}} \{C(\boldsymbol{x})\}^2$, where $C(\boldsymbol{x})$ is a cost function that accounts for the trajectory length $l_{\boldsymbol{x}}$ and the dissimilarity $D(\mathcal{S}_t, \mathcal{S}_{\boldsymbol{x}})$ between the target folded garment shape $\mathcal{S}_t$ and the simulated folded shape $\mathcal{S}_{\boldsymbol{x}}$ using the trajectory $\boldsymbol{x}$: $C(\boldsymbol{x}) = \alpha D(\mathcal{S}_t, \mathcal{S}_{\boldsymbol{x}}) + l_{\boldsymbol{x}}$. We used $\alpha = 10^3$ in our experiment. By minimizing the trajectory length as well as the dissimilarity, we aim at efficiency (i.e., reducing the time and energy for folding) as well as the correctness of the folding. We define the dissimilarity term as $D(\mathcal{S}_t, \mathcal{S}_{\boldsymbol{x}}) = \frac{1}{|\mathcal{S}_t|} \int_{\mathcal{S}_t} \|\boldsymbol{q}(\boldsymbol{y}) - \boldsymbol{y}\| dA$, where $|\mathcal{S}_t|$ is the surface area of the garment, $\boldsymbol{y} \in \mathcal{S}_t$ is a point on the target folded shape $\mathcal{S}_t$, $\boldsymbol{q}(\boldsymbol{y}) \in \mathcal{S}_{\boldsymbol{x}}$ is the corresponding point on the simulated folded shape, and dA is the area measure. For the optimization, we used a secant version of Levenberg-Marquardt algorithm [37]. With our optimized folding trajectory, we can let the robot to follow the trajectory and perform robotic folding in the real world, as in Figure 7.

5. Concluding Remarks

In this paper, we reviewed and introduced a series of our past work done from the perspective of how understanding the physics can help to develop various computing tools for predictive simulations, designing functional man-made objects, and developing smarter robots. Especially, we have focused on optics and mechanics, which are important for predicting the objects' appearance and motion, as well as designing functional objects that can shape the light and remain static under gravitational and external forces. We have also shown how optics and mechanics can be incorporated to robotic folding.

The graphics field focuses on such physics problems more at a real world scale in a resolution of our perception level. The need for the ultimate realism may sometimes ask for more efficient and accurate methods than those available in other fields. With no surprise, many state-of-the-art methods have been developed in the graphics field, involving various interesting mathematical problems, like the ones we encountered: statistics, partial differential equations, geometric flow, discrete differential geometry, combinatoric problem, randomized approach, feature registration, and so on. However, there is still a big gap between what we can perceive in the real world and what we can simulate on a computer, and there is still a long and exciting research journey for us to embark. We believe that graphics can serve as a playground to test and develop computational techniques, and its advancement would be beneficial for other fields as well.

Acknowledgments

I would like to deeply appreciate my closest collaborators (listed in the alphabetical order of the last name) Peter K. Allen, Christopher Batty, Shih-Fu Chang, Bing-Yu Chen, Yu-Huan Chung, Yoshinori Dobashi, Eitan Grinspun, Xiuhan Hu, Chun-Kai Huang, Sei Imai, Kei Iwasaki, Yinxiao Li, Sheng-Jie Luo, Tomoyuki Nishita, Breannan Smith, Yan Wang, Danfei Xu, and Changxi Zheng, with whom we have worked together for the work presented in this review paper. The work was supported in part by Grant-in-Aid for JSPS Fellows (20-7968), JSPS Grant-in-Aid for Research Activity Start-up (23800011), JSPS Postdoctoral Fellowships for Research Abroad, NSF (Grants IIS-13-19483, CMMI-11-29917, CAREER-1453101, IIS-1217904), NSERC (Grant RGPIN-04360-2014), Ministry of Science and Technology (MOST103-2218-E-002-030, MOST103-2221-E-002-158-MY3), National Science Council of Taiwan (NSC99-2221-E-002-016), Intel, The Walt Disney Company, Autodesk, Side Effects, NVIDIA, Adobe, The Foundry, MTM Corporation (Japan), and Takktile LLC.

References

[1] Y. Yue, K. Iwasaki, B.-Y. Chen, Y. Dobashi, and T. Nishita. Unbiased, adaptive stochastic sampling for rendering inhomogeneous participating media. *ACM Transactions on Graphics (Proc. of SIGGRAPH ASIA 2010)*, 29(6):177:1–177:8, 2010.

[2] Y. Yue, K. Iwasaki, B.-Y. Chen, Y. Dobashi, and T. Nishita. Toward optimal space partitioning for unbiased, adaptive free path sampling of inhomogeneous participating media. *Computer Graphics Forum (Proc. of Pacific Graphics 2011)*, 30(7):1911–1919, 2011.

[3] Y. Yue, B. Smith, C. Batty, C. Zheng, and E. Grinspun. Continuum foam: A material point method for shear-dependent flows. *ACM Transactions on Graphics (Presented at SIGGRAPH Asia 2015)*, 34(5):160:1–160:20, 2015.

[4] J. T. Kajiya. The rendering equation. *Computer Graphics (Proc. of SIGGRAPH 86)*, 20(4):143–150, 1986.

[5] E. Veach. *Robust Monte Carlo methods for light transport simulation.* PhD thesis, Stanford University, 1998.

[6] M. Pauly, T. Kollig, and A. Keller. Metropolis light transport for participating media. In *Proc. of EUROGRAPHICS Workshop on Rendering 2000*, pages 11–22, 2000.

[7] K. Perlin and E. M. Hoffert. Hypertexture. *Computer Graphics (Proc. of SIGGRAPH 89)*, 23(3):253–262, 1989.

[8] E.R. Woodcock, T. Murphy, P.J. Hemmings, and T.C. Longworth. Techniques used in the GEM code for Monte Carlo neutronics calculations in reactors and other systems of complex geometry. In *Proc. of Conference on the Application of Computing Methods to Reactor Problems, ANL-7050*, pages 557–579, 1965.

[9] M. Galtier, S. Blanco, C. Caliot, C. Coustet, J. Dauchet, M. El Hafi, V. Eymet, R. Fournier, J. Gautrais, A. Khuong, B. Piaud, and G. Terre. Integral formulation of null-collision Monte Carlo algorithms. *Journal of Quantitative Spectroscopy and Radiative Transfer*, 125:57 – 68, 2013.

[10] P. Kutz, R. Habel, Y. K. Li, and J. Novák. Spectral and decomposition tracking for rendering heterogeneous volumes. *ACM Transactions on Graphics (Proc. of SIGGRAPH 2017)*, 36(4):111:1–111:16, 2017.

[11] L.L. Carter, E.D. Cashwell, and W.M.Taylor. Monte Carlo sampling with continuously varying cross sections along flight paths. *Nuclear Science and Engineering*, 48:403–411, 1972.

[12] A. Aggarwal and S. Suri. Fast algorithms for computing the largest empty rectangle. In *Proc. of Third Annual Symposium on Computational Geometry*, pages 278–290, 1987.
[13] D. Weaire and S. Hutzler. *The Physics of Foams*. Oxford University Press, 2001.
[14] G. Ovarlez, K. Krishan, and S. Cohen-Addad. Investigation of shear banding in three-dimensional foams. *EPL (Europhysics Letters)*, 91(6):68005:1–68005:6, 2010.
[15] D. Sulsky, S.-J. Zhou, and H. L. Schreyer. Application of a particle-in-cell method to solid mechanics. *Computer Physics Communications*, 87(1-2):236–252, 1995.
[16] A. Stomakhin, C. Schroeder, L. Chai, J. Teran, and A. Selle. A material point method for snow simulation. *ACM Transactions on Graphics (Proc. of SIGGRAPH 2013)*, 32(4):102:1–102:9, 2013.
[17] A. Stomakhin, C. Schroeder, C. Jiang, L. Chai, J. Teran, and A. Selle. Augmented MPM for phase-change and varied materials. *ACM Transactions on Graphics (Proc. of SIGGRAPH 2014)*, 33(4):138:1–138:11, 2014.
[18] J. C. Simo and T. J. R. Hughes. *Computational Inelasticity*. Springer, 1998.
[19] W. H. Herschel and R. Bulkley. Konsistenzmessungen von Gummi-Benzollösungen. *Kolloid-Zeitschrift & Zeitschrift für Polymere*, 39(4):291–300, 1926.
[20] J. C. Simo. A framework for finite strain elastoplasticity based on maximum plastic dissipation and the multiplicative decomposition: Part I. continuum formulation. *Computer Methods in Applied Mechanics and Engineering*, 66(2):199–219, 1988.
[21] Y. Yue, K. Iwasaki, B.-Y. Chen, Y. Dobashi, and T. Nishita. Poisson-based continuous surface generation for goal-based caustics. *ACM Transactions on Graphics (Presented at SIGGRAPH 2014)*, 33(3):31:1–31:7, 2014.
[22] S.-J. Luo, Y. Yue, C.-K. Huang, Y.-H. Chung, S. Imai, T. Nishita, and B.-Y. Chen. Legolization: Optimizing LEGO designs. *ACM Transactions on Graphics (Proc. of SIGGRAPH ASIA 2015)*, 34(6):222:1–222:12, 2015.
[23] Y. Yu, K. Zhou, D. Xu, X. Shi, H. Bao, B. Guo, and H.-Y. Shum. Mesh editing with poisson-based gradient field manipulation. *ACM Transactions on Graphics (Proc. of SIGGRAPH 2004)*, 23(3):644–651, 2004.
[24] R. Gower, A. Heydtmann, and H. Petersen. LEGO: automated model construction. In *Proc. of 32nd European Study Group with Industry*, pages 81–94, 1998.
[25] S. Mueller, T. Mohr, K. Guenther, J. Frohnhofen, and P. Baudisch. faBrickation: fast 3D printing of functional objects by integrating construction kit building blocks. In *Proc. of SIGCHI 2014*, pages 3827–3834, 2014.
[26] P. Petrovic. Solving the LEGO brick layout problem using evolutionary algorithms. In *Proc. of NIK 2001*, 2001.
[27] R. Testuz, Y. Schwartzburg, and M. Pauly. Automatic generation of constructable brick sculptures. In *Eurographics 2013 Short papers*, pages 81–84, 2013.
[28] L. van Zijl and E. Smal. Cellular automata with cell clustering. In *Proc. of Automata 2008*, pages 425–440, 2008.
[29] Y. Li, D. Xu, Y. Yue, Y. Wang, S.-F. Chang, E. Grinspun, and P. K. Allen. Regrasping and unfolding of garments using predictive thin shell modeling. In *Proc. of the IEEE International Conference on Robotics and Automation (ICRA) 2015*, pages 1382–1388. IEEE, 2015.
[30] Y. Li, X. Hu, D. Xu, Y. Yue, E. Grinspun, and P. K. Allen. Multi-sensor surface analysis for robotic ironing. In *Proc. of the IEEE International Conference on Robotics and Automation (ICRA) 2016*, 2016.
[31] Y. Li, Y. Yue, D. Xu, E. Grinspun, and P. K. Allen. Folding deformable objects using predictive simulation and trajectory optimization. In *Proc. of the 27th IEEE/RSJ International Conference on Intelligent Robots and Systems (IROS) 2015*, 2015.
[32] M. Cusumano-Towner, A. Singh, S. Miller, J. F. O'Brien, and P. Abbeel. Bringing clothing into desired configurations with limited perception. In *Proc. of IEEE International Conference on Robotics and Automation (ICRA) 2011*, pages 3893–3900, 2011.
[33] A. Doumanoglou, A. Kargakos, T. K. Kim, and S. Malassiotis. Autonomous active recognition and unfolding of clothes using random decision forests and probabilistic planning. In *Proc. of IEEE International Conference on Robotics and Automation (ICRA) 2014*, pages 987–993, 2014.
[34] F. Osawa, H. Seki, and Y. Kamiya. Unfolding of massive laundry and classification types of dual manipulator. *Journal of Advanced and Intelligent Informatics*, 11(5), 2007.
[35] S. Miller, M. Fritz, T. Darrel, and P. Abbeel. Parametrized shape models for clothing. In *Proc. of IEEE International Conference on Robotics and Automation (ICRA) 2011*, 2011.
[36] S. Miller, J. van den Berg, M. Fritz, T. Darrell, K. Goldberg, and P. Abbeel. A geometric approach to robotic laundry folding. *The International Journal of Robotics Research*, 31(2):249–267, 2012.
[37] K. Madsen, H. B. Nielsen, and O. Tingleff. Methods for non-linear least squares problems (2nd ed.), 2004.

Data-driven diagnosis for compressed sensing algorithms

Y Nakanishi-Ohno[1,2] and K Hukushima[1]

[1]Graduate School of Arts and Sciences, The University of Tokyo, 3-8-1, Komaba, Meguro, Tokyo, 153-8902, Japan
[2]PRESTO, Japan Science and Technology Agency, 4-1-8, Honcho, Kawaguchi, Saitama, 332-0012, Japan

E-mail: nakanishi@phys.c.u-tokyo.ac.jp

Abstract. A data-driven diagnosis approach is developed for compressed sensing algorithms to objectively detect the failure of each algorithm and the insufficiency of data. Compressed sensing enables us to reduce the amount of data to a threshold; however, the problem is that no one can tell where the threshold is in practice. The approach is based on the statistical technique of cross-validation, in which part of the data is set aside as a validation dataset to assess the results of the analysis of the remaining data with an objective algorithm. The relation between cross-validation error and the size ratio of the whole dataset and validation dataset is evaluated and compared with a power-law relation to determine whether data analysis was successfully conducted. After summarizing our recent theoretical work on the first application of the approach to the basis pursuit method, it is shown by numerical experiments that the approach is universally applicable to several other algorithms.

1. Introduction

Data-driven science is growing as a tree trunk of science, which stands on the foundation of information science to support the many branches of natural science. Every branch commonly suffers from high-dimensional data but studying data handling exclusively within itself can sometimes goes astray. In data-driven science, scientists from various disciplines collaborate on data-related issues, such as data acquisition, data analysis, data interpretation, and data re-acquisition, in order to create a driving force for sustaining and accelerating the whole cycle of science. In this paper, "data-driven diagnosis" is discussed as a kind of data interpretation. It is interesting to consider what happens if you come across a situation where the results of data analysis with different algorithms are vague or ambiguous. One can speculate that some algorithms are going the wrong way, while others may doubt whether there is enough data to be interpreted as something scientifically significant in the first place. Then, it is essential to objectively detect the failure of each algorithm and the insufficiency of data. What is important here is that the procedure of data-driven diagnosis should not involve any hypothesis to be tested. Researchers from information science will probably develop algorithms based on unconfirmed facts about natural phenomena; in contrast, those from natural science will misuse algorithms beyond the scope of their applications if they work separately. Therefore, we cannot make too much of the culture of data-driven science.

Here, we develop a data-driven diagnosis framework for compressed sensing. The purpose of compressed sensing is to improve the efficiency and availability of experiments [1, 2]. A key technology is sparse modeling, where one assumes that an object is described by a parsimoniously small number of explanatory factors. Owing to the affinity of sparse modeling with natural science, there are many applications in various fields: exploration geophysics [3, 4], magnetic resonance imaging [5, 6], ghost imaging [7, 8], nuclear magnetic resonance spectroscopy [9, 10], astronomical radio interferometry [11, 12], and scanning tunneling microscopy/spectroscopy [13]. On the other hand, in information science, there are various compressed sensing algorithms with sparse modeling, which include:

- orthogonal least squares (OLS) [14]
- orthogonal matching pursuit (OMP) [15]
- homotopy method for basis pursuit (Homotopy-BP) [16]
- least angle regression (LARS) [17]
- iterative hard thresholding (IHT) [18, 19]
- compressive sampling matching pursuit (CoSaMP) [20]
- subspace pursuit (SP) [21]
- hard thresholding pursuit (HTP) [22]

The algorithm that should be used remains confusing to users although many studies have evaluated and compared these algorithms in terms of factors such as the necessary and sufficient conditions for success, computational complexity, number of tuning parameters, etc. It makes sense that a tractable algorithm should be chosen and more importantly that a reliable algorithm is used. However, the conditions for success are often described using the sparsity of what one wants to know as pointed out in [23], and this is against the rule that one should not employ a hypothesis. This is nothing but a data-driven diagnosis problem.

In our most recent work [24], we developed a cross-validation approach to data-driven diagnosis and theoretically verified that it works well for compressed sensing with the basis pursuit method. Cross-validation is a classical statistical technique for parameter tuning and model selection [25–27], but we will attempt to use it in an unconventional way. We have found that when cross-validation error is plotted against the size ratio of the training data and whole data, it behaves in a power-law fashion at the phase boundary of the success and failure of compressed sensing; this suggests that a critical factor for data-driven diagnosis can be obtained by monitoring the relation. In the current work, further progress in data-driven diagnosis will be made by investigating the applicability of our approach to the algorithms listed above.

The rest of the paper is organized as follows: Section 2 formulates the compressed sensing problem and the cross-validation procedure. Section 3 summarizes our recent theoretical work on the basis pursuit method. Section 4 discusses the results of numerical experiments to examine the applicability of our approach to several algorithms, and concludes the paper.

2. Formulation

2.1. Compressed sensing

The general problem of sensing, not restricted to compressed sensing, is formulated as finding the representation of an object, $\mathbf{x} = (x_i)_{i \in [N]}$, from a set of data $\mathbf{y} = (y_\mu)_{\mu \in [M]}$, which is given by

$$\mathbf{y} = \mathbf{A}\mathbf{x}^0. \quad (1)$$

Matrix $\mathbf{A} = (a_{\mu i})_{\mu \in [M], i \in [N]}$ represents a linear measurement process and each item of data y_μ is acquired by a measurement denoted by a corresponding row vector $\mathbf{a}_\mu = (a_{\mu i})_{i \in [N]}$. The object has a true representation $\mathbf{x}^0$. It is noted that $[N]$ is the set of all positive integers not exceeding N, namely, $[N] = \{1, 2, \ldots, N\}$. For convenience, we introduce a parameter $\alpha = M/N$ to denote the amount of data. If the dimension of $\mathbf{y}$ is greater than or equal to that of $\mathbf{x}$, that is, $\alpha \geq 1$, one can reconstruct $\mathbf{x}^0$ from $\mathbf{y}$ as follows:

$$\mathbf{x}^0 = \mathbf{A}^+\mathbf{y}, \tag{2}$$

Matrix $\mathbf{A}^+$ is the Moore–Penrose pseudoinverse of $\mathbf{A}$, and it coincides with the inverse $\mathbf{A}^{-1}$ when $\alpha = 1$. In other words, $\mathbf{x}^0$ is obtained by using the method of least squares as follows:

$$\mathbf{x}^0 = \arg\min_{\mathbf{x}} \frac{1}{2M} \sum_{\mu} (\mathbf{a}_\mu \mathbf{x} - y_\mu)^2. \tag{3}$$

Another problem with *compressed* sensing is solving an underdetermined equation system due to $\alpha < 1$. In the underdetermined case, there are many solution candidates satisfying $\mathbf{A}\mathbf{x} = \mathbf{y}$. If you multiply the pseudoinverse of $\mathbf{A}$ to $\mathbf{y}$ in the same way as the case of $\alpha \geq 1$, you are supposed to calculate

$$\mathbf{A}^+\mathbf{y} = \arg\min_{\mathbf{x}} ||\mathbf{x}||_2 \quad \text{subject to} \quad \mathbf{A}\mathbf{x} = \mathbf{y} \tag{4}$$

as derived from the definition of pseudoinverse in the case of $\alpha < 1$, where the l_2-norm $||\cdot||_2$ is defined as $||\mathbf{x}||_2 = \sqrt{\sum_i x_i^2}$. However, the l_2-norm minimization technique does not give the true representation $\mathbf{x}^0$ in the underdetermined case [35]. An effective approach to this problem is sparse modeling, which assumes that $\mathbf{x}^0$ has a sparse representation composed of a small number of non-zero elements. The number of non-zeros that $\mathbf{x}$ has is denoted by the l_0-norm $||\mathbf{x}||_0$, which is used to measure sparsity. The efficacy of sparse modeling is intuitively understood as follows: If the assumption is true, then the number of unknowns can be sufficiently reduced so that $M > ||\mathbf{x}||_0$, and the ill-posed problem can be solved.

To simplify the discussion, we will consider the following case throughout the study: Each element of $\mathbf{x}^0$ is generated from a mixture distribution $p(x_i^0) = \rho_0 \mathcal{N}(0, 1) + (1 - \rho_0)\delta(x_i^0)$ in an independent and identically distributed (iid) way. The distribution $\mathcal{N}(\mu, \sigma^2)$ is a Gaussian distribution with mean μ and variance σ^2. The function $\delta(\cdot)$ is the Dirac delta function. The parameter ρ_0 represents the sparsity of the true representation since

$$\rho_0 = \int d\mathbf{x}^0 \left(\prod_i p(x_i^0) \right) \frac{||\mathbf{x}^0||_0}{N} \equiv \left\langle \frac{||\mathbf{x}^0||_0}{N} \right\rangle_{\mathbf{x}^0}. \tag{5}$$

The angle brackets $\langle\cdot\rangle$ denote the expectation with respect to the subscript in this way. Each element of $\mathbf{A}$ is generated from $\mathcal{N}(0, N^{-1})$ in an iid way. In this setting, each element of $\mathbf{y}$ is acquired without any noise according to equation (1); in other words, generated from $p(y_\mu) = \delta(y_\mu - \mathbf{a}_\mu \mathbf{x}^0)$.

The remaining question in compressed sensing is which algorithm successfully finds the support set of $\mathbf{x}^0$, namely, $S_0 = \{i | x_i^0 \neq 0\}$. Although an exhaustive search should be conducted, in the high-dimensional case, it is intractable owing to its exponentially increasing computational complexity. As listed in the Introduction, there are various kinds of algorithms, but one often becomes confused about which algorithm to use for data analysis. Therefore, a data-driven diagnosis framework is required.

2.2. Cross-validation

Cross-validation is a general but simple statistical technique for evaluating generalization error, which measures the predictability of an algorithm on unseen data [25–27]. This error is called cross-validation error in the context of cross-validation. Cross-validation error is a frequently used criterion for parameter tuning and model selection in various fields of science. The procedure of calculating cross-validation error has three main steps. First, the available dataset $\mathbf{y}$ is divided into two non-empty subsets: a training dataset $(y_\mu)_{\mu\in\mathcal{T}}$ and validation dataset $(y_\mu)_{\mu\in\mathcal{V}}$, where $\mathcal{V} = [M] \setminus \mathcal{T}$. For convenience, a parameter $k = M/|\mathcal{V}|$ $(k > 1)$, where $|\mathcal{V}|$ is the number of elements in $\mathcal{V}$, is introduced to represent the size ratio of the validation dataset and whole dataset, and it will also play an important role later in data-driven diagnosis. Second, an estimation algorithm is applied to the training dataset to obtain an outcome $\hat{\mathbf{x}}_\mathcal{T}$. Finally, how accurately the outcome $\hat{\mathbf{x}}_\mathcal{T}$ describes the validation dataset $(y_\mu)_{\mu\in\mathcal{V}}$ is measured by cross-validation error introduced as follows:

$$\mathrm{CVE} = \frac{1}{2|\mathcal{V}|}\sum_{\mu\in\mathcal{V}}(\mathbf{a}_\mu\hat{\mathbf{x}}_\mathcal{T} - y_\mu)^2. \tag{6}$$

If cross-validation error has a small value close to zero, it means that the *unseen* data $(y_\mu)_{\mu\in\mathcal{V}}$ acquired from *future* measurements $(\mathbf{a}_\mu)_{\mu\in\mathcal{V}}$ are well *predicted* by the outcome $\hat{\mathbf{x}}_\mathcal{T}$ estimated from the *already seen* data $(y_\mu)_{\mu\in\mathcal{T}}$ acquired from *past* measurements $(\mathbf{a}_\mu)_{\mu\in\mathcal{T}}$ with an algorithm of interest. The cross-validation error fluctuates according to how the data is divided. Then, the calculation of cross-validation error is often repeated with many different partitions of the same dataset to obtain a typical value of cross-validation error. There are various ways of partitioning data, such as leave-one-out cross-validation, k-fold cross-validation, and Monte Carlo cross-validation. In the repetitive calculations, every item of the data plays a role of training and validation data by turns. What is used as training data in a calculation is used as validation data in another calculation, and vice versa. This is why the statistical technique is called *cross*-validation.

Cross-validation is a key ingredient in our data-driven diagnosis approach because of two points. One is that the calculation of cross-validation does not involve a true representation of $\mathbf{x}^0$, its support set S_0, or its sparsity ρ_0 at all. The other point is that cross-validation is applicable to whatever algorithm is used for estimation in compressed sensing, since the procedure does not assume the use of a specific algorithm to obtain $\hat{\mathbf{x}}_\mathcal{T}$. Therefore, a cross-validation approach meets the need for objectivity and is expected to be universally applicable.

3. Application to basis pursuit

This section summarizes our previous work on the initial application of our approach to compressed sensing with the basis pursuit method [24].

Basis pursuit is formulated as the following optimization problem [28]:

$$\hat{\mathbf{x}} = \arg\min_{\mathbf{x}} \|\mathbf{x}\|_1 \quad \text{subject to} \quad \mathbf{A}\mathbf{x} = \mathbf{y}. \tag{7}$$

The l_1-norm $\|\cdot\|_1$ is defined as $\|\mathbf{x}\|_1 = \sum_i |x_i|$. As from a linear programming algorithm [29, 30], efficient algorithms such as Homotopy-BP [16], a Bregman iterative algorithm [31], an approximate message-passing algorithm [32], and a primal-dual algorithm [33] have been developed for carrying out basis pursuit. The performance of basis pursuit in compressed sensing has been well studied and representative results are shown in figures 1(a) and 1(b) [34–36]. Figure 1(a) is a phase diagram of basis pursuit and the yellow line divides the two phases of success and failure. The success phase of basis pursuit is broader than that of the pseudoinverse and this fact has expanded the possibilities of compressed sensing. However, the result of the

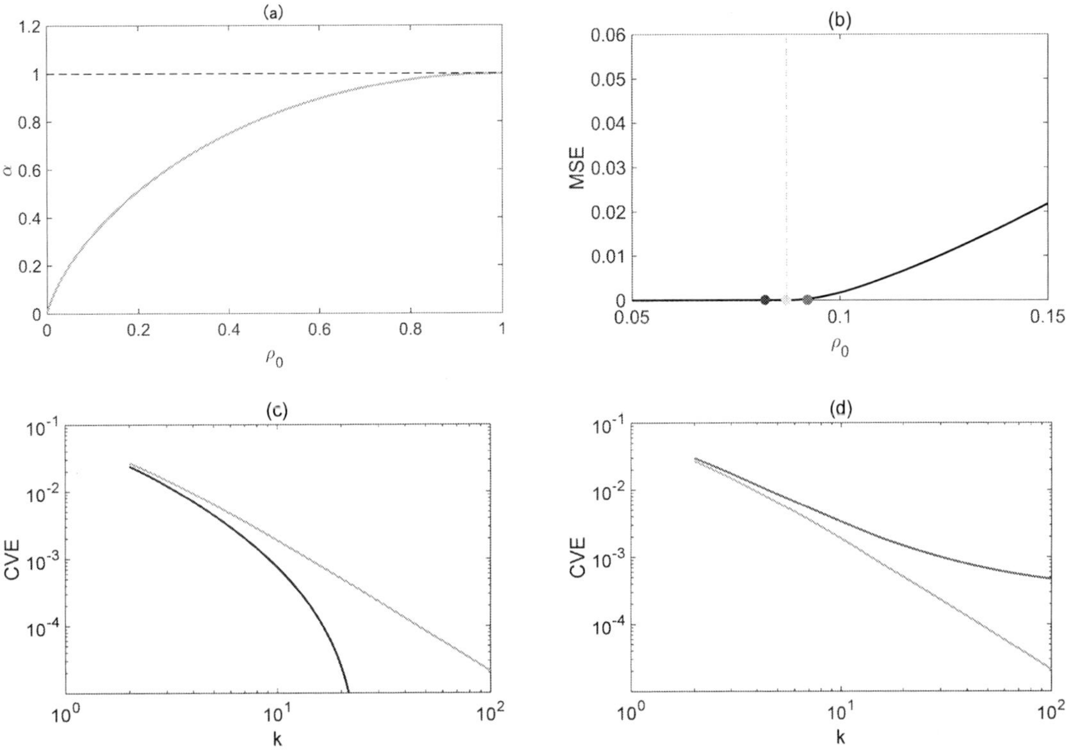

Figure 1. (a) Phase diagram on the success and failure in compressed sensing. Horizontal and vertical axes represent true sparsity ρ_0 and data amount α, respectively. The yellow line is the phase boundary of basis pursuit. The dashed line is the phase boundary of the pseudoinverse. For each method, the top-left region and bottom-right regions are the success phase and failure phase, respectively. (b) Plots of mean squared error against ρ_0 with respect to basis pursuit. $\alpha = 0.3$. (c), (d) Plots of cross-validation error against k with respect to basis pursuit. $\alpha = 0.3$. $\rho_0 = 0.0822$ (blue), 0.0872 (yellow), 0.0922 (red). These points of ρ_0 are indicated by corresponding colored circles in (b), and the yellow one is at the phase boundary. (a) and (b) are derived from calculations in [35]. (c) and (d) are derived from calculations in [24].

phase diagram has a critical defect in practice. For example, consider the case of $\alpha = 0.3$ as shown in figure 1(b). Figure 1(b) plots the typical value of the following mean squared error,

$$\mathrm{MSE} = \frac{1}{2N}\sum_i (\hat{x}_i - x_i^0)^2, \tag{8}$$

against ρ_0. In the left success phase, the mean squared error is equal to zero, while in the right failure phase, it takes a non-zero value. However, without the true sparsity ρ_0, no one knows in which phase the outcome of basis pursuit is.

To discuss data-driven diagnosis in [24], we analyzed the typical cross-validation error values of basis pursuit, defined as

$$\mathrm{CVE}_{\mathrm{BP}} = \left\langle \left\langle \frac{1}{2|\mathcal{V}|}\sum_{\mu\in\mathcal{V}} (\mathbf{a}_\mu \mathbf{x} - y_\mu)^2 \right\rangle_{\mathbf{x}} \right\rangle_{\mathbf{y},\mathbf{A},\mathbf{x}^0}, \tag{9}$$

where the use of basis pursuit with a training dataset is reflected in the probability

$$p(\mathbf{x}) \propto e^{-\beta\|\mathbf{x}\|_1} \prod_{\mu \in \mathcal{T}} \delta(\mathbf{a}_\mu \mathbf{x} - y_\mu). \tag{10}$$

Comparison with equation (7) shows that the Boltzmann factor $e^{-\beta\|\mathbf{x}\|_1}$ in the limit $\beta \to +\infty$ corresponds to the l_1-norm minimization and the delta functions represent the equality constraint. We applied the replica method, which was developed in the field of spin-glass theory [37], to obtain the results for a replica-symmetric ansatz. In our theoretical analysis, the limit $N \to +\infty$ was dealt with because the use of the saddle-point method was justified. Figures 1(c) and 1(d) plot the typical values of cross-validation error against k, the size ratio of the validation dataset and whole dataset, when $\alpha = 0.3$. The blue line in figure 1(c) represents a success case of $\rho_0 = 0.0822$, while the red line in figure 1(d) represents a failure case of $\rho_0 = 0.0922$. For reference, the yellow lines in both figures represent the case of $\rho_0 = 0.0872$ at the phase boundary.

Surprisingly, a glance at figures 1(c) and 1(d) shows that there is a qualitative difference between the success and failure phases. In the success phase, the cross-validation error decreases in an exponential fashion as k increases; in the failure phase, it decreases to a finite value for large k. At the phase boundary between them, a power-law relation is seen in the asymptotic behavior of the cross-validation error. Our intuitive understanding of this is as follows: Imagine that your dataset is actually in the success phase in figure 1(a) although you do not know it, and that you try setting aside a part of the data as validation data. If the amount of validation data is small, your training dataset remains in the success phase and the cross-validation error disappears for large k, as shown in figure 1(c). In the opposite case, if your whole dataset is in the failure phase, your training dataset is also within the failure phase and the cross-validation error takes a finite value for any k, as shown in figure 1(d). This may be the reason why the behavior of cross-validation error distinguishes the success or failure of basis pursuit.

Next, we have to consider what should be done if a failure of basis pursuit is successfully detected. Although data re-acquisition may be an option, it is also promising to perform data analysis with another algorithm. According to the intuitive explanation above, our approach is expected to be universally applicable to other algorithms. In the following section, this applicability is examined.

4. Application to algorithms

In this section, we compare various aspects of well-known compressed sensing algorithms, including data-driven diagnosis by numerical experiments. We examined the eight algorithms, which are listed in Introduction and are explained in detail in Appendix A. We set the parameter ϵ to stop the iteration of each algorithm at 1×10^{-9}. There are some variants of these algorithms with respect to initialization, stopping criterion, and so forth, each of which is an interesting target for data-driven diagnosis as well.

4.1. Conventional work

We examined the performance of the algorithms in terms of mean squared error and success rate among 100 instances of dataset $\mathbf{y}$, measurement matrix $\mathbf{A}$, and true representation of an object $\mathbf{x}^0$. Figures 2(a) and 2(b) show the results for mean squared error and success rate, respectively. The success rate is defined as the number of success instances divided by the total number of instances. The parameters are as follows: $N = 1000$, $\rho_0 = 0.05, 0.06, \ldots, 0.15$, $\alpha = 0.3$. We can see that numerical mean squared error of Homotopy-BP is consistent with the theoretical results of basis pursuit shown in figure 1(b), which confirms our theoretical work. Although Homotopy-BP estimates a representation with relatively low mean squared error, it also has

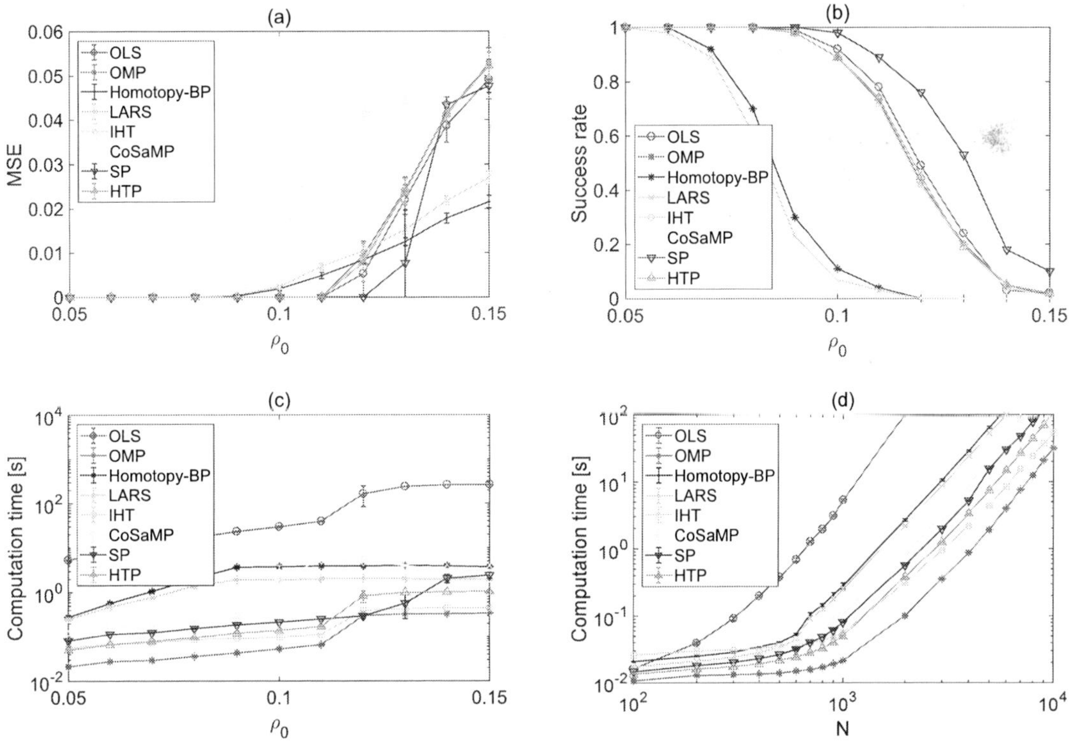

Figure 2. Performance of algorithms. (a) Plots of mean squared error against ρ_0. $N = 1000$. $\alpha = 0.3$. (b) Plots of success rate against ρ_0. $N = 1000$. $\alpha = 0.3$. (c) Plots of computation time against ρ_0. $N = 1000$. $\alpha = 0.3$. (d) Plots of computation time against N. $\rho_0 = 0.05$. $\alpha = 0.3$. Each mark and its error bar represent the median value and its standard error, respectively, evaluated for over 100 instances of $(\mathbf{y}, \mathbf{A}, \mathbf{x}^0)$ using the bootstrap method.

the lowest success rate. Thus, the basis pursuit method is considered to be unsuitable for the correct estimation of the support set. Among the algorithms studied, SP achieved the highest performance in success rate regardless of ρ_0, and all but LARS were superior to the basis pursuit method. From the viewpoint of data-driven diagnosis, it is worthwhile to mention that every algorithm as well as basis pursuit exhibited a phase transition of success and failure, and our approach is expected to be applicable to the algorithms.

We also examined the computation time of the algorithms among 100 instances. Figures 2(c) and 2(d) show the results. The parameters are as follows: (c) is the same as (a) and (b), that is, $N = 1000$, $\rho_0 = 0.05, 0.06, \ldots, 0.15$, $\alpha = 0.3$. (d) $N = 100, 200, \ldots, 1000, 2000, \ldots, 10000$, $\rho_0 = 0.05$, $\alpha = 0.3$. We find that OLS had the longest computation time, while OMP had the shortest time, regardless of N and ρ_0. Unexpectedly, in addition to its good performance in terms of success rate, the computation time of SP was the fourth shortest when it correctly estimated the true representation. As shown in figure 2(c), computation time increases as ρ_0 increases because many iterations are needed to find the true representation composed of many non-zeros. Besides, it may be noted that the relation between sparsity ρ_0 and computation time in the failure phase is different from that in the success phase. Except for Homotopy-BP and LARS, computation time greatly increased around the transition point of ρ_0. Since it can be measured without knowing $\mathbf{x}^0$ or ρ_0, computation time is perhaps a promising quantity for future work on data-driven diagnosis.

Can conventional work answer the question of what algorithm should be used? Indeed one

IOP Conf. Series: Journal of Physics: Conf. Series **1036** (2018) 012014 doi:10.1088/1742-6596/1036/1/012014

can recommend the use of SP because it achieved the best success rate at the fourth shortest computation time. However, it is not guaranteed that every result of SP will be correct unless its success rate is exactly one in all cases. Thus, what is really important is data-driven diagnosis to interpret whether the algorithm actually succeeds or fails in data analysis.

4.2. Data-driven diagnosis

Our approach to data-driven diagnosis is based on cross-validation, which can be conducted without the true representation $\mathbf{x}^0$ or the true sparsity ρ_0. In our approach, one only has to monitor the relation between the cross-validation error and parameter $k = M/|\mathcal{V}|$, which represents the size ratio of the whole dataset and validation dataset. If our approach works well, cross-validation error decreases asymptotically in an exponential fashion in the case of success, and remains a finite value for large k in the case of failure.

Figure 3 shows the results of data-driven diagnosis for the algorithms. The parameters are as follows: $N = 300$, $\rho_0 = 0.1$, $\alpha = 0.3$. Cross-validation is plotted at various points of $k = 2, 3, \ldots, 90$. The maximum k that can be examined is equal to the size of dataset $M = \alpha N$ because at least one item of data has to be set aside as validation data. Each plotted value of cross-validation error is the median among 100 trials of cross-validation with random partitions of data into training and validation data. In each panel of the algorithm, the results of 10 instances of $\mathbf{y}$, $\mathbf{A}$, and $\mathbf{x}^0$ are shown. The blue lines represent success instances and red lines represent failure instances in light of the mean squared error.

On the whole, every red instance has a finite value at $k = 90$, and in most blue instances, cross-validation error rapidly decreased to zero. The relation between the cross-validation error and parameter k can successfully distinguish the success or failure in a data-driven way. We note that some blue instances remained a finite value even at maximum k. This indicates that the training data does not give a correct representation, unlike the whole data. The reason for this is because the whole amount of data is barely above the phase boundary between success and failure, and it enters the failure phase by setting aside a small amount of validation data. Then, our approach sometimes gives conservative results. However, it is better to temporarily delay the interpretation of the results of data analysis rather than cause confusion with an inaccurate analysis.

As shown in figure 3, the cross-validation error value fluctuates, which could lead to a mistake in data-driven diagnosis in more serious situations. In our numerical experiments, only 100 trials of cross-validation were conducted to obtain a typical value, but the number of possible ways of data partition is a binomial coefficient $\binom{M}{k}$ for k. A precise and efficient method of calculating a typical value of cross-validation error needs to be developed. Originally, cross-validation was a time-consuming process, but the use of parallel computing, along with the invention of approximation methods [38], is worth discussing.

Consequently, we claim that our data-driven diagnosis approach performs well not only for the basis pursuit method but also for various compressed sensing algorithms. Our data-driven diagnosis approach provides a novel criterion to determine what algorithm should be used. Conventional works just recommend a specific algorithm according to its typical performance, but the present work enables us to select an algorithm that provides reliable results for specific data, or tentatively withhold an interpretation owing to insufficient data. There are many potential applications of data-driven diagnosis in natural science. Compressed sensing with *noisy* measurements is a practically important one. Moreover, the scope of our approach is not restricted to the formulation of linear regression, and can be expanded to include other formulations such as matrix completion [39], affine rank minimization [40], and demixing problems [41]. This is because cross-validation, on which our approach is based, is a versatile statistical tool.

In conclusion, we developed a cross-validation approach to data-driven diagnosis and

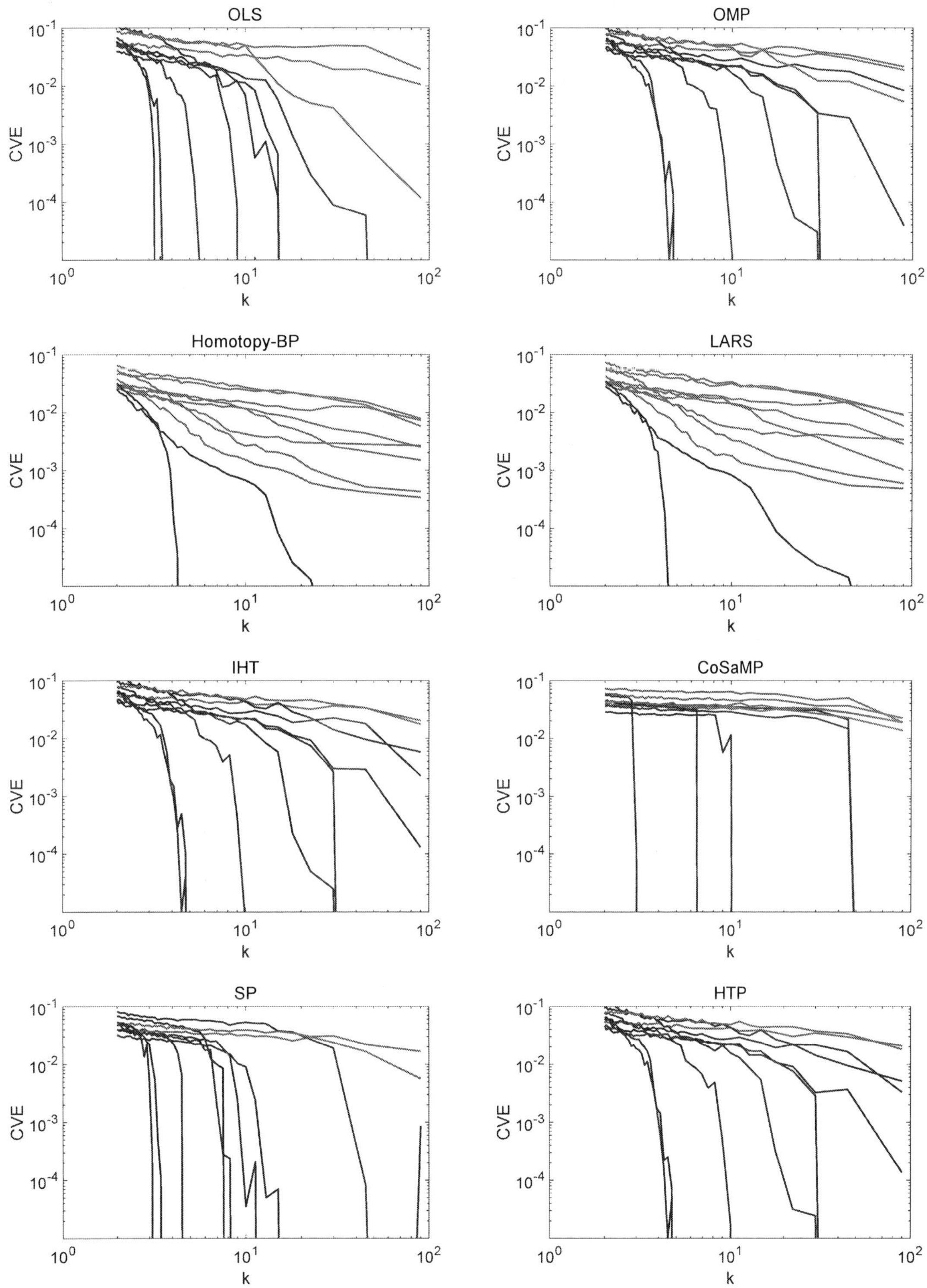

Figure 3. Relation between cross-validation error and parameter k for 10 instances of $(\mathbf{y}, \mathbf{A}, \mathbf{x}^0)$. $N = 300$. $\rho_0 = 0.1$. $\alpha = 0.3$. The blue lines represent success instances and red lines represent failure instances.

demonstrated that the approach works well for various compressed sensing algorithms. Our approach is useful for identifying an algorithm that provides reliable information and that definitely states when there is insufficient data. Data-driven diagnosis should have a significant impact on data-driven science because it solves the general problem of what algorithm should be used for data analysis in natural science among a variety of methods in information science.

Acknowledgments
This work was supported by JSPS KAKENHI Grant Numbers JP17K12749 (YN-O) and JP25120010 (KH), and JST PRESTO Grant Number JPMJPR1773 (YN-O).

Appendix A. Algorithms of compressed sensing

The appendix explains algorithms examined in this study.

Notation

- $\mathbf{A}_S$: The submatrix of a matrix $\mathbf{A}$ with columns indexed by a support S.
- $\mathbf{x}_S$: The subvector of a vector $\mathbf{x}$ with elements indexed by a support S.
- $\mathbf{A}^\dagger$: The adjoint matrix of a matrix $\mathbf{A}$.
- $\overset{+}{\min}\{\cdot\}$: The minimum taken only over positive arguments.
- $\mathrm{sign}(\cdot)$: Sign function. It is defined element-wise when the argument is a vector.
- $L_s(\cdot)$: An index set of s largest vector elements in the sense of absolute value.
- $H_s(\cdot)$: Hard thresholding operator. It sets all but s largest vector elements in the sense of absolute value to zero.

Appendix A.1. OLS

The OLS algorithm for estimating $\mathbf{x}$, given a vector $\mathbf{y}$, a matrix $\mathbf{A}$, and a small positive scalar ϵ, is described as follows:

(i) $S = \emptyset$, $r = \frac{(\|\mathbf{y}\|_2)^2}{2M}$.
(ii) repeat the following (1) and (2) while $r > \epsilon$:
   (1) $[r, j_{\mathrm{add}}] = [\min_{j \notin S}, \arg\min_{j \notin S}] \frac{1}{2M} \left(\|\mathbf{y} - \mathbf{A}_{S \cup \{j\}} (\mathbf{A}_{S \cup \{j\}})^+ \mathbf{y}\|_2\right)^2$.
   (2) $S = S \cup \{j_{\mathrm{add}}\}$.
(iii) $\mathbf{x} = \mathbf{0}$, $\mathbf{x}_S = (\mathbf{A}_S)^+ \mathbf{y}$.

OLS takes an index j_{add} in the support S at each iteration of (ii), and the additional index j_{add} is determined by using the method of least squares at (ii-1). Be careful not to confuse OLS with the method of *ordinary* least squares.

Appendix A.2. OMP

The OMP algorithm for estimating $\mathbf{x}$, given a vector $\mathbf{y}$, a matrix $\mathbf{A}$, and a small positive scalar ϵ, is described as follows:

(i) $S = \emptyset$, $\mathbf{x} = \mathbf{0}$, $\mathbf{r} = \mathbf{y}$, $r = \frac{(\|\mathbf{y}\|_2)^2}{2M}$.
(ii) repeat the following (1)–(5) while $r > \epsilon$:
   (1) $j_{\mathrm{add}} = \arg\max_{j \notin S} |(\mathbf{A}_{\{j\}})^\dagger \mathbf{r}|$.
   (2) $S = S \cup \{j_{\mathrm{add}}\}$.
   (3) $\mathbf{x}_S = (\mathbf{A}_S)^+ \mathbf{y}$.

(4) $\mathbf{r} = \mathbf{y} - \mathbf{A}\mathbf{x}$.
(5) $r = \frac{(\|\mathbf{r}\|_2)^2}{2M}$.

OMP as well as OLS takes an index j_{add} in the support S at each iteration of (ii), but the additional index j_{add} is determined by minimizing the absolute inner product between each column vector in $\mathbf{A}$ and the residual $\mathbf{r}$ at (ii-1). Owing to the replacement for the method of least squares, one can reduce computation time since the calculation of the pseudoinverse needs to be done only once per iteration at (ii-3).

Appendix A.3. Homotopy-BP

The Homotopy-BP algorithm for estimating $\mathbf{x}$, given a vector $\mathbf{y}$, a matrix $\mathbf{A}$, and a small positive scalar ϵ, is described as follows:

(i) $\mathbf{x} = \mathbf{0}$, $\mathbf{r} = \mathbf{y}$, $r = \frac{(\|\mathbf{y}\|_2)^2}{2M}$.
(ii) $[\lambda, j_{\text{add}}] = [\max_j, \arg\max_j] |(\mathbf{A}_{\{j\}})^\dagger \mathbf{r}|$.
(iii) $S = \{j_{\text{add}}\}$.
(iv) repeat the following (1)–(9) while $r > \epsilon$:
   (1) $\mathbf{d} = \mathbf{0}$, $\mathbf{d}_S = ((\mathbf{A}_S)^\dagger \mathbf{A}_S)^{-1} \text{sign}((\mathbf{A}_S)^\dagger \mathbf{r})$.
   (2) $[\gamma_{\text{add}}, j_{\text{add}}] = [\overset{+}{\min_{j \notin S}}, \arg \overset{+}{\min_{j \notin S}}] \left\{ \frac{\lambda - (\mathbf{A}_{\{j\}})^\dagger \mathbf{r}}{1 - (\mathbf{A}_{\{j\}})^\dagger \mathbf{A}_S \mathbf{d}_S}, \frac{\lambda + (\mathbf{A}_{\{j\}})^\dagger \mathbf{r}}{1 + (\mathbf{A}_{\{j\}})^\dagger \mathbf{A}_S \mathbf{d}_S} \right\}$.
   (3) $[\gamma_{\text{rem}}, j_{\text{rem}}] = [\overset{+}{\min_{j \in S}}, \arg \overset{+}{\min_{j \in S}}] \left\{ -\frac{\mathbf{x}_{\{j\}}}{\mathbf{d}_{\{j\}}} \right\}$
   (4) $\gamma = \min\{\gamma_{\text{add}}, \gamma_{\text{rem}}\}$
   (5) $\mathbf{x} = \mathbf{x} + \gamma \mathbf{d}$.
   (6) $\mathbf{r} = \mathbf{y} - \mathbf{A}\mathbf{x}$.
   (7) $r = \frac{(\|\mathbf{r}\|_2)^2}{2M}$.
   (8) $\lambda = \lambda - \gamma$.
   (9) if $\gamma = \gamma_{\text{add}}$, $S = S \cup \{j_{\text{add}}\}$; or if $\gamma = \gamma_{\text{rem}}$, $S = S \setminus \{j_{\text{rem}}\}$.

Homotopy-BP is a solver of basis pursuit with a homotopy method. The homotopy method solves a convergent series of optimization problems for an original problem and takes the limit of solutions with respect to the parameter of the series as follows:

$$\hat{\mathbf{x}} = \lim_{\lambda \to +0} \hat{\mathbf{x}}_\lambda, \tag{A.1}$$

where in the case of basis pursuit $\hat{\mathbf{x}}_\lambda$ is given by

$$\hat{\mathbf{x}}_\lambda = \arg\min_{\mathbf{x}} \left\{ \frac{1}{2} (\|\mathbf{y} - \mathbf{A}\mathbf{x}\|_2)^2 + \lambda \|\mathbf{x}\|_1 \right\}. \tag{A.2}$$

The optimization problem in equation (A.2) is called basis pursuit denoising [42] and equivalently, lasso (least absolute shrinkage and selection operator) [43]. The Homotopy-BP algorithm follows the solution $\hat{\mathbf{x}}_\lambda$ from $\mathbf{0}$ to $\hat{\mathbf{x}}$, and the solution path is piecewise linear as shown in (iv-5). Then, it is enough to find the ends of the linear pieces in the process of (iv-2)–(iv-4).

Appendix A.4. LARS

The LARS algorithm for estimating $\mathbf{x}$, given a vector $\mathbf{y}$, a matrix $\mathbf{A}$, and a small positive scalar ϵ, is described as follows:

(i) $\mathbf{x} = \mathbf{0}$, $\mathbf{r} = \mathbf{y}$, $r = \frac{(\|\mathbf{y}\|_2)^2}{2M}$.

(ii) $[\lambda, j_{\text{add}}] = [\max_j, \arg\max_j]|(\mathbf{A}_{\{j\}})^\dagger \mathbf{r}|$.

(iii) $S = \{j_{\text{add}}\}$.

(iv) repeat the following (1)–(7) while $r > \epsilon$:

(1) $\mathbf{d} = \mathbf{0}$, $\mathbf{d}_S = ((\mathbf{A}_S)^\dagger \mathbf{A}_S)^{-1}\text{sign}((\mathbf{A}_S)^\dagger \mathbf{r})$.

(2) $[\gamma, j_{\text{add}}] = [\overset{+}{\min_{j\notin S}}, \arg\overset{+}{\min_{j\notin S}}]\left\{\dfrac{\lambda - (\mathbf{A}_{\{j\}})^\dagger \mathbf{r}}{1 - (\mathbf{A}_{\{j\}})^\dagger \mathbf{A}_S \mathbf{d}_S}, \dfrac{\lambda + (\mathbf{A}_{\{j\}})^\dagger \mathbf{r}}{1 + (\mathbf{A}_{\{j\}})^\dagger \mathbf{A}_S \mathbf{d}_S}\right\}$.

(3) $\mathbf{x} = \mathbf{x} + \gamma \mathbf{d}$.

(4) $\mathbf{r} = \mathbf{y} - \mathbf{A}\mathbf{x}$.

(5) $r = \frac{(\|\mathbf{r}\|_2)^2}{2M}$.

(6) $\lambda = \lambda - \gamma$.

(7) $S = S \cup \{j_{\text{add}}\}$.

LARS takes an index j_{add} in the support S at each iteration of (iv-7) when the j_{add}-th column vector in $\mathbf{A}$ gets as close to the residual vector $\mathbf{r}$ in angle as the closest vector in the span of $\mathbf{A}_S$ is. LARS is the same as Homotopy-BP except that the former does not remove an index once it is selected.

Appendix A.5. IHT

The IHT algorithm for estimating $\mathbf{x}$, given a vector $\mathbf{y}$, a matrix $\mathbf{A}$, and a small positive scalar ϵ, is described as follows:

(i) $s = 0$, $S = \emptyset$, $\mathbf{x} = \mathbf{0}$, $\mathbf{r} = \mathbf{y}$, $r = \frac{(\|\mathbf{y}\|_2)^2}{2M}$.

(ii) repeat the following (1)–(5) while $r > \epsilon$:

(1) $s = s + 1$, $S_{\text{old}} = S^0(\neq \emptyset)$.

(2) repeat the following (a)–(d) while $S \neq S_{\text{old}}$:

(a) $S_{\text{old}} = S$.

(b) $\mathbf{x} = H_s(\mathbf{x} + \mathbf{A}^\dagger \mathbf{r})$

(c) $S = \{j | x_j \neq 0\}$.

(d) $\mathbf{r} = \mathbf{y} - \mathbf{A}\mathbf{x}$.

(3) $\mathbf{x} = \mathbf{0}$, $\mathbf{x}_S = (\mathbf{A}_S)^+ \mathbf{y}$.

(4) $\mathbf{r} = \mathbf{y} - \mathbf{A}\mathbf{x}$.

(5) $r = \frac{(\|\mathbf{r}\|_2)^2}{2M}$.

IHT is based on gradient descent for the method of least squares, which is modified by using a hard thresholding operator, as shown in (ii-2-b).

Appendix A.6. CoSaMP

The CoSaMP algorithm for estimating $\mathbf{x}$, given a vector $\mathbf{y}$, a matrix $\mathbf{A}$, and a small positive scalar ϵ, is described as follows:

(i) $s = 0$, $S = \emptyset$, $\mathbf{x} = \mathbf{0}$, $\mathbf{r} = \mathbf{y}$, $r = \frac{(\|\mathbf{y}\|_2)^2}{2M}$.

(ii) repeat the following (1)–(6) while $r > \epsilon$:

(1) $s = s + 1$, $\Delta r = \Delta r^0(> 0)$.

(2) repeat the following (a)–(g) while $\Delta r > 0$:

(a) $S_{\text{old}} = S$, $r_{\text{old}} = r$.

(b) $S = S_{\text{old}} \cup L_{2s}(\mathbf{A}^\dagger \mathbf{r})$.

(c) $\mathbf{x} = \mathbf{0}$, $\mathbf{x}_S = H_s((\mathbf{A}_S)^+ \mathbf{y})$.

(d) $S = \{j | x_j \neq 0\}$.

(e) $\mathbf{r} = \mathbf{y} - \mathbf{A}\mathbf{x}$.

(f) $r = \frac{(\|\mathbf{r}\|_2)^2}{2M}$.
(g) $\Delta r = r^{\text{old}} - r$.
(3) $S = S_{\text{old}}$.
(4) $\mathbf{x} = \mathbf{0}$, $\mathbf{x}_S = (\mathbf{A}_S)^+\mathbf{y}$.
(5) $\mathbf{r} = \mathbf{y} - \mathbf{A}\mathbf{x}$.
(6) $r = \frac{(\|\mathbf{r}\|_2)^2}{2M}$.

CoSaMP merges an index set of $2s$ largest elements in absolute value into the current support at (ii-2-b), and prunes some of them by using the method of pseudoinverse and hard thresholding operator at (ii-2-c).

Appendix A.7. SP

The SP algorithm for estimating $\mathbf{x}$, given a vector $\mathbf{y}$, a matrix $\mathbf{A}$, and a small positive scalar ϵ, is described as follows:

(i) $s = 0$, $S = \emptyset$, $\mathbf{x} = \mathbf{0}$, $\mathbf{r} = \mathbf{y}$, $r = \frac{(\|\mathbf{y}\|_2)^2}{2M}$.
(ii) repeat the following (1)–(6) while $r > \epsilon$:
(1) $s = s + 1$, $\Delta r = \Delta r^0 (> 0)$.
(2) repeat the following (a)–(g) while $\Delta r > 0$:
(a) $S_{\text{old}} = S$, $r_{\text{old}} = r$.
(b) $S = S_{\text{old}} \cup L_s(\mathbf{A}^\dagger \mathbf{r})$.
(c) $\mathbf{x} = \mathbf{0}$, $\mathbf{x}_S = H_s((\mathbf{A}_S)^+\mathbf{y})$.
(d) $S = \{j | x_j \neq 0\}$.
(e) $\mathbf{r} = \mathbf{y} - \mathbf{A}\mathbf{x}$.
(f) $r = \frac{(\|\mathbf{r}\|_2)^2}{2M}$.
(g) $\Delta r = r^{\text{old}} - r$.
(3) $S = S_{\text{old}}$.
(4) $\mathbf{x} = \mathbf{0}$, $\mathbf{x}_S = (\mathbf{A}_S)^+\mathbf{y}$.
(5) $\mathbf{r} = \mathbf{y} - \mathbf{A}\mathbf{x}$.
(6) $r = \frac{(\|\mathbf{r}\|_2)^2}{2M}$.

SP can be regarded as a variant of CoSaMP. A slight difference between them is the size of an index set merged at (ii-2-b).

Appendix A.8. HTP

The HTP algorithm for estimating $\mathbf{x}$, given a vector $\mathbf{y}$, a matrix $\mathbf{A}$, and a small positive scalar ϵ, is described as follows:

(i) $s = 0$, $S = \emptyset$, $\mathbf{x} = \mathbf{0}$, $\mathbf{r} = \mathbf{y}$, $r = \frac{(\|\mathbf{y}\|_2)^2}{2M}$.
(ii) repeat the following (1) and (6) while $r > \epsilon$:
(1) $s = s + 1$, $\Delta r = \Delta r^0 (> 0)$.
(2) repeat the following (a)–(g) while $\Delta r > 0$:
(a) $S_{\text{old}} = S$, $r_{\text{old}} = r$.
(b) $S = L_s(\mathbf{x} + \mathbf{A}^\dagger \mathbf{r})$.
(c) $\mathbf{x} = \mathbf{0}$, $\mathbf{x}_S = (\mathbf{A}_S)^+\mathbf{y}$.
(d) $S = \{j | x_j \neq 0\}$.
(e) $\mathbf{r} = \mathbf{y} - \mathbf{A}\mathbf{x}$.
(f) $r = \frac{(\|\mathbf{r}\|_2)^2}{2M}$.
(g) $\Delta r = r_{\text{old}} - r$.
(3) $S = S_{\text{old}}$.

(4) $\mathbf{x} = \mathbf{0}$, $\mathbf{x}_S = (\mathbf{A}_S)^{+}\mathbf{y}$.
(5) $\mathbf{r} = \mathbf{y} - \mathbf{A}\mathbf{x}$.
(6) $r = \frac{(\|\mathbf{r}\|_2)^2}{2M}$.

HTP can be regarded as a variant of IHT. A slight difference between them is that the pseudoinverse is used at (ii-2-c) on a support which is selected by using the hard thresholding operator.

References

[1] Donoho D L 2006 *IEEE Trans Inf Theory* **52** 1289–1206.
[2] Candès E J and Wakin M B 2008 *IEEE Signal Process Mag* **25** 21–30.
[3] Claerbout J F and Muir F 1973 *Geophysics* **38** 826–844.
[4] Taylor H L, Banks S C, and McCoy J F 1979 *Geophysics* **44** 39–52.
[5] Lustig M, Donoho D, and Pauly J M 2007 *Magn Reson Med* **58** 1182–1195.
[6] Lustig M, Donoho D L, Santos J M, and Pauly J M 2008 *IEEE Signal Process Mag* **25** 72–82.
[7] Katz O, Bromberg Y, and Siberberg Y 2009 *Appl. Phys. Lett.* **95** 131110.
[8] Zhao C, Gong W, Chen M, Li E, Wang H, Zu W, Han S 2012 *Appl. Phys. Lett,* **101** 141123.
[9] Kazimierczuk K and Orekhov V Y 2011 *Angew Chem Int Ed* **50** 5556–5559.
[10] Holland D J, Bostock M J, Gladden L F, and Nietlispach D 2011 *Angew Chem Int Ed* **50** 6548–6681.
[11] Honma M, Akiyama K, Uemura M, and Ikeda S 2014 *Publ Astron Soc Jpn* **66** 95.
[12] Ikeda S, Tazaki F, Akiyama K, Hada K, and Honma M 2016 *Publ Astron Soc Jpn* **68** 45.
[13] Nakanishi-Ohno Y, Haze M, Yoshida Y, Hukushima K, Hasegawa Y, and Okada M 2016 *J Phys Soc Jpn* **85** 093702.
[14] Chen S, Billings S A, and Luo W 1989, *Int J Control* **50** 1873–1896.
[15] Pati Y C, Rezaiifar R, and Krishnaprasad P S 1993 in *Proceedings of 27th Asilomar Conference on Signals, Systems and Computers* 40–44.
[16] Osborne M R, Presnell B, and Turlach B A 2000 *IMA J Numer Anal* **20** 389–403.
[17] Efron B, Hastie T, Johnstone I, and Tibshirani R 2004 *Ann Stat* **32** 407–499.
[18] Blumensath T and Davies M E 2008 *J Fourier Anal Appl* **14** 629–654.
[19] Blumensath T and Davies M E 2009 *Appl Comput Harmon Anal* **27** 265–274.
[20] Needell D and Tropp J A 2009 *Appl Comput Harmon Anal* **26** 301–321.
[21] Dai W and Milenkovic O 2009 *IEEE Trans Inf Theory* **55** 2230–2249.
[22] Foucart S 2011 *SIAM J Numer Anal* **49** 2543–2563.
[23] Lopes M E 2013 in *Proceedings of Machine Learning Research* **28** 217–225.
[24] Nakanishi-Ohno Y and Hukushima K in preparation.
[25] Mosier C I 1951 *Educ Psychol Meas* **11** 5–11; Cureton E E *Ibid* 12–15; Katzell R A *Ibid* 16–22; Wherry R J *Ibid* 23–28.
[26] Efron B and Gong G 1983 *Am Stat* **37** 36–48.
[27] Kohavi R 1995 in *Proceedings of 14th International Joint Conference on Artificial Intelligence* **2** 1137–1143.
[28] Chen S and Donoho D 1994 in *Proceedings of 28th Asilomar Conference on Signals, Systems and Computers* 41–44.
[29] Bloomfield P and Steiger W 1983 *Least Absolute Deviations: Theory, Applications, and Algorithms* Basel Birkhäuser
[30] Candès E J and Tao T 2005 *IEEE Trans Inf Theory* **51** 4203–4215.
[31] Yin W, Osher S, Goldfarb D, and Darbon J 2008 *SIAM J Imaging Sci* **1** 143–168.
[32] Donoho D L, Maleki A, and Montanari A 2009 *Proc Natl Acad Sci* **106** 18914–18919.
[33] Chambolle A and Pock T 2011 *J Math Imaging Vis* **40** 120–145.
[34] Donoho D L 2006 *Discrete Comput Geom* **35** 617–652.
[35] Kabashima Y, Wadayama T, and Tanaka T 2009 *J Stat Mech* **2009** L09003.
[36] Amelunxen D, Lotz M, McCoy M B, and Tropp J A 2014 *Inf Inference* **3** 224–294.
[37] Nishimori H 2001 *Statistical Physics of Spin Glasses and Information Processing: An Introduction*
[38] Obuchi T and Kabashima Y 2016 *J Stat Mech* **2016** 053304.
[39] Candès E J and Recht B 2009 *Found Comput Math* **9** 717–772.
[40] Recht B, Fazel M, and Parrilo P A 2010 *SIAM Rev* **52** 471–501.
[41] McCoy M B and Tropp J A 2014 *Found Comput Math* **14** 503–567.
[42] Chen S S, Donoho D L, and Saunders M A 2001 *SIAM Rev* **43** 129–159.
[43] Tibshirani R 1996 *J R Stat Soc Ser B* **58** 267–288.

Fast Randomized Semi-Supervised Clustering

Alaa Saade[1], Florent Krzakala[1,2], Marc Lelarge[3], Lenka Zdeborová[4]

1 Laboratoire de Physique Statistique (CNRS UMR-8550), PSL Universités and École Normale Supérieure, 75005 Paris (e-mail: alaa.saade,florent.krzakala@ens.fr)
2 Sorbonne Universités, UPMC Univ. Paris 06
3 INRIA and École Normale Supérieure Paris, France (e-mail: marc.lelarge@ens.fr)
4 Institut de Physique Théorique CEA Saclay and CNRS, France (e-mail: lenka.zdeborova@gmail.com)

Abstract. We consider the problem of clustering partially labeled data from a minimal number of randomly chosen pairwise comparisons between the items. We introduce an efficient local algorithm based on a power iteration of the non-backtracking operator and study its performance on a generative model. For the case of two clusters, we give bounds on the classification error and show that a small error can be achieved from $O(n)$ randomly chosen measurements, where n is the number of items in the dataset. Our algorithm is therefore efficient both in terms of time and space complexities. We also investigate numerically the performance of the algorithm on synthetic and real-world data.

1. Introduction

Similarity-based clustering aims at classifying data points into homogeneous groups based on some measure of their resemblance. The problem can be stated formally as follows: given n items $\{x_i\}_{i\in[n]} \in \mathcal{X}^n$, and a symmetric similarity function $s : \mathcal{X}^2 \to \mathbb{R}$, the aim is to cluster the dataset from the knowledge of the pairwise similarities $s_{ij} := s(x_i, x_j)$, for $1 \leq i < j \leq n$. This information is usually represented in the form of a weighted similarity graph $G = (V, E)$ where the nodes represent the items of the dataset and the edges are weighted by the pairwise similarities. Popular choices for the similarity graph are the fully-connected graph, or the k-nearest neighbors graph, for a suitable k (see e.g. [1] for a review). Both choices, however, require the computation of a large number of pairwise similarities, typically $O(n^2)$. For large datasets, with n in the millions or billions, or large-dimensional data, where computing each similarity s_{ij} is costly, the complexity of this procedure is often prohibitive, both in terms of computational and memory requirements.

It is then natural to ask whether it is really required to compute as many as $O(n^2)$ similarities to accurately cluster the data. In the absence of additional information on the data, a reasonable alternative is to compare each item to a small number of other items in the dataset, chosen uniformly at random. Random subsampling methods are a well-known means of reducing the complexity of a problem, and they have been shown to yield substantial speed-ups in clustering [2] and low-rank approximation [3, 4]. In particular, [5] recently showed that an unsupervised spectral method based on the principal eigenvectors of the non-backtracking operator of [6] can cluster the data better than chance from only $O(n)$ similarities.

In this paper, we build upon previous work by considering two variations motivated by real-world applications. The first question we address is how to incorporate the knowledge of the

IOP Conf. Series: Journal of Physics: Conf. Series **1036** (2018) 012015 doi:10.1088/1742-6596/1036/1/012015

labels of a small fraction of the items to aid clustering of the whole dataset, resulting in a more efficient algorithm. This question, referred to as semi-supervised clustering, is of broad practical interest [7, 8]. For instance, in a social network, we may have pre-identified individuals of interest, and we might be looking for other individuals sharing similar characteristics. In biological networks, the function of some genes or proteins may have been determined by costly experiments, and we might seek other genes or proteins sharing the same function. More generally, efficient human-powered methods such as crowdsourcing can be used to accurately label part of the data [9, 10], and we might want to use this knowledge to cluster the rest of the dataset at no additional cost.

The second question we address is the number of randomly chosen pairwise similarities that are needed to achieve a given classification error. Previous work has mainly focused on two related, but different questions. One line of research has been interested in exact recovery, i.e. how many measurements are necessary to exactly cluster the data. Note that for exact recovery to be possible, it is necessary to choose at least $O(n \log n)$ random measurements for the similarity graph to be connected with high probability. On simple models, [11, 12, 13] showed that this scaling is also sufficient for exact recovery. On the sparser end of the spectrum, [14, 15, 16, 5] have focused on the detectability threshold, i.e. how many measurements are needed to cluster the data better than chance. On simple models, this threshold is typically achievable with $O(n)$ measurements only. While this scaling is certainly attractive for large problems, it is important for practical applications to understand how the expected classification error decays with the number of measurements.

To answer these two questions, we introduce a highly efficient, local algorithm based on a power iteration of the non-backtracking operator. For the case of two clusters, we show on a generative model that the classification error decays exponentially with the number of measured pairwise similarities, thus allowing the algorithm to cluster data to arbitrary accuracy while being efficient both in terms of time and space complexities. We demonstrate the good performance of this algorithm on both synthetic and real-world data, and compare it to the popular label propagation algorithm [8].

2. Algorithm and guarantee

2.1. Algorithm for 2 clusters

Consider n items $\{x_i\}_{i\in[n]} \in \mathcal{X}^n$ and a symmetric similarity function $s : \mathcal{X}^2 \to \mathbb{R}$. The choice of the similarity function is problem-dependent, and we will assume one has been chosen. For concreteness, s can be thought of as a decreasing function of a distance if $\mathcal{X}$ is an Euclidean space. The following analysis, however, applies to a generic function s, and our bounds will depend explicitly on its statistical properties. We assume that the true labels $(\sigma_i = \pm 1)_{i\in\mathcal{L}}$ of a subset $\mathcal{L} \subset [n]$ of items is known. Our aim is to find an estimate $(\sigma_i)_{i\in[n]}$ of the labels of all the items, using a small number of similarities. More precisely, let E be a random subset of all the $\binom{n}{2}$ possible pairs of items, containing each given pair $(ij) \in [n]^2$ with probability α/n, for some $\alpha > 0$. We compute only the $O(\alpha n)$ similarities $(s_{ij} := s(x_i, x_j))_{(ij)\in E}$ of the pairs thus chosen.

From these similarities, we define a weighted similarity graph $G = (V, E)$ where the vertices $V = [n]$ represent the items, and each edge $(ij) \in E$ carries a weight $w_{ij} := w(s_{ij})$, where w is a weighting function. Once more, we will consider a generic function w in our analysis, and discuss the performance of our algorithm as a function of the choice of w. In particular, we show in section 2.2 that there is an optimal choice of w when the data is generated from the symmetric model defined in (2) below. However, in practice, the main purpose of w is to center the similarities, i.e. we will take in our numerical simulations $w(s) = s - \bar{s}$, where $\bar{s}$ is the empirical mean of the observed similarities. The necessity to center the similarities is discussed in the following. Note that the graph G is a weighted version of an Erdős-Rényi random graph with average degree α, which controls the sampling rate: a larger α means more pairwise similarities are computed, at the expense of an increase in complexity. Algorithm 1 describes our clustering procedure for the case of 2 clusters. We denote by ∂i the set of neighbors of node i in the graph G, and by $\vec{E}$ the set of directed edges of G.

Algorithm 1 Non-backtracking local walk (2 clusters)

Input: $n\,, \mathcal{L}\,, (\sigma_i = \pm 1)_{i \in \mathcal{L}}\,, E\,, (w_{ij})_{(ij) \in E}\,, k_{\max}$
Output: Cluster assignments $(\hat{\sigma}_i)_{i \in [n]}$

Initialize the messages $v^{(0)} = (v^{(0)}_{i \to j})_{(i \to j) \in \vec{E}}$
  for all $(i \to j) \in \vec{E}$ **do**
    if $i \in \mathcal{L}$ **then** $v^{(0)}_{i \to j} \leftarrow \sigma_i$
    else $v^{(0)}_{i \to j} \leftarrow \pm 1$ uniformly at random
Iterate for $k = 1, \dots, k_{\max}$
  for $(i \to j) \in \vec{E}$ **do** $v^{(k)}_{i \to j} \leftarrow \sum_{l \in \partial i \setminus j} w_{il} v^{(k-1)}_{l \to i}$
Pool the messages
  for $i \in [n]$ **do** $\hat{v}_i \leftarrow \sum_{l \in \partial i} w_{il} v^{(k_{\max})}_{l \to i}$
Output the assignments
  for $i \in [n]$ **do** $\hat{\sigma}_i \leftarrow \mathrm{sign}(\hat{v}_i)$

This algorithm can be thought of as a linearized version of a belief propagation algorithm, that iteratively updates messages on the directed edges of the similarity graph, by assigning to each message the weighted sum of its incoming messages. More precisely, Algorithm 1 can be observed to approximate the leading eigenvector of the non-backtracking operator B, whose elements are defined, for $(i \to j), (k \to l) \in \vec{E}$, by

$$\mathrm{B}_{(i \to j),(k \to l)} := w_{kl}\, \mathbf{1}(i = l) \mathbf{1}(k \neq j)\,. \tag{1}$$

It is therefore close in spirit to the unsupervised spectral methods introduced by [6, 16, 5], which rely on the computation of the principal eigenvector of B. On sparse graphs, methods based on the non-backtracking operator are known to perform better than traditional spectral algorithms based e.g. on the adjacency matrix, or random walk matrix of the sparse similarity graph, which suffer from the appearance of large eigenvalues with localized eigenvectors (see e.g. [6, 17]). In particular, we will see on the numerical experiments of Section 3 that Algorithm 1 outperforms the label propagation algorithm, based on an iteration of the random walk matrix.

However, in contrast with the past spectral approaches based on B, our algorithm is local, in that the estimate $\hat{\sigma}_i$ for a given item $i \in [n]$ depends only on the messages on the edges that are at most $k_{\max} + 1$ steps away from i in the graph G. This fact will prove essential in our analysis. Indeed, we will show that in our semi-supervised setting, a finite number of iterations (independent of n) is enough to ensure a low classification error. On the other hand, in the unsupervised setting, we expect local algorithms not to be able to find large clusters in a graph, a limitation that has already been highlighted on the related problems of finding large independent sets on graphs [18] and community detection [19]. On the practical side, the local nature of Algorithm 1 leads to a drastic improvement in running time. Indeed, in order to compute the leading eigenvector of B, a number of iterations k scaling with the number n of items is required [20]. Here, on the contrary, the number of iterations stays independent of the size n of the dataset.

2.2. Model and guarantee

To evaluate the performance of Algorithm 1, we consider the following semi-supervised variant of the labeled stochastic block model [15]. This model extends the popular stochastic block model by allowing generic edge-weights and is used as a benchmark for a theoretical analysis of graph clustering [12]. We assign n items to 2 predefined clusters of the same average size $n/2$, by drawing for each item $i \in [n]$ a cluster label $\sigma_i \in \{\pm 1\}$ with uniform probability 1/2. We choose uniformly at random ηn items to form a subset $\mathcal{L} \subset [n]$ of items whose label is revealed, so that η is the fraction of labeled data. Next, we choose which pairs of items will be compared by generating an Erdős-Rényi random graph $G = (V = [n], E) \in \mathcal{G}(n, \alpha/n)$, for some constant $\alpha > 0$, independent of n. We will assume that the similarity s_{ij} between two items i and j is a random variable depending only on the labels of the items i and j. More precisely, we consider the symmetric model

$$\mathbb{P}(s_{ij} = s|\sigma_i, \sigma_j) = \begin{cases} p_{\text{in}}(s) & \text{if } \sigma_i = \sigma_j\,, \\ p_{\text{out}}(s) & \text{if } \sigma_i \neq \sigma_j\,, \end{cases} \tag{2}$$

where p_{in} (resp. p_{out}) is the distribution of the similarities between items within the same cluster (resp. different clusters). The properties of the weighting function w will determine the performance of our algorithm through the two following quantities. Define $2\Delta(w) := \mathbb{E}\left[w_{ij}|\sigma_j = \sigma_j\right] - \mathbb{E}\left[w_{ij}|\sigma_j \neq \sigma_j\right]$, the difference in expectation between the weights inside a cluster and between different clusters. Define also $\Sigma(w)^2 := \mathbb{E}\left[w^2\right]$, the second moment of the weights. Our first result (proved in section 4) is concerned with what value of α is required to improve the initial labeling with Algorithm 1.

Theorem 1. *Assume a similarity graph G with n items and a labeled set $\mathcal{L}$ of size ηn to be generated from the symmetric model (2) with 2 clusters. Define $\tau(\alpha, w) := \frac{\alpha \Delta(w)^2}{\Sigma(w)^2}$. If $\Delta(w) > 0$, then there exists a constant $C > 0$ such that the estimates $\hat{\sigma}_i$ from k iterations of Algorithm 1 achieve*

$$\frac{1}{n}\sum_{i=1}^{n} \mathbb{P}(\sigma_i \neq \hat{\sigma}_i) \leq 1 - r_{k+1} + C\frac{\alpha^{k+1}\log n}{\sqrt{n}}\,, \tag{3}$$

where $r_0 = \eta^2$ and for $0 \leq l \leq k$,

$$r_{l+1} = \frac{\tau(\alpha, w) r_l}{1 + \tau(\alpha, w) r_l}\,. \tag{4}$$

To understand the content of this bound, we consider the limit of a large number of items $n \to \infty$, so that the last term of (3) vanishes. Note first that if $\tau(\alpha, w) > 1$, then starting from any positive initial condition, r_k converges to $(\tau(\alpha, w) - 1)/\tau(\alpha, w) > 0$ in the limit where the number of iterations $k \to \infty$. A random guess on the unlabeled points yields an asymptotic error of $\lim_{n\to\infty} \frac{1}{n}\sum_{i=1}^{n} \mathbb{P}(\sigma_i \neq \hat{\sigma}_i) = (1-\eta)/2$, so that a sufficient condition for Algorithm 1 to improve the initial partial labeling, after a certain number of iterations $k(\tau(\alpha, w), \eta)$ independent of n, is

$$\tau(\alpha, w) > \frac{2}{1-\eta}\,. \tag{5}$$

Note that there is an optimal choice of weighting function w which maximizes $\tau(\alpha, w)$, namely (consistently with [14]):

$$\begin{aligned} w^*(s) &:= \frac{p_{\text{in}}(s) - p_{\text{out}}(s)}{p_{\text{in}}(s) + p_{\text{out}}(s)} \\ \Longleftrightarrow \tau(\alpha, w^*) &= \frac{\alpha}{2}\int \mathrm{d}s\; \frac{(p_{\text{in}}(s) - p_{\text{out}}(s))^2}{p_{\text{in}}(s) + p_{\text{out}}(s)}\,, \end{aligned} \tag{6}$$

which, however, requires knowing the parameters of the model. It is then informative to compare our bound (5) to known optimal asymptotic bounds in the unsupervised setting. In the limit of vanishing supervision $\eta \to 0$, the bound (5) guarantees improving the initial labeling if $\tau(\alpha, w^*) > 2 + O(\eta)$. In the unsupervised setting ($\eta = 0$), it has been shown by [14] that if $\tau(\alpha, w^*) < 1$, no algorithm, either local or global, can cluster the graph better than random guessing. If $\tau(\alpha, w^*) > 1$, on the other hand, then a global spectral method based on the principal eigenvectors of the non-backtracking operator improves over random guessing [5]. This suggests that, in the limit of vanishing supervision $\eta \to 0$, the bound (5) is close to optimal, but off by a factor of 2.

Note, however, that Theorem 1 applies to a generic weighting function w. In particular, while the optimal choice (6) is not practical, theorem 1 guarantees that Algorithm 1 retains the ability to improve the initial labeling from a small number of measurements, as long as $\Delta(w) > 0$. With the choice $w(s) = s - \bar{s}$ advocated for in section 2.1, we have $2\Delta(w) = \mathbb{E}\left[s_{ij}|\sigma_j = \sigma_j\right] - \mathbb{E}\left[s_{ij}|\sigma_j \neq \sigma_j\right]$. Therefore Algorithm 1 improves over random guessing for α large enough if the similarity between items in the same cluster is larger in expectation than the similarity between items in different clusters, which is a reasonable requirement. Note that the hypotheses of Theorem 1 do not require the weighting function w to be centered. However, it is easy to check that if $\mathbb{E}[w] \neq 0$, defining a new weighting function by $w'(s) := w(s) - \mathbb{E}[w]$, we have $\tau(\alpha, w') > \tau(\alpha, w)$, so that the bound (3) is improved.

While Theorem 1 guarantees improving the initial clustering from a small sampling rate α, it provides a rather loose bound on the expected error when α becomes larger. The next theorem addresses this regime. A proof is given in Section 5.

Theorem 2. *Assume a similarity graph G with n items and a labeled set $\mathcal{L}$ of size ηn to be generated from the symmetric model (2) with 2 clusters. Assume further that the weighting function w is bounded: $\forall s \in \mathbb{R}, |w(s)| \leq 1$. Define $\tau(\alpha, w) := \frac{\alpha\Delta(w)^2}{\Sigma(w)^2}$. If $\alpha\Delta(w) > 1$ and $\alpha\Sigma(w)^2 > 1$, then there exists a constant $C > 0$ such that the estimates $\hat{\sigma}_i$ from k iterations of Algorithm 1 achieve*

$$\frac{1}{n}\sum_{i=1}^{n}\mathbb{P}(\sigma_i \neq \hat{\sigma}_i) \leq \exp\left[-\frac{q_{k+1}}{4}\min\left(1, \frac{\Sigma(w)^2}{\Delta(w)}\right)\right] + C\frac{\alpha^{k+1}\log n}{\sqrt{n}}, \tag{7}$$

where $q_0 = 2\eta^2$ and for $0 \leq l \leq k$,

$$q_{l+1} = \frac{\tau(\alpha, w)q_l}{1 + 3/2\max(1, q_l)}. \tag{8}$$

Note that by linearity of Algorithm 1, the condition $\forall s, |w(s)| \leq 1$ can be relaxed to w bounded. It is once more instructive to consider the limit of large number of items $n \to \infty$. Starting from any initial condition, if $\tau(\alpha, w) < 5/2$, then $q_k \underset{k\to\infty}{\longrightarrow} 0$ so that the bound (7) is uninformative. On the other hand, if $\tau(\alpha, w) > 5/2$, then starting from any positive initial condition, $q_k \underset{k\to\infty}{\longrightarrow} \frac{2}{3}(\tau(\alpha, w) - 1) > 0$. This bound therefore shows that on a model with a given distribution of similarities (2) and a given weighting function w, an error smaller than ϵ can be achieved from $\alpha n = O(n \log 1/\epsilon)$ measurements, in the limit $\epsilon \to 0$, with a finite number of iterations $k(\tau(\alpha, w), \eta, \epsilon)$ independent of n. We note that this result is the analog, for a weighted graph, of the recent results of [21] who show that in the stochastic block model, a local algorithm similar to Algorithm 1 achieves an error decaying exponentially as a function of a relevant signal to noise ratio.

IOP Conf. Series: Journal of Physics: Conf. Series **1036** (2018) 012015 doi:10.1088/1742-6596/1036/1/012015

2.3. More than 2 clusters

Algorithm 2 gives a natural extension of our algorithm to $q > 2$ clusters. In this case, we expect the non-backtracking operator B defined in equation (1) to have $q - 1$ large eigenvalues, with eigenvectors correlated with the types σ_i of the items (see [5]). We use a deflation-based power iteration method [22] to approximate these eigenvectors, starting from informative initial conditions incorporating the knowledge drawn from the partially labeled data. Numerical simulations illustrating the performance of this algorithm are presented in Section 3. Note that each deflated matrix B_c for $c \geq 2$ is a rank-$(c-1)$ perturbation of a sparse matrix, so that the power iteration can be done efficiently using sparse linear algebra routines. In particular, both Algorithms 1 and 2 have a time and space complexities linear in the number of items n.

Algorithm 2 Non-backtracking local walk (q clusters)

Input: $n\,, q\,, \mathcal{L}\,, (\sigma_i \in [q])_{i \in \mathcal{L}}\,, E, (w_{ij})_{(ij)\in E}\,, k_{\max}$
Output: Cluster assignments $(\hat{\sigma}_i)_{i\in[n]}$
$\mathrm{B}_1 \leftarrow \mathrm{B}$ where B is the matrix of equation (1)
for $c = 1, \cdots, q-1$ **do**
  Initialize the messages $v^{(0)} = (v^{(0)}_{i\to j})_{(ij)\in E}$
    for all $(i \to j) \in \vec{E}$ **do**
      if $i \in \mathcal{L}$ and $\sigma_i = c$ **then** $v^{(0)}_{i\to j} \leftarrow 1$
      else if $i \in \mathcal{L}$ and $\sigma_i \neq c$ **then** $v^{(0)}_{i\to j} \leftarrow -1$
      else $v^{(0)}_{i\to j} \leftarrow \pm 1$ uniformly at random
  Iterate for $k = 1, \ldots, k_{\max}$
    $v^{(k)} \leftarrow \mathrm{B}_c v^{(k-1)}$
  Pool the messages in a vector $\hat{v}_c \in \mathbb{R}^n$ with
  entries $(\hat{v}_{i,c})_{i\in[n]}$
    for $i \in [n]$ **do** $\hat{v}_{i,c} \leftarrow \sum_{l\in\partial i} w_{il} v^{(k_{\max})}_{l\to i}$
  Deflate B_c

$$\mathrm{B}_{c+1} \leftarrow \mathrm{B}_c - \frac{\mathrm{B}_c v^{(k_{\max})} v^{(k_{\max})\intercal} \mathrm{B}_c}{v^{(k_{\max})\intercal} \mathrm{B}_c v^{(k_{\max})}}$$

Concatenate $\hat{V} \leftarrow [\hat{v}_1 | \cdots | \hat{v}_{q-1}] \in \mathbb{R}^{n\times(q-1)}$
Output the assignments $(\hat{\sigma}_i)_{i\in[n]} \leftarrow \mathrm{kmeans}(\hat{V})$

---

3. Numerical simulations

In addition to the theoretical guarantees presented in the previous section, we ran numerical simulations on two toy datasets consisting of 2 and 4 Gaussian blobs (Figure 1), and two subsets of the MNIST dataset [23] consisting respectively of the digits in $\{0, 1\}$ and $\{0, 1, 2\}$ (Figure 2). We also considered the 20 Newsgroups text dataset [24], consisting of text documents organized in 20 topics, of which we selected 2 for our experiments of Figure 3. All three examples differ considerably from the model we have studied analytically. In particular, the random similarities are not identically distributed conditioned on the true labels of the items, but depend on latent variables, such as the distance to the center of the Gaussian, in the case of Figure 1. Additionally, in the case of the MNIST dataset of Figure 2, the clusters have different sizes (e.g. 6903 for the 0's and 7877 for the 1's). Nevertheless, we find that our algorithm performs well, and outperforms the popular label propagation algorithm [8] in a wide range of values of the sampling rate α.

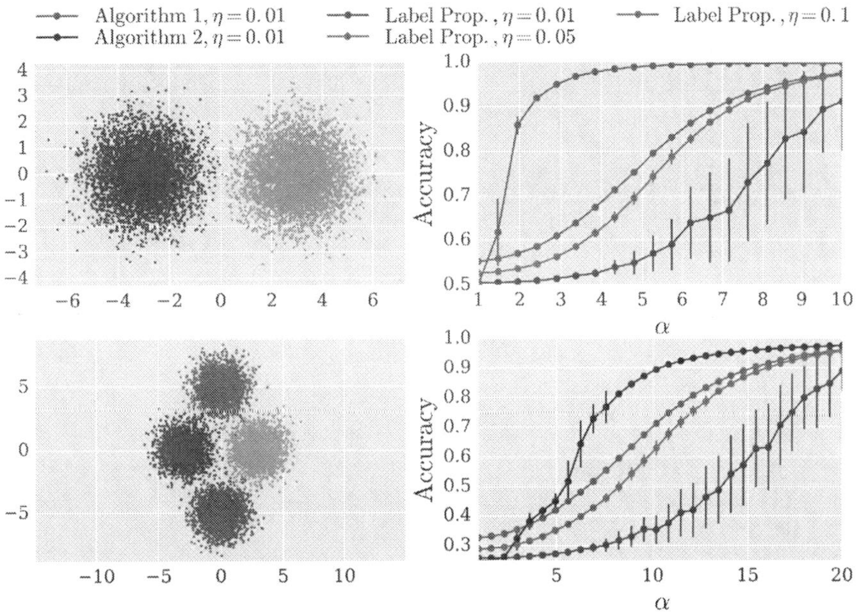

Figure 1. Performance of algorithms 1 and 2 compared to label propagation on a toy dataset in two dimensions. The left panel shows the data, composed of $n = 10^4$ points, with their true labels. The right panel shows the clustering performance on random subsamples of the complete similarity graph. Each point is averaged over 100 realizations. The accuracy is defined as the fraction of correctly labeled points. We set the maximum number of iterations of our algorithms to $k_{\max} = 30$. α is the average degree of the Erdős-Rényi random graph G, and η is the fraction of labeled data. For all methods, we used the same similarity function $s_{ij} = \exp -d_{ij}^2/\sigma^2$ where d_{ij} is the Euclidean distance between points i and j and σ^2 is a scaling factor which we set to the empirical mean of the observed squared distances d_{ij}^2. For algorithms 1 and 2, we used the weighting function $w(s) := s - \bar{s}$ (i.e. we simply center the similarities, see text). Label propagation is run on the random similarity graph G. We note that Algorithm 1 has the same preformance as label propagation with ten times fewer labeled data and with only a number of similarities of the order of three times the number of points.

In all cases, we find that the accuracy achieved by Algorithms 1 and 2 is an increasing function of α, rapidly reaching a plateau at a limiting accuracy. This limiting accuracy is also reached by label propagation and depends on the choice of the similarity function. But the most important observation is the rate of convergence of the accuracy to this limiting value, as a function of α. Indeed, on these simple datasets, it is enough to compute, for each item, their similarity with a few randomly chosen other items to reach an accuracy within a few percents of the limiting accuracy allowed by the quality of the similarity function. As a consequence, similarity-based clustering can be significantly sped up. For example, we note that the semi-supervised clustering of the 0's and 1's of the MNIST dataset (representing $n = 14780$ points in dimension 784), from 1% labeled data, and to an accuracy greater than 96% requires $\alpha \approx 6$ (see figure 2), and runs on a laptop in 2 seconds, including the computation of the randomly chosen similarities. Additionally, in contrast with our algorithms, we find that in the strongly undersampled regime (small α), the performance of label propagation depends strongly on the fraction η of available labeled data. We find in particular that Algorithms 1 and 2 outperform label propagation even starting from ten times fewer labeled data.

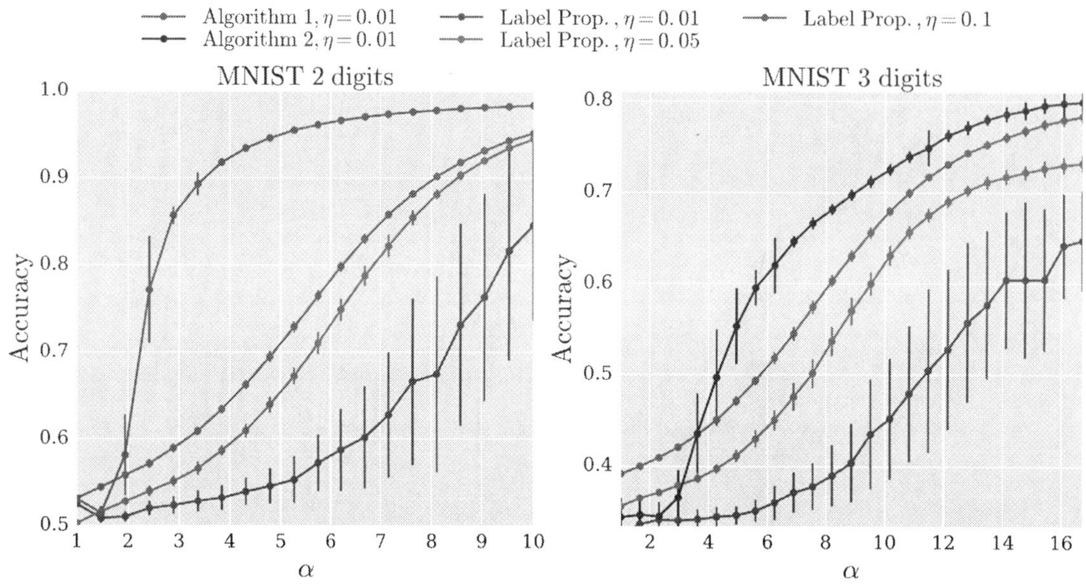

Figure 2. Performance of algorithms 1 and 2 compared to label propagation on a subset of the MNIST dataset. The left panel corresponds the set of 0's and 1's ($n = 14780$ samples) while the right panel corresponds the 0's,1's and 2's ($n = 21770$). All the parameters are the same as in the caption of figure 1, except that we used the Cosine distance in place of the Euclidean one.

4. Proof of Theorem 1

Consider the model introduced in Section 2 for the case of two clusters. We will bound the probability of error on a randomly chosen node, and (3) will follow. Denote by I an integer drawn uniformly at random from $[n]$, and by $\hat{\sigma}_I^{(k)}$ the decision variable after k iterations of Algorithm 1. We are interested in the probability of error at node I conditioned on the true label of node I. As noted previously, the algorithm is local in the sense that $\hat{\sigma}_I^{(k)}$ depends only on the messages in the neighborhood of I consisting of all the nodes and edges of G that are at most $k+1$ steps aways from I. Thanks to the total variation distance bound between the law of this neighborhood and a weighted Galton-Watson branching process (see prop. 31 in [20]), we can use a standard coupling argument to reduce the analysis of our algorithm on G to its analysis on a branching process and this will introdce the error term $C\frac{\alpha^{k+1}\log n}{\sqrt{n}}$ in (3). Using the recursive structure of the branching process allows us to determine the law of $\hat{v}_I = \hat{v}_\sigma^{(k)}$ conditioned on the type of node I being $\sigma = \pm 1$,

$$\hat{v}_\sigma^{(k)} \overset{\mathcal{D}}{=} \sum_{i=1}^{d_1} w_{i,\text{in}} v_{i,\sigma}^{(k)} + \sum_{i=1}^{d_2} w_{i,\text{out}} v_{i,-\sigma}^{(k)}, \tag{9}$$

where $\overset{\mathcal{D}}{=}$ denotes equality in distribution. The random variables $v_{i,\sigma}^{(k)}$ for $\sigma = \pm 1$ have the same distribution as the message $v_{i\to j}^{(k)}$ after k iterations of the algorithm, for a randomly chosen edge $(i \to j)$, conditioned on the type of node i being σ. They are i.i.d. copies of a random variable

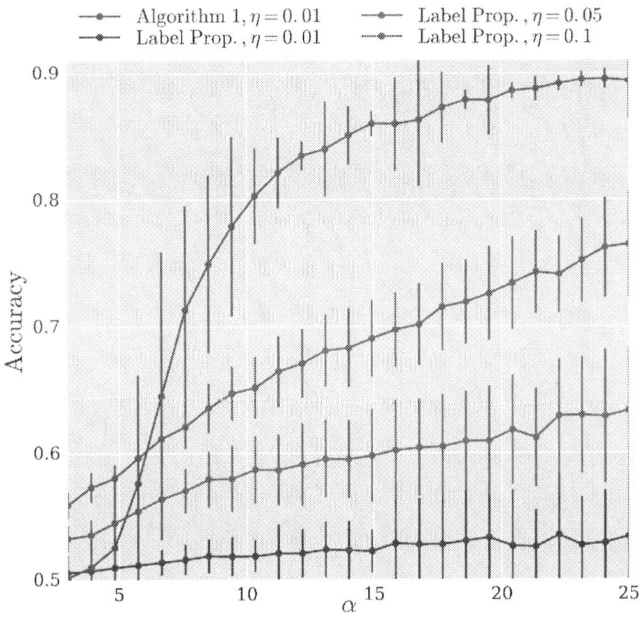

Figure 3. Performance of Algorithm 1 compared to label propagation on a subset of the 20 Newsgroups dataset with $q = 2$ clusters. We consider the topics "misc.forsale" (975 documents) and " soc.religion.christian" (997 documents), which are relatively easy to distinguish, to illustrate the efficiency of our subsampling approach. The resulting dataset consists of $n = 1972$ text documents, to which we applied a standard "tf-idf" transformation (after stemming, and using word unigrams) to obtain a vector representation of the documents. We used the same similarity (based on the Cosine distance) and weighting functions as in figure 2.

$v_\sigma^{(k)}$ whose distribution is defined recursively, for $l \geq 0$ and $\sigma = \pm 1$, through

$$v_\sigma^{(l+1)} \overset{\mathcal{D}}{=} \sum_{i=1}^{d_1} w_{i,\text{in}} v_{i,\sigma}^{(l)} + \sum_{i=1}^{d_2} w_{i,\text{out}} v_{i,-\sigma}^{(l)} . \tag{10}$$

In equations (9), d_1 and d_2 are two independent random variables with a Poisson distribution of mean $\alpha/2$, and $w_{i,\text{in}}$ (resp. $w_{i,\text{out}}$) are i.i.d copies of w_{in} (resp. w_{out}) whose distribution is the same as the weights w_{ij}, conditioned on $\sigma_i = \sigma_j$ (resp. $\sigma_i \neq \sigma_j$). Note in particular that $\hat{v}_\sigma^{(k)}$ has the same distribution as $v_\sigma^{(k+1)}$.

Theorem 1 will follow by analyzing the evolution of the first and second moments of the distribution of $v_{+1}^{(k+1)}$ and $v_{-1}^{(k+1)}$. Equations (10) can be used to derive recursive formulas for the first and second moments. In particular, the expected values verify the following linear system

$$\begin{pmatrix} \mathbb{E}\left[v_{+1}^{(l+1)}\right] \\ \mathbb{E}\left[v_{-1}^{(l+1)}\right] \end{pmatrix} = \frac{\alpha}{2} \begin{pmatrix} \mathbb{E}[w_{\text{in}}] & \mathbb{E}[w_{\text{out}}] \\ \mathbb{E}[w_{\text{out}}] & \mathbb{E}[w_{\text{in}}] \end{pmatrix} \begin{pmatrix} \mathbb{E}\left[v_{+1}^{(l)}\right] \\ \mathbb{E}\left[v_{-1}^{(l)}\right] \end{pmatrix} . \tag{11}$$

The eigenvalues of this matrix are $\mathbb{E}[w_{\text{in}}] + \mathbb{E}[w_{\text{out}}]$ with eigenvector $(1,1)^\intercal$, and $\mathbb{E}[w_{\text{in}}] - \mathbb{E}[w_{\text{out}}]$ with eigenvector $(1,-1)^\intercal$. With the assumption of our model, we have $\mathbb{E}\left[v_{+1}^{(0)}\right] = \eta = -\mathbb{E}\left[v_{-1}^{(0)}\right]$

which is proportional to the second eigenvector. Recalling the definition of $\Delta(w)$ from section 2, we therefore have, for any $l \geq 0$,

$$\mathbb{E}\left[v_{+1}^{(l+1)}\right] = \alpha\Delta(w)\mathbb{E}\left[v_{+1}^{(l)}\right], \tag{12}$$

and $\mathbb{E}\left[v_{-1}^{(l)}\right] = -\mathbb{E}\left[v_{+1}^{(l)}\right]$. With the additional observation that $\mathbb{E}\left[\left(v_{+1}^{(0)}\right)^2\right] = \mathbb{E}\left[\left(v_{-1}^{(0)}\right)^2\right] = 1$, a simple induction shows that for any $l \geq 0$, $\mathbb{E}\left[\left(v_{+1}^{(l)}\right)^2\right] = \mathbb{E}\left[\left(v_{-1}^{(l)}\right)^2\right]$. Recalling the definition of $\Sigma(w)^2$ from Section 2, we have the recursion

$$\begin{aligned}\mathbb{E}\left[\left(v_{+1}^{(l+1)}\right)^2\right] &= \alpha^2\Delta(w)^2\,\mathbb{E}\left[v_{+1}^{(l)}\right]^2 \\ &\quad + \alpha\,\Sigma(w)^2\,\mathbb{E}\left[\left(v_{+1}^{(l)}\right)^2\right].\end{aligned} \tag{13}$$

Noting that since $\Delta(w) > 0$, we have $\sigma\,\mathbb{E}\left[v_\sigma^{(k+1)}\right] > 0$ for $\sigma = \pm 1$, the proof of Theorem 1 is concluded by invoking Cantelli's inequality

$$\mathbb{P}\left(\sigma\, v_\sigma^{(k+1)} \leq 0\right) \leq 1 - r_{k+1}, \tag{14}$$

with, for $l \geq 0$,

$$r_l := \mathbb{E}\left[v_\sigma^{(l)}\right]^2 \mathbb{E}\left[\left(v_\sigma^{(l)}\right)^2\right]^{-1}, \tag{15}$$

where r_l is independent of σ, and is shown to verify the recursion (4) by combining (12) and (13).

5. Proof of theorem 2

The proof is adapted from a technique developed by [9]. We show that the random variables $v_\sigma^{(l)}$ are sub-exponential by induction on l. A random variable X is said to be sub-exponential if there exist constants $K > 0, a, b$ such that for $|\lambda| < K$

$$\mathbb{E}[e^{\lambda X}] \leq \exp\left(\lambda a + \lambda^2 b\right). \tag{16}$$

Define $f_\sigma^{(l)}(\lambda) := \mathbb{E}\left[e^{\lambda v_\sigma^{(l)}}\right]$ for $l \geq 0$ and $\sigma = \pm 1$. We introduce two sequences $(a_l)_{l\geq 0}, (b_l)_{l \geq 0}$ defined recursively by $a_0 = \eta, b_0 = 1/2$ and for $l \geq 0$

$$\begin{aligned} a_{l+1} &= \alpha\Delta(w)a_l, \\ b_{l+1} &= \alpha\Sigma(w)^2\left(b_l + \frac{3}{2}\max(a_l^2, b_l)\right). \end{aligned} \tag{17}$$

Note that since we assume that $\alpha\,\Delta(w) > 1$ and $\alpha\,\Sigma(w)^2 > 1$, both sequences are positive and increasing. In the following, we show that

$$f_\sigma^{(k+1)}(\lambda) \leq \exp\left(\sigma\lambda a_{k+1} + \lambda^2 b_{k+1}\right), \tag{18}$$

for $|\lambda| \leq \left(2\max\left(a_k, \sqrt{b_k}\right)\right)^{-1}$. Theorem 2 will follow from the Chernoff bound applied at

$$\lambda_\sigma^* = -\sigma\frac{a_{k+1}}{2b_{k+1}}\min\left(1, \frac{\Sigma(w)^2}{\Delta(w)}\right). \tag{19}$$

The fact that $|\lambda_\sigma^*| \le \left(2\max\left(a_k, \sqrt{b_k}\right)\right)^{-1}$ follows from (17). Noting that $\sigma\lambda_\sigma^* < 0$ for $\sigma = \pm 1$, the Chernoff bound allows to show

$$\begin{aligned}\mathbb{P}\left(\sigma v_\sigma^{(k+1)} \le 0\right) &\le f_\sigma^{(k+1)}(\lambda_\sigma^*) \\ &\le \exp\left[-\frac{q_{k+1}}{4}\min\left(1, \frac{\Sigma(w)^2}{\Delta(w)}\right)\right],\end{aligned} \tag{20}$$

where $q_{k+1} := a_{k+1}^2/b_{k+1}$ is shown using (17) to verify the recursion (8). We are left to show that $f_\sigma^{(k+1)}(\lambda)$ verifies (18). First, with the choice of initialization in Algorithm 1, we have for any $\lambda \in \mathbb{R}$

$$\begin{aligned}f_{+1}^{(0)}(\lambda) = f_{-1}^{(0)}(-\lambda) &= \frac{1+\eta}{2}\exp(\lambda) + \frac{1-\eta}{2}\exp(-\lambda) \\ &\le \exp\left(\eta\lambda + \lambda^2/2\right),\end{aligned} \tag{21}$$

where we have used the inequality for $x \in [0,1], \lambda \in \mathbb{R}$

$$xe^{\lambda} + (1-x)e^{-\lambda} \le \exp\left((2x-1)\lambda + \lambda^2/2\right). \tag{22}$$

Therefore we have shown $f_\sigma^{(0)}(\lambda) \le \exp\left(\sigma\lambda a_0 + \lambda^2 b_0\right)$. Next, let us assume that for some $l \ge 0$ and for any λ such that $|\lambda| \le \left(2\max\left(a_{l-1}, \sqrt{b_{l-1}}\right)\right)^{-1}$,

$$f_\sigma^{(l)}(\lambda) \le \exp\left(\sigma\lambda a_l + \lambda^2 b_l\right), \tag{23}$$

with the convention $a_{-1} = b_{-1} = 0$ so that the previous statement is true for any $\lambda \in \mathbb{R}$ if $l = 0$. The density evolution equations (10) imply the following recursion on the moment-generating functions, for any $\lambda \in \mathbb{R}, \sigma = \pm 1$,

$$\begin{aligned}f_\sigma^{(l+1)}(\lambda) = \exp\Big(-\alpha + \frac{\alpha}{2}\Big(&\mathbb{E}_{w_{\text{in}}}\left[f_\sigma^{(l)}(\lambda w_{\text{in}})\right] \\ &+ \mathbb{E}_{w_{\text{out}}}\left[f_{-\sigma}^{(l)}(\lambda w_{\text{out}})\right]\Big)\Big).\end{aligned} \tag{24}$$

We claim that for $|\lambda| \le \left(2\max\left(a_l, \sqrt{b_l}\right)\right)^{-1}$ and for $\sigma = \pm 1$,

$$\begin{aligned}&\frac{1}{2}\left(\mathbb{E}_{w_{\text{in}}}\left[f_\sigma^{(l)}(\lambda w_{\text{in}})\right] + \mathbb{E}_{w_{\text{out}}}\left[f_{-\sigma}^{(l)}(\lambda w_{\text{out}})\right]\right) \\ &\le 1 + \sigma a_l\Delta(w)\lambda + \lambda^2\Sigma(w)^2\left(b_l + \frac{3}{2}\max(a_l^2, b_l)\right)\end{aligned} \tag{25}$$

Injecting equation (25) in the recursion (24) yields $f_\sigma^{(l+1)}(\lambda) \le \exp\left(\sigma\lambda\, a_{l+1} + \lambda^2\, b_{l+1}\right)$, for any λ such that $|\lambda| \le \left(2\max\left(a_l, \sqrt{b_l}\right)\right)^{-1}$, with a_{l+1}, b_{l+1} defined by (17). The proof is then concluded by induction on $0 \le l \le k$. To show (25), we start from the following inequality: for $|a| \le 3/4$,

$$\exp(a) \le 1 + a + (2/3)a^2. \tag{26}$$

With $|w| \le 1$ as per the assumption of theorem 2, for $|\lambda| \le \left(2\max\left(a_l, \sqrt{b_l}\right)\right)^{-1}$, we have for $\sigma = \pm 1$ that $|\sigma\lambda w a_l + \lambda^2 w^2 b_l| \le 3/4$. Additionally, since a_l and b_l are non-decreasing in l, we also have that $|\lambda| \le \left(2\max\left(a_{l-1}, \sqrt{b_{l-1}}\right)\right)^{-1}$, so that by our induction hypothesis, for $\sigma = \pm 1$

$$\begin{aligned}f_\sigma^{(l)}(\lambda w) &\le \exp\left(\sigma\lambda w a_l + \lambda^2 w^2 b_l\right) \\ &\le 1 + \sigma\lambda w a_l + \lambda^2 w^2 b_l + \frac{2}{3}\left(\sigma\lambda w a_l + \lambda^2 w^2 b_l\right)^2 \\ &\le 1 + \sigma\lambda w a_l + \lambda^2 w^2 b_l + \frac{2}{3}\lambda^2 w^2\left(a_l + |\lambda| b_l\right)^2 \\ &\le 1 + \sigma\lambda w a_l + \lambda^2 w^2 b_l + \frac{3}{2}\lambda^2 w^2 \max(a_l^2, b_l),\end{aligned}$$

where we have used in the last inequality that $(a_l + |\lambda|b_l)^2 \leq 9/4 \max(a_l^2, b_l)$. The claim (25) follows by taking expectation, and the proof is completed.

Acknowledgments

This work has been supported by the ERC under the European Union ' s FP7 Grant Agreement 307087-SPARCS and by the French Agence Nationale de la Recherche under reference ANR-11-JS02-005-01 (GAP project).

References

[1] Ulrike Luxburg. A tutorial on spectral clustering. *Statistics and Computing*, 17(4):395, 2007.
[2] Petros Drineas, Alan M Frieze, Ravi Kannan, Santosh Vempala, and V Vinay. Clustering in large graphs and matrices. In *SODA*, volume 99, pages 291–299. Citeseer, 1999.
[3] Dimitris Achlioptas and Frank McSherry. Fast computation of low-rank matrix approximations. *Journal of the ACM (JACM)*, 54(2):9, 2007.
[4] Nathan Halko, Per-Gunnar Martinsson, and Joel A Tropp. Finding structure with randomness: Probabilistic algorithms for constructing approximate matrix decompositions. *SIAM review*, 53(2):217–288, 2011.
[5] A. Saade, M. Lelarge, F. Krzakala, and L. Zdeborov. Clustering from sparse pairwise measurements. pages 780–784, July 2016.
[6] F. Krzakala, C. Moore, E. Mossel, J. Neeman, A. Sly, L. Zdeborová, and P. Zhang. Spectral redemption in clustering sparse networks. *Proc. Natl. Acad. Sci.*, 110(52):20935–20940, 2013.
[7] Sugato Basu, Arindam Banerjee, and Raymond Mooney. Semi-supervised clustering by seeding. In *In Proceedings of 19th International Conference on Machine Learning (ICML-2002*. Citeseer, 2002.
[8] Xiaojin Zhu and Zoubin Ghahramani. Learning from labeled and unlabeled data with label propagation. Technical report, Citeseer, 2002.
[9] David R Karger, Sewoong Oh, and Devavrat Shah. Iterative learning for reliable crowdsourcing systems. In *Advances in neural information processing systems*, pages 1953–1961, 2011.
[10] David R Karger, Sewoong Oh, and Devavrat Shah. Efficient crowdsourcing for multi-class labeling. In *ACM SIGMETRICS Performance Evaluation Review*, volume 41, pages 81–92. ACM, 2013.
[11] Emmanuel Abbe, Afonso S Bandeira, Annina Bracher, and Amit Singer. Decoding binary node labels from censored edge measurements: Phase transition and efficient recovery. *arXiv:1404.4749*, 2014.
[12] Se-Young Yun and Alexandre Proutiere. Optimal cluster recovery in the labeled stochastic block model. *ArXiv e-prints*, 5, 2015.
[13] Bruce Hajek, Yihong Wu, and Jiaming Xu. Achieving exact cluster recovery threshold via semidefinite programming: Extensions. *arXiv preprint arXiv:1502.07738*, 2015.
[14] M. Lelarge, L. Massoulie, and Jiaming Xu. Reconstruction in the labeled stochastic block model. In *Information Theory Workshop (ITW), 2013 IEEE*, pages 1–5, Sept 2013.
[15] Simon Heimlicher, Marc Lelarge, and Laurent Massoulié. Community detection in the labelled stochastic block model. 09 2012.
[16] Alaa Saade, Florent Krzakala, Marc Lelarge, and Lenka Zdeborová. Spectral detection in the censored block model. *IEEE International Symposium on Information Theory (ISIT2015), to appear*, 2015.
[17] Pan Zhang. Robust spectral detection of global structures in the data by learning a regularization. *arXiv preprint arXiv:1609.02906*, 2016.
[18] David Gamarnik and Madhu Sudan. Limits of local algorithms over sparse random graphs. In *Proceedings of the 5th conference on Innovations in theoretical computer science*, pages 369–376. ACM, 2014.
[19] Varun Kanade, Elchanan Mossel, and Tselil Schramm. Global and local information in clustering labeled block models. 2014.
[20] Charles Bordenave, Marc Lelarge, and Laurent Massoulié. Non-backtracking spectrum of random graphs: community detection and non-regular ramanujan graphs. arXiv:1501.06087, 2015.
[21] T Tony Cai, Tengyuan Liang, and Alexander Rakhlin. Inference via message passing on partially labeled stochastic block models. *arXiv preprint arXiv:1603.06923*, 2016.
[22] Nguyen Duc Thang, Young-Koo Lee, Sungyoung Lee, et al. Deflation-based power iteration clustering. *Applied intelligence*, 39(2):367–385, 2013.
[23] Yann LeCun, Léon Bottou, Yoshua Bengio, and Patrick Haffner. Gradient-based learning applied to document recognition. *Proceedings of the IEEE*, 86(11):2278–2324, 1998.
[24] Ken Lang. Newsweeder: Learning to filter netnews. In *Proceedings of the 12th international conference on machine learning*, pages 331–339, 1995.

IOP Conf. Series: Journal of Physics: Conf. Series **1036** (2018) 012016 doi:10.1088/1742-6596/1036/1/012016

Cross-validation estimates of the number of modules in higher-order networks

Tatsuro Kawamoto

Artificial Intelligence Research Center, National Institute of Advanced Industrial Science and Technology, 2-3-26 Aomi, Koto-ku, Tokyo, Japan

E-mail: `kawamoto.tatsuro@aist.go.jp`

Abstract. A graph clustering algorithm detects macroscopic patterns of connectivity hidden in network data. We regard each vertex in the network as belonging to a module, and the vertices in the same module have a statistically equivalent connection pattern. While a network with binary labels—e.g., similar or unrelated—is often considered as the simplest example, network data with finer information about the relationships—e.g., similar, unrelated, and dissimilar—are common in practice. For this reason, we consider the graph clustering of networks with labeled edges. In graph clustering, selecting the number of modules that are considered to be statistically significant is a crucial step. In other words, it determines the level of coarse graining. Here, we propose efficient cross-validation estimates for the number of modules based on the prediction errors. While the straightforward implementation of cross-validation is typically computationally demanding, we show that leave-one-out cross-validation has a good affinity with the belief propagation algorithm, and thus can be measured efficiently.

1. Introduction

Suppose that you collect relational data among the people in an international conference so that you can build a social network. Each vertex represents a participant of the conference, and each edge represents the relationship between a pair of participants. You ask a participant whether a pair of other participants are friends with each other (and believe his/her information); the corresponding vertices are connected via an undirected edge if they are friends and are not connected otherwise. You conduct this survey to multiple participants until you obtain a sufficient number of edges. From the resulting social network, we expect that a graph clustering algorithm detects social groups among the participants.

However, when the number of participants N is very large, it is difficult to obtain the relational data of $O(N^2)$. Thus, the resulting network will typically be very sparse. When the network is sparse, because we do not have sufficient information on each vertex, it is a difficult task to infer the social groups accurately. A simple solution is to introduce a new edge type, "enemy," and distinguish the unconnected pairs into "enemy" and "unknown," so that we can increase the number of edges. We can also ask for more details when we collect the relationships and add more edges of different types.

To detect the module structure from sparse networks, we consider the statistical inference of modules using the so-called stochastic blockmodel (SBM) [1]. The SBM is a random graph model with a module structure. In the SBM, each vertex has a planted module assignment, and the vertices in the same label have a statistically equivalent connection pattern. In particular,

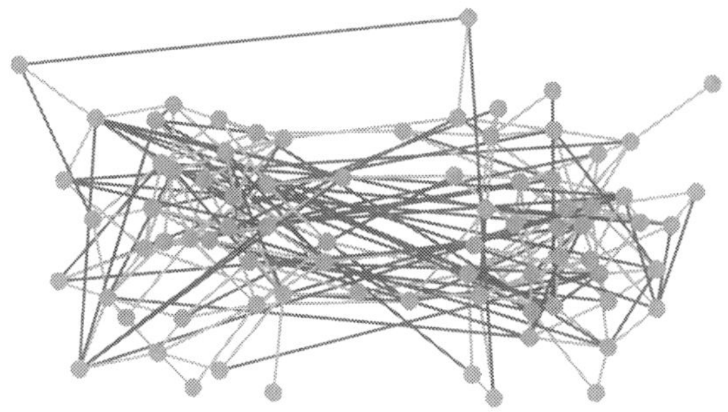

Figure 1. An instance of the labeled SBM with two planted modules. The red edges represent the assortative edges, and the black edges represent the disassortative edges.

we consider a variant of the SBM that has labeled edges; it is called the labeled SBM. An instance of the labeled SBM is shown in Fig. 1. Note that if we increase the number of modules q to which vertices can belong, the SBM becomes more flexible so that it is easier to fit the real network. However, we obviously do not want to set each vertex to a different module—i.e., $q = N$. Therefore, we wish to find the parsimonious number of modules q^* such that the corresponding SBM efficiently summarizes the network structure.

Here, we extend the cross-validation estimate for the number of modules that we proposed [6] for the networks with binary edges (edges and nonedges) to the ones with labeled edges. Our cross-validation estimate is closely related to the inference algorithm. We consider the expectation-maximization (EM) algorithm with belief propagation (BP) for its E-step. By utilizing the quantities we obtained as a result of the algorithm, we can efficiently perform leave-one-out cross-validation (LOOCV).

The rest of the article is organized as follows. First, we describe the generation of the labeled SBM in detail (Sec. 2) and the specific inference algorithm (Sec. 3). We then explain the method of cross-validation estimate for the networks with labeled edges (Sec. 4), and investigate the performance of the proposed estimates (Sec. 5). We confirm the performance of the cross-validation estimates through numerical experiments (Sec. 6). The last section is devoted to summary and discussion (Sec. 7).

2. Labeled stochastic blockmodel

We consider a network with vertex set V and edge set E. The number of vertices is N, and the total number of edges is L. The edge set comprises p types of edge sets, so $E = \cup_{\alpha=1}^{p} E_\alpha$, and we denote the number of α-edges as L_α (i.e., $\sum_\alpha L_\alpha = L$). We also refer to the nonedges as the edges with $\alpha = 0$. We denote the adjacency matrix of a network as A, i.e., $A_{ij} = \alpha$ if vertex pair i and j is connected via an α-edge. We denote the average degree of α-edges as c_α and $c \equiv \sum_\alpha c_\alpha$ as the average degree of all edges.

The labeled SBM of q planted modules is generated as follows. We denote the module label as $\sigma \in \{1, \ldots, q\}$. For each vertex in the vertex set V, we assign a module assignment σ_i $(i \in V)$ independently and randomly with probability $p(\sigma_i|\gamma) = \gamma_{\sigma_i}$, where $\sum_{\sigma=1}^{q} \gamma_\sigma = 1$. We then generate edges independently based on the matrices called the affinity matrices $\{\boldsymbol{c}^\alpha\}$. Each affinity matrix $\boldsymbol{c}^\alpha$ represents the connection pattern between/within modules. For example, vertices i and j are connected via an α-edge $(\alpha > 0)$ with probability $c^\alpha_{\sigma_i\sigma_j}/N$. Therefore, the likelihood function of the labeled SBM is

$$p(A, \boldsymbol{\sigma}|\boldsymbol{\gamma}, \{\boldsymbol{c}^\alpha\}) = \prod_{i=1}^{N} \gamma_{\sigma_i} \prod_{i<j} \prod_{\alpha=0}^{p} \left(\frac{c^\alpha_{\sigma_i \sigma_j}}{N} \right)^{\delta_{A_{ij},\alpha}}. \tag{1}$$

Note that because each vertex pair is connected via a nonedge or one of the α-edges, the elements of the affinity matrices are normalized as $\sum_{\alpha=0}^{p} c^\alpha_{\sigma\sigma'}/N = 1$ for any σ and σ'. We consider that each element of $\boldsymbol{c}^\alpha$ is of $O(1)$ except for $\alpha = 0$, so the resulting network is sparse.

3. EM algorithm with BP

We infer the module assignments using the EM algorithm [3]. For the statistical inference of the labeled SBM, we aim to infer the posterior marginal probability $p(\sigma_i|A, \boldsymbol{\gamma}, \{\boldsymbol{c}^\alpha\})$ of the module assignment for each vertex and learn the model parameters so that the marginal log-likelihood $\log p(A|\boldsymbol{\gamma}, \{\boldsymbol{c}^\alpha\})$ is maximized (i.e., we perform the point estimate for the model parameters). Using the variational expression, the marginal log-likelihood can be written as

$$\log \sum_{\boldsymbol{\sigma}} p(A, \boldsymbol{\sigma}|\boldsymbol{\gamma}, \{\boldsymbol{c}^\alpha\}) = \mathbb{E}_\psi \left[\log \frac{p(A, \boldsymbol{\sigma}|\boldsymbol{\gamma}, \{\boldsymbol{c}^\alpha\})}{\psi_{\boldsymbol{\sigma}}} \right] + D_{\mathrm{KL}} \left[\psi_{\boldsymbol{\sigma}} || p(\boldsymbol{\sigma}|A, \boldsymbol{\gamma}, \{\boldsymbol{c}^\alpha\}) \right], \tag{2}$$

where $\psi_{\boldsymbol{\sigma}}$ is the variational distribution of the module assignments $\boldsymbol{\sigma}$, and $\mathbb{E}_\psi$ represents the average with respect to $\psi_{\boldsymbol{\sigma}}$. $D_{\mathrm{KL}}[P||Q]$ is the Kullback–Leibler divergence between P and Q. In the EM algorithm, in the E-step, we evaluate the variational distribution $\psi_{\boldsymbol{\sigma}}$ so that $D_{\mathrm{KL}} [\psi_{\boldsymbol{\sigma}} || p(\boldsymbol{\sigma}|A, \boldsymbol{\gamma}, \{\boldsymbol{c}^\alpha\})]$ vanishes (or is minimized). In the M-step, we update the model parameters so that the first term of Eq. (2) is maximized. After the model-parameter update, the variational distribution $\psi_{\boldsymbol{\sigma}}$ that minimizes $D_{\mathrm{KL}} [\psi_{\boldsymbol{\sigma}} || p(\boldsymbol{\sigma}|A, \boldsymbol{\gamma}, \{\boldsymbol{c}^\alpha\})]$ may be varied. Thus, we iterate the E-step and M-step until convergence. Hereafter, we often omit the hat notation that we use for the estimated quantities.

3.1. E-step

To perform the E-step, we use BP, which is an approximate algorithm that solves for the marginal distribution of a probabilistic model with a treelike structure [4, 5]. We expect that the performance of BP is good because the networks we consider are sparse. Following the standard formulation of BP, the estimate of the marginal probability ψ^i_σ with respect to the likelihood (1) for vertex i is expressed as

$$\psi^i_\sigma = \frac{\gamma_\sigma}{Z^i} \prod_{\ell \in [\partial i]^0} \left(1 - \sum_{\alpha>0} \sum_{\sigma_\ell} \psi^{\ell \to i}_{\sigma_\ell} \frac{c^\alpha_{\sigma_\ell \sigma}}{N} \right) \prod_{\alpha>0} \prod_{k \in [\partial i]^\alpha} \left(\sum_{\sigma_k} \psi^{k \to i}_{\sigma_k} c^\alpha_{\sigma_k \sigma} \right) \simeq \frac{\gamma_\sigma}{Z^i} \prod_{\alpha>0} \varphi^i_{\alpha,\sigma}, \tag{3}$$

$$\varphi^i_{\alpha,\sigma} = \mathrm{e}^{-N^{-1} \sum_\ell \sum_{\sigma_\ell} \psi^\ell_{\sigma_\ell} c^\alpha_{\sigma_\ell \sigma}} \prod_{k \in [\partial i]^\alpha} \left(\sum_{\sigma_k} \psi^{k \to i}_{\sigma_k} c^\alpha_{\sigma_k \sigma} \right), \tag{4}$$

where the vertex $k \in [\partial i]^\alpha$ is a neighbor of i such that $(i, k) \in E_\alpha$, Z^i is a normalization factor of ψ^i_σ, and

$$\psi^{i \to j}_\sigma = \frac{\gamma_\sigma}{Z^{i \to j}} \prod_{\ell \in [\partial i \backslash j]^0} \left(1 - \sum_{\alpha>0} \sum_{\sigma_\ell} \psi^{\ell \to i}_{\sigma_\ell} \frac{c^\alpha_{\sigma_\ell \sigma}}{N} \right) \prod_{\alpha>0} \prod_{k \in [\partial i \backslash j]^\alpha} \left(\sum_{\sigma_k} \psi^{k \to i}_{\sigma_k} c^\alpha_{\sigma_k \sigma} \right) \simeq \frac{\gamma_\sigma}{Z^{i \to j}} \prod_{\alpha>0} \varphi^{i \to j}_{\alpha,\sigma}, \tag{5}$$

$$\varphi^{i \to j}_{\alpha,\sigma} = \mathrm{e}^{-N^{-1} \sum_\ell \sum_{\sigma_\ell} \psi^\ell_{\sigma_\ell} c^\alpha_{\sigma_\ell \sigma}} \prod_{k \in [\partial i \backslash j]^\alpha} \left(\sum_{\sigma_k} \psi^{k \to i}_{\sigma_k} c^\alpha_{\sigma_k \sigma} \right). \tag{6}$$

Analogously, the vertex $k \in [\partial i \backslash j]^\alpha$ is a neighbor of i such that $(i,k) \in E_\alpha$ but $k \neq j$ and $Z^{i \to j}$ is a normalization factor of $\psi_\sigma^{i \to j}$.

The probability $\psi_\sigma^{i \to j}$ represents the marginal probability of vertex i with the missing knowledge about the edge (i,j), which is called the cavity bias. Because Eq. (5) constitutes a set of close equations with respect to the cavity biases, we iteratively update them until convergence. We then substitute the result into Eq. (3) to obtain the full marginal.

3.2. M-step

To derive the updated equation of the M-step, as we mentioned above, we take the extremum point of the first term in Eq. (2). As a result, we have the estimate $\hat{\gamma}_\sigma$ for the fraction of the module size as

$$\hat{\gamma}_\sigma = \frac{1}{N} \sum_{i=1}^{N} \psi_\sigma^i. \tag{7}$$

For the estimate $\hat{c}^\alpha_{\sigma\sigma'}$ of an element of the affinity matrix $\boldsymbol{c}^\alpha$, we have

$$\hat{c}^{\alpha(\mathrm{new})}_{\sigma\sigma'} = \frac{c_\alpha}{\hat{\gamma}_\sigma \hat{\gamma}_{\sigma'}} \frac{1}{L_\alpha} \sum_{(i,j) \in E_\alpha} \frac{\hat{c}^\alpha_{\sigma\sigma'} \left(\psi_\sigma^{i \to j} \psi_{\sigma'}^{j \to i} + \psi_{\sigma'}^{i \to j} \psi_\sigma^{j \to i} \right)}{\sum_{\sigma\sigma'} \hat{c}^\alpha_{\sigma\sigma'} \left(\psi_\sigma^{i \to j} \psi_{\sigma'}^{j \to i} + \psi_{\sigma'}^{i \to j} \psi_\sigma^{j \to i} \right)}, \tag{8}$$

where $\hat{c}^{\alpha(\mathrm{new})}_{\sigma\sigma'}$ is the updated estimate, which is written in terms of the estimate from the previous update.

We then substitute these updated values of the model parameters into the BP equations (3) and (5) to update the E-step again.

4. Cross-validation estimates of the prediction errors

To perform the EM algorithm above, we must specify the number of modules q as an input. While evaluating the saturation of the free energy (or the Bethe free energy [3]) is a choice, it is empirically known that it tends to overfit for real-world networks. For this reason, we measure the prediction errors using the leave-one-out cross-validation (LOOCV) [7]. While the straightforward implementation of cross-validation is computationally demanding, LOOCV is an exceptional case in which the prediction errors can be efficiently estimated using the results of BP.

4.1. Bayes prediction error

The prediction error that we consider is the predictability of edge information. To this end, we measure the posterior distribution of A_{ij} given the adjacency matrix except for A_{ij} (we denote it as $A^{\backslash(i,j)}$). Because the edge is generated based on the module assignments of the vertex pair in the SBM, we have

$$p(A_{ij} = \alpha | A^{\backslash(i,j)}) = \sum_{\sigma_i, \sigma_j} p(A_{ij} = \alpha | \sigma_i, \sigma_j) p(\sigma_i, \sigma_j | A^{\backslash(i,j)}). \tag{9}$$

For $\alpha > 0$, $p(A_{ij} = \alpha | \sigma_i, \sigma_j) = c^\alpha_{\sigma_i \sigma_j}/N$ by the model definition and $p(\sigma_i, \sigma_j | A^{\backslash(i,j)})$ can be estimated using the result of BP as

$$p(\sigma_i, \sigma_j | A^{\backslash(i,j)}) = p(\sigma_i | A^{\backslash(i,j)}) p(\sigma_j | A^{\backslash(i,j)}) = \psi_{\sigma_i}^{i \to j} \psi_{\sigma_j}^{j \to i}. \tag{10}$$

IOP Conf. Series: Journal of Physics: Conf. Series **1036** (2018) 012016 doi:10.1088/1742-6596/1036/1/012016

Thus, we have

$$p(A_{ij} = \alpha | A^{\backslash(i,j)}) = \sum_{\sigma_i,\sigma_j} \psi_{\sigma_i}^{i\to j} \frac{c^{\alpha}_{\sigma_i\sigma_j}}{N} \psi_{\sigma_j}^{j\to i} \tag{11}$$

and we denote it as Z^{ij}_{α}.

We measure the prediction error using the cross-entropy error function [2]. We denote the cross-entropy with respect to $p(A_{ij} = \alpha | A^{\backslash(i,j)})$ that is normalized by L as the Bayes prediction error $E_{\mathrm{Bayes}}(q)$.

$$E_{\mathrm{Bayes}}(q) = \frac{1}{L} \sum_{i<j} \sum_{\alpha=0}^{p} \delta_{A_{ij},\alpha} \left[-\log p(A_{ij} = \alpha | A^{\backslash(i,j)}) \right] \tag{12}$$

Using the fact that the network is sparse, $E_{\mathrm{Bayes}}(q)$ can be approximated as follows.

$$\begin{aligned} E_{\mathrm{Bayes}}(q) &= -\frac{1}{L} \sum_{\alpha>0} \sum_{(i,j)\in E_{\alpha}} \left(\log Z^{ij}_{\alpha} - \log(1 - \sum_{\alpha>0} Z^{ij}_{\alpha}) \right) - \frac{1}{L} \sum_{i<j} \log(1 - \sum_{\alpha>0} Z^{ij}_{\alpha}) \\ &\simeq 1 - \frac{1}{L} \sum_{\alpha>0} \sum_{(i,j)\in E_{\alpha}} \log Z^{ij}_{\alpha}, \end{aligned} \tag{13}$$

where we used the fact that $-\log(1 - \sum_{\alpha>0} Z^{ij}_{\alpha}) \simeq \sum_{\alpha>0} Z^{ij}_{\alpha}$ is of $O(1/N)$ and is the probability that an edge exists between i and j irrespective of its label.

4.2. Gibbs prediction error

As shown in Eq. (9), an edge is generated via a two-step process in the SBM; i.e., the selection of module assignments and the generation of an edge based on the selected module assignments. When we measure the prediction error, instead of using $p(A_{ij} = \alpha | A^{\backslash(i,j)})$, we can measure the cross-entropy with respect to the probability $p(A_{ij} = \alpha | \sigma_i, \sigma_j)$ with fixed module assignments. We then average over the selection of the module assignments. That is,

$$E_{\mathrm{Gibbs}}(q) = \frac{1}{L} \sum_{i<j} \sum_{\sigma_i,\sigma_j} p(\sigma_i, \sigma_j | A^{\backslash(i,j)}) \sum_{\alpha=0}^{p} \delta_{A_{ij},\alpha} \left[-\log p(A_{ij} = \alpha | \sigma_i, \sigma_j) \right] \tag{14}$$

We denote this estimate of the prediction error $E_{\mathrm{Gibbs}}(q)$ as the Gibbs prediction error. As with the Bayes prediction error $E_{\mathrm{Bayes}}(q)$, we rewrite $E_{\mathrm{Gibbs}}(q)$ using the sparse approximation, as follows.

$$\begin{aligned} E_{\mathrm{Gibbs}}(q) &= -\frac{1}{L} \sum_{\alpha=0}^{p} \sum_{(i,j)\in E_{\alpha}} \sum_{\sigma_i,\sigma_j} \psi_{\sigma_i}^{i\to j} \psi_{\sigma_j}^{j\to i} \left(\log \frac{c^{\alpha}_{\sigma_i\sigma_j}}{N} \right) \\ &\simeq 1 - \frac{1}{L} \sum_{\alpha=1}^{p} \sum_{(i,j)\in E_{\alpha}} \sum_{\sigma_i,\sigma_j} \psi_{\sigma_i}^{i\to j} \psi_{\sigma_j}^{j\to i} \log c^{\alpha}_{\sigma_i\sigma_j} + \log N. \end{aligned} \tag{15}$$

Similarly, we can also define the so-called MAP estimate of the Gibbs prediction error. We denote it as $E_{\mathrm{MAP}}(q)$. It can be obtained by replacing $\psi_{\sigma_i}^{i\to j}$ and $\psi_{\sigma_j}^{j\to i}$ with the delta functions that is peaked at $\mathrm{argmax}\{\psi_{\sigma_i}^{i\to j}\}$ and $\mathrm{argmax}\{\psi_{\sigma_j}^{j\to i}\}$, respectively.

IOP Conf. Series: Journal of Physics: Conf. Series **1036** (2018) 012016 doi:10.1088/1742-6596/1036/1/012016

4.3. Gibbs training error

Finally, we can define the LOOCV estimate of the training error. In the Gibbs prediction error $E_{\mathrm{Gibbs}}(q)$, we used the module-assignment probability $p(\sigma_i, \sigma_j|A^{\setminus(i,j)})$ without the knowledge of A_{ij}. On the other hand, it will be the training error if we utilize the actual information of A_{ij} as well. We denote it as the Gibbs training error and it is defined as follows.

$$E_{\mathrm{training}}(q) = \frac{1}{L}\sum_{i<j}\sum_{\sigma_i,\sigma_j} p(\sigma_i,\sigma_j|A)\sum_{\alpha=0}^{p}\delta_{A_{ij},\alpha}\left[-\log p(A_{ij}=\alpha|\sigma_i,\sigma_j)\right] \tag{16}$$

Using Bayes' rule, $p(\sigma_i, \sigma_j|A)$ is recast as

$$p(\sigma_i,\sigma_j|A) = p(\sigma_i,\sigma_j|A_{ij},A^{\setminus(i,j)}) = \frac{p(A_{ij},\sigma_i,\sigma_j|A^{\setminus(i,j)})}{p(A_{ij}|A^{\setminus(i,j)})} = \frac{1}{Z^{ij}_{\alpha}}\psi^{i\to j}_{\sigma_i}\frac{c^{\alpha}_{\sigma_i\sigma_j}}{N}\psi^{j\to i}_{\sigma_j}. \tag{17}$$

Again, we approximate $E_{\mathrm{Gibbs}}(q)$ using the sparse approximation as follows.

$$E_{\mathrm{training}}(q) \simeq 1 - \frac{1}{L}\sum_{\alpha=1}^{p}\sum_{(i,j)\in E_\alpha}\frac{1}{Z^{ij}_{\alpha}}\sum_{\sigma_i,\sigma_j}\psi^{i\to j}_{\sigma_i}\frac{c^{\alpha}_{\sigma_i\sigma_j}}{N}\psi^{j\to i}_{\sigma_j}\log c^{\alpha}_{\sigma_i\sigma_j} + \log N. \tag{18}$$

Note that all cross-validation estimates that we introduced are written as averages over the set of edges E. Therefore, we can also measure their standard errors, and those errors can be useful when we employ a criterion such as the "one-standard-error rule" [8] in model selection. In the numerical experiments, we include the standard errors as the error bars of the cross-validation estimates.

4.4. Prediction error for $q = 1$

To evaluate whether the network has any module structure at all, it is informative to calculate the prediction error for $q = 1$. Note that because there are no model parameters to be learned when $q = 1$, there are no distinctions among the Bayes, Gibbs, and training errors. Thus, we refer to the prediction error as $E(q = 1)$.

Because we have $Z^{ij}_{\alpha} = p(A_{ij} = \alpha|A^{\setminus(i,j)}) \simeq c_\alpha/N$, Eq. (13) yields

$$E(q=1) \simeq 1 - \frac{1}{L}\sum_{\alpha>0}\sum_{(i,j)\in E_\alpha}\log\frac{c_\alpha}{N} = 1 - \log\frac{c}{N} + S\left[P_\alpha\right], \tag{19}$$

where we define $P_\alpha \equiv c_\alpha/c$ as the fraction of α-edges, and $S\left[P_\alpha\right] \equiv -\sum_{\alpha>0} P_\alpha \log P_\alpha$ is the entropy.

5. Performance of the LOOCV estimates of the prediction errors

In this section, we investigate the performance of each prediction error. In particular, we focus on the performance near the detectability threshold. The detectability threshold is the critical point in the strength of the module structure, below which the algorithm does not perform better than chance[1].

When we discuss the performance near the detectability threshold, we assume that the sizes of the planted module are equal. In this case, below the detectability threshold, the estimated probability of module label σ is $1/q$ for any vertex; i.e., $\psi^i_\sigma = \psi^{i\to j}_\sigma = \gamma_\sigma = 1/q$. Here, we do

[1] The detectability threshold under the Nishimori condition (i.e., the planted values of model parameters are given as inputs) is derived in Ref. [9], and the algorithmic detectability threshold of the EM algorithm is derived in Ref. [10, 11]

not argue the values of model parameters at the detectability threshold themselves. Instead, we analyze how the prediction errors behave when $\psi_\sigma^{i\to j}$ is close to $1/q$. In the following, we assume that the planted model parameters are learned accurately, which is correct in the algorithmically detectable region.

To gain deeper insight about the performance of prediction errors, we focus on the case of simple community structure—i.e.,

$$c_{\sigma\sigma'}^{\alpha} = \begin{cases} c_{\text{in}}^{\alpha} & (\sigma = \sigma') \\ c_{\text{out}}^{\alpha} & (\sigma \neq \sigma') \end{cases}. \tag{20}$$

Here, we introduce $\Delta c_\alpha \equiv c_{\text{in}}^{\alpha} - c_{\text{out}}^{\alpha}$. Under this parametrization, the Bayes prediction error $E_{\text{Bayes}}(q)$ in Eq. (13) is recast as

$$E_{\text{Bayes}}(q) = E(q=1) - \sum_{\alpha>0} P_\alpha \frac{1}{L_\alpha} \sum_{(i,j)\in E_\alpha} \log\left[1 + \frac{\Delta c_\alpha}{c_\alpha} \sum_\sigma \left(\psi_\sigma^{i\to j}\psi_\sigma^{j\to i} - \gamma_\sigma^2\right)\right]. \tag{21}$$

The parenthesis in the second term indicates how much the estimate of the algorithm is correlated to the planted structure. In the case of edges that indicate an assortative structure, because the edge is more likely to connect the vertices of the same module assignment, $\psi_\sigma^{i\to j}\psi_\sigma^{j\to i}$ is greater than γ_σ^2 (this is the case of the random guess). If this tends to be true, the overall sign of the second term in Eq. (21) is negative because $\Delta c_\alpha > 0$. Conversely, in the case of edges that indicate a disassortative structure, $\psi_\sigma^{i\to j}\psi_\sigma^{j\to i}$ is smaller than γ_σ^2. If this tends to be true, the overall sign of the second term is again negative because $\Delta c_\alpha < 0$. Therefore, as long as the estimated module assignments are correlated with the planted structure, the Bayes prediction error $E_{\text{Bayes}}(q)$ is smaller than $E(q=1)$ and supports partition of the network. When the set of model parameters reaches the detectability threshold, the second term in Eq. (21) vanishes, and we have $E_{\text{Bayes}}(q) = E(q=1)$.

Let us next consider the behavior of the Gibbs prediction error $E_{\text{Gibbs}}(q)$. When the affinity matrix is given by Eq. (20), Eq. (15) reads

$$E_{\text{Gibbs}}(q) = E(q=1) - \sum_{\alpha>0}^{p} P_\alpha G_\alpha, \tag{22}$$

$$\begin{aligned} G_\alpha = &\left(1 - \left\langle \sum_\sigma \psi_\sigma^{i\to j}\psi_\sigma^{j\to i} \right\rangle_\alpha\right) \log\left(1 - \frac{\Delta c_\alpha}{c_\alpha}\boldsymbol{\gamma}^\top\boldsymbol{\gamma}\right) \\ &+ \left\langle \sum_\sigma \psi_\sigma^{i\to j}\psi_\sigma^{j\to i} \right\rangle_\alpha \log\left(1 - \frac{\Delta c_\alpha}{c_\alpha}\boldsymbol{\gamma}^\top\boldsymbol{\gamma} + \frac{\Delta c_\alpha}{c_\alpha}\right), \end{aligned} \tag{23}$$

where we define

$$\left\langle \sum_\sigma \psi_\sigma^{i\to j}\psi_\sigma^{j\to i} \right\rangle_\alpha \equiv \frac{1}{L_\alpha} \sum_{(i,j)\in E_\alpha} \sum_\sigma \psi_\sigma^{i\to j}\psi_\sigma^{j\to i} \tag{24}$$

and this is equal to $\boldsymbol{\gamma}^\top\boldsymbol{\gamma}$ at the detectability threshold. Note that $0 \leq \left\langle \sum_\sigma \psi_\sigma^{i\to j}\psi_\sigma^{j\to i} \right\rangle_\alpha \leq 1$ and $0 \leq \boldsymbol{\gamma}^\top\boldsymbol{\gamma} \leq 1$. Note also that the conditions $c_{\text{in}} > 0$ and $c_{\text{out}} > 0$ restrict the value of $\Delta c_\alpha / c_\alpha$ to

$$-\frac{1}{1-\boldsymbol{\gamma}^\top\boldsymbol{\gamma}} \leq \frac{\Delta c_\alpha}{c_\alpha} \leq \frac{1}{\boldsymbol{\gamma}^\top\boldsymbol{\gamma}}. \tag{25}$$

When the planted structure is strong enough, we can confirm that $E_{\mathrm{Gibbs}}(q)$ is smaller than $E(q=1)$ and supports partitioning of the network. For the α-edges with an assortative structure, $G_\alpha > 0$ because the first term in Eq. (23) is dominant and it gives a positive contribution. We also have $G_\alpha > 0$ for the α-edges with a disassortative structure because the second term in Eq. (23) is dominant and it again gives a positive contribution.

When the planted structure is close to the detectability threshold, however, the Gibbs prediction error $E_{\mathrm{Gibbs}}(q)$ becomes strictly greater than $E(q=1)$. To see this, we recast G_α as

$$G_\alpha = \log\left(1 - \frac{\Delta c_\alpha}{c_\alpha}\boldsymbol{\gamma}^\top\boldsymbol{\gamma}\right) + \left\langle \sum_\sigma \psi_\sigma^{i\to j}\psi_\sigma^{j\to i}\right\rangle_\alpha \log\left(1 + \frac{\frac{\Delta c_\alpha}{c_\alpha}\boldsymbol{\gamma}^\top\boldsymbol{\gamma}}{1-\frac{\Delta c_\alpha}{c_\alpha}\boldsymbol{\gamma}^\top\boldsymbol{\gamma}}\frac{1}{\boldsymbol{\gamma}^\top\boldsymbol{\gamma}}\right). \tag{26}$$

Here, we set

$$x = \frac{\frac{\Delta c_\alpha}{c_\alpha}\boldsymbol{\gamma}^\top\boldsymbol{\gamma}}{1-\frac{\Delta c_\alpha}{c_\alpha}\boldsymbol{\gamma}^\top\boldsymbol{\gamma}}, \qquad y = \frac{1}{\boldsymbol{\gamma}^\top\boldsymbol{\gamma}}, \tag{27}$$

and use the inequality $\log(1+xy) < y\log(1+x)$ $(x \geq 0, y > 1)$. We then have

$$G_\alpha < \left(1 - \frac{\left\langle \sum_\sigma \psi_\sigma^{i\to j}\psi_\sigma^{j\to i}\right\rangle_\alpha}{\boldsymbol{\gamma}^\top\boldsymbol{\gamma}}\right)\log\left(1 - \frac{\Delta c_\alpha}{c_\alpha}\boldsymbol{\gamma}^\top\boldsymbol{\gamma}\right). \tag{28}$$

At the detectability threshold ($\left\langle \sum_\sigma \psi_\sigma^{i\to j}\psi_\sigma^{j\to i}\right\rangle_\alpha = \boldsymbol{\gamma}^\top\boldsymbol{\gamma}$), $G_\alpha < 0$. Thus, $E_{\mathrm{Gibbs}}(q) > E(q=1)$.

6. Numerical experiments

We numerically confirm the performance of the cross-validation estimates for the prediction errors. Two results of the labeled SBM are shown in Fig. 2. In both cases, the labeled SBM has two types of edges ($p = 2$) whose average degrees are $c_1 = 3$ and $c_2 = 5$. Figure 2a shows the result on the network with five planted modules. The values of model parameters are $\Delta c_1/c_1 = 4.375$ and $\Delta c_2/c_2 = -0.625$. The prediction errors, the Gibbs prediction errors in particular, are minimized or saturated at $q = 5$. On the other hand, 2b shows the result on the network with three planted modules. The values of model parameters are $\Delta c_1/c_1 = 2.1$ and $\Delta c_2/c_2 = 0.075$. In this case, however, the network is in the region where the planted structure is algorithmically undetectable. According to Eq. (19), the prediction error for $q = 1$ is at $E(q=1) \simeq 6.38$. We can thus confirm that the prediction errors are minimized or already saturated at $q = 1$. Therefore, the cross-validation estimates perform reasonably on the labeled SBMs.

7. Summary and Discussion

We introduced the LOOCV estimates of the prediction errors for the labeled SBM. While the cross-validation is computationally demanding when it is implemented straightforwardly, we showed that it can be performed efficiently using BP. Our analytical expressions of the LOOCV estimates for the labeled SBM are extensions of the results for the standard (binary label) SBM [6].

Compared to the case of the standard SBM, the cross-validation estimate of the labeled SBM is given as the linear combination of the contribution from each edge type. We also analyzed the performance of the Bayes and Gibbs prediction errors near the detectability threshold. As in the case of the standard SBM, we confirmed that the Gibbs prediction error strictly underfits, whereas the Bayes prediction error supports partitioning all the way down to the detectability threshold.

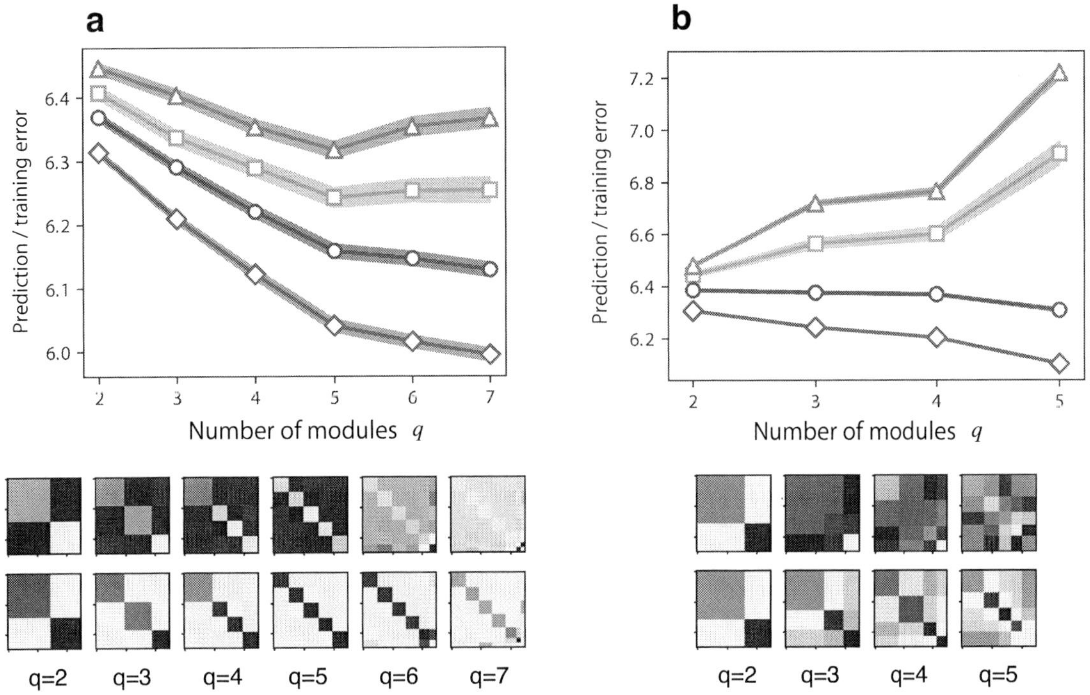

Figure 2. The cross-validation estimates of the prediction errors on the labeled SBM with two types of edges. The average degrees are $c_1 = 3$ and $c_2 = 5$. In both figures, the horizontal axis indicates the input value for the number of modules q, and the vertical axis indicates the value of the cross-validation estimates. From top to bottom, the curves represent the Gibbs prediction error E_{Gibbs} (green), MAP estimate of the Gibbs prediction error E_{MAP} (yellow), Bayes prediction error E_{Bayes} (red), and Gibbs training error E_{training} (blue). The shadow on each curve represents the standard error. In (a), the network has 1,000 vertices of five equally sized planted modules, and the model parameters are in the detectable region. On the other hand, in (b), the network consists of 900 vertices with three equally sized planted modules, and the model parameters are in the algorithmically undetectable region. For each network, the learned affinity matrices are visualized at the bottom. The block in darker color represents the densely connected components in the affinity matrix, whereas the block in the lighter color represents the sparsely connected component. The size of each module is reflected in the size of the corresponding element in the matrix.

In Ref. [6], we derived that the inequality $E_{\mathrm{training}} \leq E_{\mathrm{Bayes}} \leq E_{\mathrm{Gibbs}}$ holds in general. Moreover, when the partitions with different q constitute a hierarchical structure, we can also show the inequality $q^*_{\mathrm{Gibbs}} \leq q^*_{\mathrm{Bayes}}$, where q^*_{Bayes} and q^*_{Gibbs} are the estimated number of modules by E_{Bayes} and E_{Gibbs}, respectively. Because the derivations of these inequalities are independent of the edge type, those relations among the cross-validation errors also hold in the present case.

Acknowledgments

This work was supported by the New Energy and Industrial Technology Development Organization (NEDO).

References

[1] Holland P W, Laskey K B, and Leinhardt S 1983, *Soc.Netw.* **5**, 109
[2] Bishop C M 2006, *Pattern Recognition and Machine Learning*, (Singapore: Springer)
[3] Decelle A, Krzakala F, Moore C, and Zdeborová L 2011, *Phys. Rev. E* **84** 066106
[4] Yedidia J S, Freeman W T, and Weiss Y 2001, *Advances in neural information processing systems* 13

[5] Mézard M and Montanari A 2009, *Information, Physics, and Computation* (Oxford: Oxford University Press)
[6] Kawamoto T and Kabashima Y 2017, *Sci. Rep.* **7**, 3327
[7] Arlot S and Celisse A 2010, *Statistics surveys*, **4**, 40-79
[8] Friedman J, Hastie T, and Tibshirani R 2001, *The elements of statistical learning* (New York: Springer series in statistics)
[9] Heimlicher S, Lelarge M, and Massoulié L 2012, *arXiv preprint* arXiv:1209.2910
[10] Kawamoto T 2017, *arXiv preprint* arXiv:1710.08816
[11] Kawamoto T 2017, *arXiv preprint* arXiv:1710.08841

Exact Bayesian learning of Partition Directed Acyclic Graphs

J Pensar[1,5] and J Kohonen[2,5] and J Corander[1,3,4]

[1]Department of Mathematics and Statistics, University of Helsinki, Finland
[2]Department of Computer Science, University of Helsinki, Finland
[3]Department of Biostatistics, University of Oslo, Norway
[4]Wellcome Trust Sanger Institute, Cambridge, UK

E-mail: jukka.corander@medisin.uio.no

Abstract. It is known that directed acyclic graphs (DAGs) may hide several local features of the joint probability distribution that can be essential for some applications. To remedy this, more expressive model classes have been introduced. In addition to the restrictions implied by conditional independence, these model classes typically include some form of local structure that implies equality constraints on the node-wise conditional distribution. In particular, the concept of context-specific independence (CSI) was introduced to increase the flexibility of traditional Bayesian networks. Furthermore, in the most expressive class of generalized Bayesian networks, decision graphs were used to model arbitrary parameter restrictions. Here we formulate an alternative representation of such models called a partition DAG (PDAG), which defines the parameter restrictions using a partition-based representation of the parent outcome spaces. We establish a criterion that can identify whether an arbitrary PDAG has a CSI-consistent representation using an efficient basic graph theoretic algorithm. Based on a recursive inference algorithm for partition posteriors, an exact Bayesian learning method is introduced. We demonstrate on real data that exact learning of PDAGs can identify important relationships between variables that have not been discovered by previous graphical model learning methods.

1. Introduction

Bayesian networks have represented a standard workhorse in artificial intelligence and machine learning for more than two decades [15]. However, it has also been acknowledged that the directed acyclic graphs (DAGs) they are based upon, may hide several local features of the joint probability distribution that can be essential for some applications. To remedy this, more expressive classes of Bayesian networks have been introduced [4, 5, 9, 10, 17, 18, 20]. In particular, several of the introduced model classes [4, 9, 17, 20] have been built around the concept of context-specific independence (CSI), which is a natural generalization of conditional independence. In addition to generalized Bayesian networks, the related classes of probabilistic decision graphs (PDGs) [12] and chain event graphs (CEGs) [22] have been introduced for the purpose of capturing asymmetric model structures.

The common feature among most classes of generalized Bayesian networks is that additional local restrictions impose some form of equality constraints on the node-wise conditional

[5] JP and JK contributed equally to this work.

distributions. The equality constraints effectively partition the outcome space of the parents of a node into classes, such that parent configurations belonging to the same class induce the same distribution. The structure of the possible partitions depends on the generality of the restrictions [18]. In this work we will consider Bayesian networks with arbitrary equality constraints, that is, arbitrary partitions. This corresponds to using decision graphs for modeling the local structure between a node and its parents [5], however, we will consider an alternative representation based on the actual parent outcome partitions. In contrast to decision graphs, each distinct partition will correspond to a unique set of restrictions. We refer to the model structure containing both the global DAG structure and the local parent outcome partitions as a partition DAG (PDAG). When each partition in a PDAG has maximum cardinality, only the independence characteristics of the DAG itself are retained. With a decreasing number of classes in a partition, an increasing set of additional local constraints are introduced to the conditional distribution of the corresponding node. Some of these will typically correspond to CSIs, but in general not all of them will be interpretable as such.

The modular composition of Bayesian networks is a very important structural property in terms of model learning. In particular, for a given a DAG, the variables specifying the local structure of a node are confined to the parents of the node, enabling efficient learning of Bayesian networks with local structure. In contrast, one of the key features of models such as PDGs and CEGs is the ability to model conditioning contexts that consist of specific joint configurations of large collections of variables. Although these high-order interactions make the models extremely flexible, they also make learning of such graphs very challenging. As a result, methods developed for PDG and CEG learning have exploited Bayesian network learning as a step of the learning procedure [13, 1]. Still, numerical experiments have indicated the difficulty in learning PDG/CEG models with higher predictive accuracy than standard Bayesian networks [13, 21].

The paper is organized as follows. In Section 2 we introduce PDAGs and discuss their connection to the well-known concept of CSI. In section 3 we introduce a graph theoretical condition for determining if an arbitrary PDAG has a CSI-based dependence structure, and if not, we explain how one can modify the partitions in order to make them consistent with CSI. Section 4 describes how to evaluate the Bayesian score of a PDAG for a given set of data. Section 5 presents a recent recursive algorithm for exact learning of the optimal parent partitions. The algorithm can in the binary case efficiently identify the maximum *a posteriori* partition of a parent outcome space for up to four binary parents. In Section 6 we demonstrate using real data that exact learning of PDAGs can identify important relationships between variables that have not been earlier discovered by any other graphical model learning methods. Finally, in Section 7 we provide some additional remarks and discuss some ideas for future research.

2. Partition Directed Acyclic Graphs

A DAG is a graph $G = (V, E)$ consisting of a set of nodes $V = \{1, \ldots, d\}$ and a set of edges $E \subset V \times V$ such that $(u, v) \in E$ if there is a directed edge from node u to v. The acyclicity restriction prevents the edge set from containing directed cycles. The set of parents of node v is denoted by $pa(v) = \{u \in V : (u, v) \in E\}$. In a Bayesian network, the nodes V correspond to a set of stochastic variables $X = \{X_1, \ldots, X_d\}$. We use X_S, where $S \subseteq V$, to denote a set of variables and we use lower case letters x_S to denote a value taken by X_S. The outcome space of a set of variables X_S is denoted by $\mathcal{X}_S$, and the cardinality is denoted by $|\mathcal{X}_S|$. In this work, we assume that all considered variables are binary, $\mathcal{X}_v = \{0, 1\}$. Our results generalize to non-binary variables in a relatively straightforward manner.

A Bayesian network models the joint distribution over X_V by asserting statements of conditional independence,

$$X_A \perp X_B \mid X_S \Leftrightarrow p(X_A \mid X_B, X_S) = p(X_A \mid X_S).$$

X_1	X_2	X_3	$p(X_4 \mid X_{1,2,3})$
0	0	0	p_1
0	0	1	p_2
0	1	0	p_3
0	1	1	p_4
1	0	0	p_5
1	0	1	p_6
1	1	0	p_7
1	1	1	p_8

(a)

X_1	X_2	X_3	$p(X_4 \mid X_{1,2,3})$
0	0	0	p_1
0	0	1	p_2
0	1	0	p_3
0	1	1	p_3
1	0	0	p_4
1	0	1	p_5
1	1	0	p_3
1	1	1	p_6

(b)

X_1	X_2	X_3	$p(X_4 \mid X_{1,2,3})$
0	0	0	p_1
0	0	1	p_2
0	1	0	p_2
0	1	1	p_3
1	0	0	p_4
1	0	1	p_2
1	1	0	p_2
1	1	1	p_5

(c)

Table 1. CPT of node 4 when $pa(4) = \{1, 2, 3\}$: (a) no regularities, (b) CSI-based regularities, and (c) arbitrary regularities.

The conditional independence statements assumed by a Bayesian network are compactly encoded by its DAG, where edges correspond to direct dependencies and lack of edges corresponds to conditional independencies. More specifically, the dependence structure is characterized by the local directed Markov property, which implies an explicit factorization of the joint distribution according to

$$p(X_1, \ldots, X_d) = \prod_{v=1}^{d} p(X_v \mid X_{pa(v)}). \tag{1}$$

In (1), each factor corresponds to a collection of conditional probability distributions (CPDs). By definition, each collection of CPDs implicitly assumes full dependence of v on its parents, i.e. separate parameters are needed to specify the CPD for all combinations in $\mathcal{X}_{pa(v)}$. In general, this requires

$$(|\mathcal{X}_v| - 1) \cdot \prod_{u \in pa(v)} |\mathcal{X}_u|$$

free parameters. In other words, under the assumption of binary variables, the number of parameters needed to specify all CPDs over X_v is $2^{|pa(v)|}$. The standard approach for representing the CPDs associated with a node is to use a conditional probability table (CPT). An example of a traditional CPT with distinct CPDs is shown in Table 1(a).

A limitation of traditional Bayesian networks is that the number of model parameters associated with a node grows exponentially with the number of parents. However, in many situations, there may exist regularities in form of identical CPDs within a CPT. One of the earliest and most well-known approaches for capturing such regularities introduced the concept of context-specific independence (CSI) [4]:

$$X_A \perp X_B \mid x_C, X_S \Leftrightarrow p(X_A \mid X_B, x_C, X_S) = p(X_A \mid x_C, X_S).$$

CSI is a natural generalization of conditional independence that only holds in part of the outcome space, as specified by the context $X_C = x_C$. In terms of CPTs, we consider local CSIs of the form

$$X_v \perp X_u \mid x_{pa(v)\setminus u}, \text{ where } u \in pa(v),$$

since such statements will imply identical CPDs within the CPT:

$$p(X_v \mid x_u, x_{pa(v)\setminus u}) = p(X_v \mid x'_u, x_{pa(v)\setminus u}) \text{ for all } x_u, x'_u \in \mathcal{X}_u.$$

In the binary case, a local CSI thus corresponds to a restriction forcing exactly two values in the CPT of v to be identical. As an example, consider the CPT in Table 1(b), where the parent configurations $(0,1,0),(0,1,1),(1,1,0)$ all induce the same conditional probability p_3. A closer examination reveals that this particular CPT structure can be explained by the CSIs

$$X_3 \perp X_4 \mid X_1 = 0, X_2 = 1 \text{ and } X_1 \perp X_4 \mid X_2 = 1, X_3 = 0.$$

Rather than specifying the same distribution multiple times, the above parent configuration can then be merged into a single class of configurations. Hence, the set of restrictions imposed by a collection of CSIs defines a partition of the parent outcome space into classes, where all elements in the same class induce an identical CPD.

Note, however, that the converse is not true, that is, an arbitrary partition of the parent outcome space does not necessarily correspond to a collection of CSIs. In this work, we go beyond CSI-based regularities and consider arbitrary regularities of the form

$$p(X_v \mid x_{pa(v)}) = p(X_v \mid x'_{pa(v)}).$$

As an example, consider the CPT in Table 1(c), where $(0,0,1),(0,1,0),(1,0,1),(1,1,0)$ all induce the same conditional probability p_2. In contrast to the previous example (Table 1(b)), these regularities cannot be fully explained by CSI and we need to allow arbitrary partitions to compactly represent the CPT structure. We discuss this more in detail in the next section, where we also introduce a criterion for determining whether the regularities in a CPT can be explained by CSI or not.

In general, Bayesian networks with parent outcome partitions enjoy similar flexibility in terms of parameter restrictions as probabilistic decision graphs, but are based on a different representation that enables efficient learning algorithms to be developed. To formally represent the structure of Bayesian networks with arbitrary parent outcome partitions, we introduce the concept of partition DAGs.

Definition 1. *Partition DAG*
Let $G = (V, E)$ be a DAG for the stochastic variables $\{X_1, \ldots, X_d\}$. For all $v \in V$, let S_v be a partition of the set $\mathcal{X}_{pa(v)}$ into k classes $s_1, \ldots, s_k$ such that the conditional probabilities $p(x_v \mid x_{pa(v)})$ are equal for all $x_{pa(v)} \in s_c, c = 1, \ldots, k$ and unrestricted otherwise. The graph G together with the collection of partitions $\mathcal{S} = \{S_v : v \in V\}$ define a Partition DAG (PDAG), denoted by $G_{\mathcal{S}}$.

For notational simplicity, indexing of the partition S_v and the number of classes k in it with respect to the node is not explicitly shown for each class of S_v when the meaning is unambiguous. Note that k may freely vary across the nodes $v \in V$.

3. CSI-consistent partition

In [17], a parent outcome partition is referred to as CSI-consistent if it can be constructed according to a collection of local CSIs. As discussed in the previous section, an arbitrary partition would typically be expected to reside beyond the scope of CSI, however, then the following questions arise: what restrictions must a partition satisfy in order to be CSI-consistent, and if a partition is not CSI-consistent, how can it be modified to become CSI-consistent? CSI-consistency can be a useful property. For example, previous research has shown that CSI can be exploited to improve the efficiency of inference algorithms [19, 20].

To construct a procedure for determining if a partition is CSI-consistent we introduce the following graphical criterion.

IOP Conf. Series: Journal of Physics: Conf. Series **1036** (2018) 012017 doi:10.1088/1742-6596/1036/1/012017

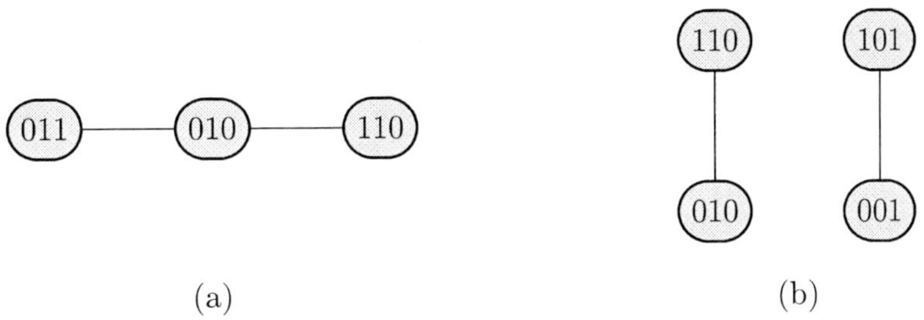

Figure 1. The class connection graph of the non-singleton class in the partition obtained from the CPT in (a) Table 1(b) and (b) Table 1(c). For clarity, the nodes are labeled with the corresponding parent configurations.

Definition 2. *Class connection graph*
A class connection graph $G_c = (N, E)$ *for a class* c *for* $v \in V$ *is an undirected graph over the class elements* $s_c = \{x_{pa(v)}^{(l)}\}_{l=1}^{q}$ *such that* $q = |s_c|$ *and*

$$N = s_c \text{ and } \{l, r\} \in E \text{ if } d(x_{pa(v)}^{(l)}, x_{pa(v)}^{(r)}) = 1,$$

where $d(\cdot, \cdot)$ *is the Hamming distance between the two configurations.*

Theorem 1. *A partition* $S_v = \{s_1, \ldots, s_k\}$ *is CSI-consistent if and only if the corresponding class connection graphs* $G_1, \ldots, G_k$ *are connected.*

Proof. See Appendix A. □

Theorem 1 basically reformulates the problem of checking for CSI-consistency to a well-known graph connectivity problem for which there already exist several efficient methods such as the depth-first and breadth-first algorithms [11].

To illustrate the criterion, consider the partition obtained from the CPT in Table 1(b). The class connection graph of the non-singleton class is shown in Figure 1(a). Note that the class connection graph of a singleton class is connected by definition. Since the graph in Figure 1(a) is connected, the partition is CSI-consistent. On the other hand, consider the partition obtained from the CPT in Table 1(c). The class connection graph of the non-singleton class is shown in Figure 1(b). In this case, the graph is not connected and, consequently, the partition is not CSI-consistent. However, if we would split the class

$$\{(0,0,1), (0,1,0), (1,0,1), (1,1,0)\}$$

into two new classes

$$\{(0,0,1), (1,0,1)\} \text{ and } \{(0,1,0), (1,1,0)\},$$

the partition would become CSI-consistent, since all class connection graphs would then be connected. The new representation would no longer be minimal in terms of the number of parameters, since the same distribution would have to be defined twice. Still, the criterion provides a straightforward way of making an arbitrary partition CSI-consistent by removing as few equality constraints as possible.

IOP Conf. Series: Journal of Physics: Conf. Series **1036** (2018) 012017 doi:10.1088/1742-6596/1036/1/012017

4. Bayesian score for PDAGs

A very useful property of PDAGs and similar models is that the marginal likelihood can still be evaluated by a closed-form expression [5]. Let $\mathbf{X} = \{\mathbf{x}_i\}_{i=1}^{n}$ denote a set of training data consisting of n observations $\mathbf{x}_i = (x_{i1}, \dots x_{id})$, assuming that $\mathbf{X}$ contains no missing values. We let θ_{vc} denote the probability of $X_v = 1$ given that $X_{pa(v)}$ is assigned a value in class c of the partition S_v. For a given PDAG $G_{\mathcal{S}}$, we let $\theta_{G_{\mathcal{S}}}$ denote collectively all the unknown parameters in the CPTs. The likelihood of a PDAG can then be written as

$$p(\mathbf{X} \mid \theta_{G_{\mathcal{S}}}, G_{\mathcal{S}}) = \prod_{v=1}^{d} \prod_{c=1}^{k_v} \theta_{vc}^{n(v,c,1)} (1 - \theta_{vc})^{n(v,c,0)},$$

where $n(v, c, 1)$ denotes the total count of observations with $x_v = 1$ and $x_{pa(v)}$ falling into the class c of S_v. Further, $n(v, c, 0)$ is the corresponding count with $x_v = 0$ and $n(v, c) = n(v, c, 0) + n(v, c, 1)$. We use a prior for $\theta_{G_{\mathcal{S}}}$ which factorizes over the DAG and the classes, such that the marginal likelihood (or evidence) of the data can be calculated analytically. Under the standard conjugate Beta prior, we obtain the following expression for the marginal likelihood given the PDAG:

$$p(\mathbf{X} \mid G_{\mathcal{S}}) = \prod_{v=1}^{d} \prod_{c=1}^{k_v} \frac{\Gamma(\alpha)}{\Gamma(n(v,c) + \alpha)} \prod_{l=0}^{1} \frac{\Gamma(n(v,c,l) + \alpha_l)}{\Gamma(\alpha_l)}, \tag{2}$$

where $\alpha = \alpha_0 + \alpha_1$.

There are several options in terms of specifying the hyperparameters in (2), see e.g. [15]. In all the numerical experiments reported here we have used the uniform prior with $\alpha_1 = \alpha_0 = 1$. Following the Bayesian approach of structure learning, we aim at maximizing $\log p(\mathbf{X} \mid G_{\mathcal{S}}) + \log p(G_{\mathcal{S}})$, where $p(G_{\mathcal{S}})$ is a product prior over the d partitions. Each prior factor in the product is assumed to be defined by the uniform distribution on the number of classes k_v in the partition S_v. Further discussion about the choice of priors is found in the final section of the paper.

5. Exact PDAG learning algorithm

Learning PDAGs from training data is in general extremely challenging. Even for a fixed ordering of the variables specified by a candidate DAG, there exist extremely many possible partitions of the parent outcome space for each single node. For example, for a node with 3 (binary) parents, there are 4140 ways to partition the 8 possible parent configurations, whereas for a node with 4 parents, the corresponding number is already 10,480,142,147. In general, the number of unordered partitions of $|\mathcal{X}_{pa(v)}|$ entries is given by the Bell number of $|\mathcal{X}_{pa(v)}|$. While direct complete enumeration of all possible partitions is in practice tedious for cases with more than two parents, the recursive algorithm for exact posterior inference about clustering introduced in [14] is sufficiently fast for exact learning of partitions in a PDAG with at most 4 parents for any node.

Let $S(\mathcal{X}_{pa(v)}, k_v)$ be the Stirling number of the second kind, i.e. the number of possible partitions of $\mathcal{X}_{pa(v)}$ into exactly k_v non-overlapping and non-empty subsets. We set the prior probability of a partition S_v equal to $|\mathcal{X}_{pa(v)}|^{-1} S(\mathcal{X}_{pa(v)}, k_v)^{-1}$, which corresponds to the uniform distribution for the number of classes k_v for node v. To enable use of the recursive algorithm for posterior inferences, we consider now *ordered partitions*, each of which is a tuple of disjoint nonempty subsets, whose union is $\mathcal{X}_{pa(v)}$.

The posterior probability of an ordered partition S_v with k_v classes in the PDAG model equals

$$p(S_v \mid \mathbf{X}) = \frac{p(S_v) p(\mathbf{X} \mid S)}{p(\mathbf{X})} = Z \cdot \prod_{c=1}^{k_v} f(s_{v,c}),$$

IOP Conf. Series: Journal of Physics: Conf. Series **1036** (2018) 012017 doi:10.1088/1742-6596/1036/1/012017

where Z is the normalizing constant and $f(s_{v,c})$ is the marginal likelihood term for class c in (2). The posterior probability of k clusters equals under the above formulation

$$p(k_v \mid \mathbf{X}) = Z \cdot \sum_{\substack{S_v \in \mathcal{S}_v \\ |S_v|=k}} \prod_{c=1}^{k_v} f(s_{v,c}), \tag{3}$$

where $\mathcal{S}_v$ is the space of ordered partitions for node v. This sum of products can be conveniently expressed by subset convolution. Given two real-valued functions f and g defined on the subsets of $\mathcal{X}_{\Pi_v}$, their *subset convolution* is the function

$$(f * g)(Y) = \sum_{A \subseteq Y} f(A) \cdot g(Y \setminus A), \quad \text{for all } Y \subseteq \mathcal{X}_{pa(v)}. \tag{4}$$

Expressed in a more symmetric form we get

$$(f * g)(Y) = \sum_{\substack{A,B \subseteq X \\ A+B=Y}} f(A) \cdot g(B),$$

where $A + B = Y$ represents disjoint union. Convolution is associative, and iterative application hence yields

$$(f_1 * \ldots * f_{k_v})(Y) = \sum_{\substack{A_1,\ldots,A_{k_v} \subseteq Y \\ A_1+\ldots+A_{k_v}=Y}} \prod_{c=1}^{k_v} f_c(A_c).$$

The k_v-fold convolution expresses summation over ordered k_v-partitions of a set Y. Writing (3) in terms of iterated convolution yields

$$p(k_v \mid \mathbf{X}) = Z \cdot f^{(k_v)}(\mathcal{X}_{pa(v)}), \tag{5}$$

where $f^{(k_v)} = (f * \ldots * f)$ denotes the convolution of k_v copies of f.

The full convolution table for $f*g$ can be obtained in $O(3^{|\mathcal{X}_{pa(v)}|})$ operations, which is still very fast for the case with 4 parent nodes. In practice these calculations can be done in approximately 1 minute for such a node of a PDAG model, whereas 3 parent nodes can be handled in a fraction of a second. When the summation in (3) is replaced with maximization, one obtains the maximum posterior probability among k_v-partitions. This can be computed using a variant of subset convolution, where the summation is replaced with maximization such that the subset convolution is performed over the max-product semiring, instead of the sum-product ring. This yields an $O(|\mathcal{X}_{pa(v)}|3^{|\mathcal{X}_{pa(v)}|})$ algorithm for finding the maximum *a posteriori* (MAP) partition for any node-parent combination in a PDAG.

6. Experiments

6.1. *Heart disease data*

A classical data set used for benchmarking various graph learning algorithms is the coronary heart disease risk factor data set with 1841 observations on 6 binary variables [8, 23, 17]. Definitions of the variables are given in Table 2. The data set has been considered also in numerous other articles about graphical model learning and a consistent finding is that the Family anamnesis of coronary heart disease is independent of all the remaining variables.

Figure 2 shows the DAG with the parent sets of nodes being determined by exact learning such that at most four parent candidates were considered simultaneously and the available partial

Variable	Outcomes
X_1: Smoking	No $:= 0$, Yes $:= 1$
X_2: Strenuous mental work	No $:= 0$, Yes $:= 1$
X_3: Strenuous physical work	No $:= 0$, Yes $:= 1$
X_4: Systolic blood pressure	$< 140 := 0, > 140 := 1$
X_5: Ratio of β and α lipoproteins	$< 3 := 0, > 3 := 1$
X_6: Family anamnesis of CHD	No $:= 0$, Yes $:= 1$

Table 2. Description of the variables in coronary heart disease (CHD) data.

X_{1356}	#0	#1	$\bar{X}_4$
0001	26	15	0.3659
1000	107	45	0.2961
1100	168	76	0.3115
class 1	301	136	0.3112
class 2	753	651	0.4637

(a)

X_{1256}	#0	#1	$\bar{X}_4$
0011	16	9	0.3600
1000	185	79	0.2992
1100	90	42	0.3182
class 1	291	130	0.3088
class 2	763	657	0.4627

(b)

X_{156}	#0	#1	$\bar{X}_4$
100	275	121	0.3056
class 1	275	121	0.3056
class 2	779	666	0.4609

(c)

Table 3. MAP partition for X_4 with parent set (a) 1356, (b) 1256, and (c) 156. Details of class 2 have been omitted. For each combination of parent values, #0 and #1 are the counts of $X_4 = 0$ and $X_4 = 1$, respectively, and the last column is the proportion of ones.

ordering of the variables was used to exclude certain candidate nodes from the set of possible parents. In practice, nodes 4 and 5 had each four possible parents, whereas node 3 had two and node 2 only one. Nodes 1 and 6 had no possible parents given the imposed partial ordering information. For a detailed discussion of these data see, e.g. [23].

In the more comprehensive analysis reported below, we concentrate on variable X_4 (systolic blood pressure) and its parent structure. As at most four parent candidates were considered at a time, there are $\binom{5}{4} = 5$ different candidate sets of parents, in shorthand notation: 2356, 1356, 1256, 1236 and 1235.

For each candidate set we obtained the MAP partition for the 16 parent value combinations as discussed before. The log marginal likelihoods of the MAP partitions (relative to the trivial partition) are +0.0, +13.5, +13.4, +0.0 and +10.5, respectively. The two highest likelihoods (with almost a tie) occur for the candidate sets 1356 and 1256, which are highly similar to each other. In both cases the MAP partition is a 2-partition where the smaller class contains three parent value combinations, as illustrated in Tables 3(a)–(b).

In both tables, we can see that the smaller class contains parent configurations, for which X_4 has a low probability of high blood pressure. In both cases, the first row, which contains relatively few observations overall, seems somewhat odd and might be a random artefact. The remaining two rows, 1000 and 1100 in either case, appear more plausible, and they would seem to indicate an interesting phenomenon in the data: if $(X_1, X_5, X_6) = (1, 0, 0)$, then the probability of $X_4 = 1$ is much lower than otherwise.

This dependence, where just one particular combination of three binary variables has a joint effect on a fourth variable, is not explicitly caught by the structure of ordinary DAG models. As mentioned above, despite of numerous previous analyses of this data with various Bayesian and Markov network models, any significant association between the family anamnesis of CHD

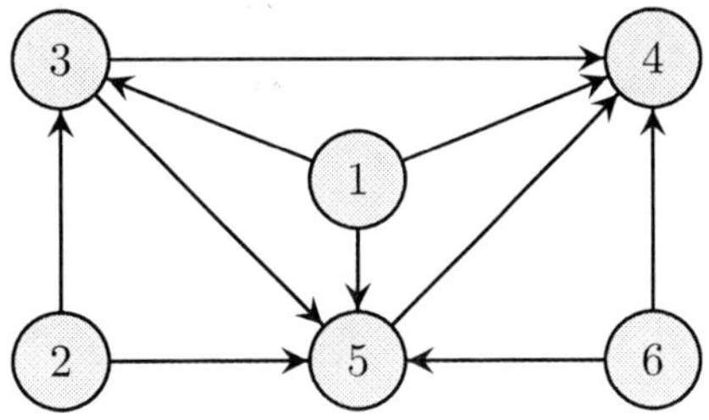

Figure 2. DAG structure for the heart disease data.

and the remaining variables has never been detected. However, it has been remarked that the lack of an association between node 6 and the two biologically measured variables (nodes 4 and 5) is surprising, given that CHD among relatives is a known risk factor.

To explore the association a bit further, we also examine the optimal parent set with only 3 nodes. This contains the nodes 156 and the MAP partition of the 8 parent configurations is shown in Table 3(c). Again, we obtain a 2-partition, but now the smaller class contains just the one parent configuration 100, while the seven other combinations are in the other class. The finding can be interpreted such that individuals that smoke and have a low lipoprotein ratio as well as no family anamnesis of CHD, have a lower probability of high blood pressure than others. The reason why the connection between the family history variable (node 6) and the other CHD risk factors is not revealed in ordinary pairwise association or network analysis, is the context-specific nature of their dependence, which is masked by the distribution of the other conditional probabilities in Table 3(c). All these probabilities are within 3 percentage points from the mean of class 2 (0.46), whereas only the configuration 100 corresponds to much smaller probability (0.31), still based on substantial amount of data (396 observations).

6.2. *Simulated data*

To gauge the behavior of the learning algorithm, we performed a simulated data study using the CSI-based model with 10 binary variables introduced in [17]. Due to space limitations, we focus here on variable X_5, which has three parents X_1, X_4 and X_8, and satisfies the CSI

$$X_5 \perp X_4, X_8 \mid X_1 = 0.$$

In other words, there are 8 parent configurations which are partitioned into 5 classes such that one class contains the four configurations where $X_1 = 0$, and the four remaining classes contain a single configuration.

In the simulation, we generated 100 random replicates of data with varying sizes according to the given model, and then identified the MAP partition using the subset convolution algorithm for each replicate. The MAP partition was compared to the generating partition using two metrics: adjusted Rand index, and zero-one-loss (one if the partitions are identical, zero otherwise). The results of both metrics are shown in Figure 3.

As expected, the more data we have, the closer the learned partition is to the structure in the generating model. We note that fairly large sample sizes are required to obtain very accurate results. This is reasonable, since the PDAG model class is extremely flexible and e.g. 1000 observations split over the 16 value combinations (two child node outcomes for each combination of parental values) means on average only 60 samples per category and the variation around this may also be substantial. It is important to notice the difference between the generating distribution used for the simulations and the structure of the real CHD data analyzed previously. As shown for the CHD data, the conditional probabilities were highly similar across

IOP Conf. Series: Journal of Physics: Conf. Series **1036** (2018) 012017 doi:10.1088/1742-6596/1036/1/012017

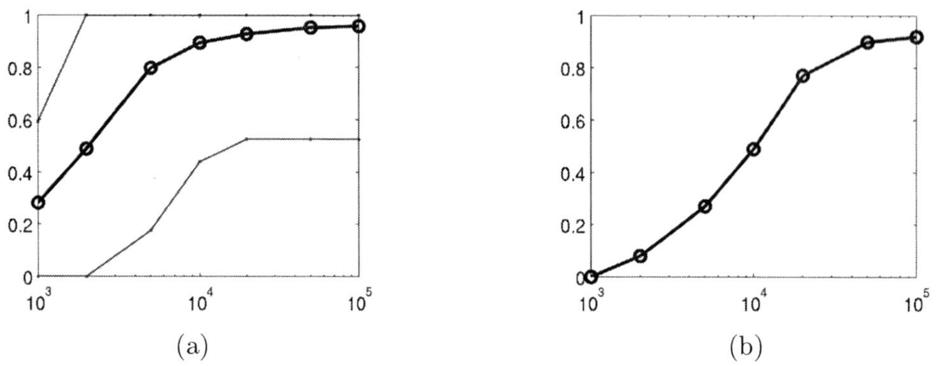

Figure 3. Simulated data study with data sizes ranging from 1000 to 100000. Vertical axis: (a) adjusted Rand index between inferred and the generating partition (black: average over 100 replications; blue: 5% and 95% quantiles), (b) proportion of inferred partitions equal to the structure of the generating model in 100 replications.

the parental combinations of node 4, except in one case. In contrast, the conditional probabilities in the synthetic model have been sampled from a prior distribution (when not forced equal), which means that they are more uniformly spread across the interval $(0, 1)$, such that any two probabilities may coincidentally be fairly close to each other. This would typically require more data for accurate learning of the underlying partition.

7. Discussion

Our approach has focused on the representability and interpretability of partitions of CPT elements to encode local restrictions on the parameters of a DAG model. The exact Bayesian learning method we introduced uses recursive computation to identify the posterior optimal partition for a node given any candidate set of parents. However, direct brute-force approach still necessitates enumeration across such candidate sets, which is in general infeasible. It would therefore be interesting to explore how the recursive computation over subsets of parental outcomes could be combined with an approach based on computational logic to identify the optimal parent sets. There has recently been a considerable interest in use of computational logic algorithms for learning DAGs and Markov networks, such as integer programming, answer-set programming and maximum satisfiability [7, 2, 6, 3, 16]. In particular, the maximum satisfiability approach introduced in [3] scales surprisingly well with respect to the number of nodes and could potentially leverage a solution to more scalable exact PDAG learning in combination with the subset convolution.

The prior we use for partitions of CPT elements is a uniform distribution for the number of classes in the partition. This prior penalizes most partitions that have cardinality approximately in the middle of the range from a single class to maximum cardinality, as the Stirling number of the second kind is maximized for them on a given set of parent configurations. The rationale for our current prior can be explained as follows. When the number of classes in a partition is small, the model for the conditional distribution of the node is fairly simple as the node is almost contextually independent of its parents except in a limited set of cases. Since there are only few different such models for a candidate set of parents, the prior penalty needs not to be as large. Similarly, if there are many classes, i.e. the partition has almost maximal cardinality, the model is again *structurally* relatively simple since nearly all combinations of the values of parents have their respective unique conditional distributions. When the partition cardinality is neither small nor large, the resulting model has a more diverse set of dependencies and local independencies.

Since there are many such models, some of them may coincidentally provide a fairly good fit to the data, and therefore, the additional penalty from the uniform prior on the number of classes can be a useful way to prevent too flexible models to overfit the data. However, more research is needed to establish the advantages and disadvantages of different types of possible priors for this kind of flexible models.

The partition-based formulation chosen here to represent parameter constraints may also prove particularly useful for inference compared with decision graphs. For instance, [19] concluded that rule-based representations may be more efficient than tree-based and [24] analyzed potential computational advantages resulting from CSIs. Additionally, [20] improved the approach of [19] by using a combination of contexts and tables, leading to contextual belief networks. As noted in [17], the underlying CSIs of a CSI-consistent partition directly correspond to the parent contexts of a contextual belief network. By making the partitions of a PDAG CSI-consistent, the inference method in [20] can thus be applied as such. Whether it is possible to utilize the structure of PDAGs for inference even more efficiently is an open problem.

Acknowledgments
The authors would like to thank professor Teemu Roos for useful discussions on context-specific graphical models.

Appendix A. Proof of Theorem 1

Proof. A partition is CSI-consistent if the structure of all of its classes can be explained by local CSI statements. That is, for each class s_c and each distinct pair of elements $(x^{(l_1)}_{pa(v)}, x^{(l_m)}_{pa(v)})$ within that class, the implied regularity

$$p(x_v \mid x^{(l_1)}_{pa(v)}) = p(x_v \mid x^{(l_m)}_{pa(v)})$$

must be explained by a collection of local CSI statements. This is equivalent to saying that there must exist a sequence of elements $(x^{(l_i)}_{pa(v)})^{m-1}_{i=2}$ in the class s_c for which

$$d(x^{(l_i)}_{pa(v)}, x^{(l_{i+1})}_{pa(v)}) = 1 \quad \text{for } i = 1, \ldots, m-1, \tag{A.1}$$

where $d(\cdot, \cdot)$ denotes the Hamming distance. If (A.1) holds, the equality restrictions in the chain

$$p(x_v \mid x^{(l_1)}_{pa(v)}) = p(x_v \mid x^{(l_2)}_{pa(v)}) = \ldots = p(x_v \mid x^{(l_{m-1})}_{pa(v)}) = p(x_v \mid x^{(l_m)}_{pa(v)})$$

can all be explained by local CSIs of the form

$$X_v \perp X_u \mid x^{(l_i)}_{pa(v)\setminus u},$$

where $u \in pa(v)$ is the parent with different values in configurations l_i and l_{i+1}. On the other hand, if

$$d(x^{(l_i)}_{pa(v)}, x^{(l_{i+1})}_{pa(v)}) > 1,$$

the same restriction can no longer be explained be such a statement. It could still be explained by a local CSI of the form

$$X_v \perp X_U \mid x_{pa(v)\setminus U},$$

where $U \subset pa(v)$ and $|U| > 1$. However, for such a CSI to be consistent with the class, there must exist a sequence satisfying (A.1) between the considered elements.

The (non-)existence of a sequence satisfying (A.1) between a pair of elements in a class s_c is equivalent to the corresponding node pair being (dis)connected in the class connection graph G_c. For all pairs in the class to be connected, the corresponding class connection graph must be connected. Consequently, all classes in a partition are CSI-consistent if and only if all the corresponding class connection graphs are connected. □

References

[1] L.M. Barclay, J.L. Hutton and J.Q. Smith. Refining a Bayesian Network using a Chain Event Graph. *International Journal of Approximate Reasoning*, 54:1300–1309, 2013.

[2] M. Bartlett and J. Cussens. Advances in Bayesian network learning using integer programming. In *Proceedings of the 29th Conference on Uncertainty in Artificial Intelligence*, 182–191, 2013.

[3] J. Berg, M. Järvisalo and B. Malone. Learning optimal bounded treewidth Bayesian networks via maximum satisfiability. In *Proceedings of the 17th Conference on Artificial Intelligence and Statistics*, 86–95, 2014.

[4] C. Boutilier, N. Friedman, M. Goldszmidt and D. Koller. Context-specific independence in Bayesian networks. In *Proceedings of the Twelfth Annual Conference on Uncertainty in Artificial Intelligence*, 115–123, 1996.

[5] D.M. Chickering, D. Heckerman and C. Meek. A Bayesian approach to learning Bayesian networks with local structure. In *Proceedings of the Thirteenth Conference on Uncertainty in Artificial Intelligence*, 1997.

[6] J. Corander, T. Janhunen, J. Rintanen, H. Nyman and J. Pensar. Learning chordal Markov networks by constraint satisfaction. In *Advances in Neural Information Processing Systems 26*, 1349–1357, 2013.

[7] J. Cussens. Bayesian network learning by compiling to weighted MAX-SAT. In *Proceedings of the Conference on Uncertainty in Artificial Intelligence*, 105–112, 2008.

[8] D. Edwards and H. Tomá. A fast procedure for model search in multidimensional contingency tables. *Biometrika*, 72:339–351, 1985.

[9] N. Friedman and M. Goldszmidt. Learning Bayesian networks with local structure. In *Proceedings of the Twelfth Annual Conference on Uncertainty in Artificial Intelligence*, 252–262, 1996.

[10] D. Geiger and D. Heckerman. Knowledge representation and inference in similarity networks and Bayesian multinets. *Artificial Intelligence*, 82:45–74, 1996.

[11] M.C. Golumbic. *Algorithmic Graph Theory and Perfect graphs: Second Edition.* Elsevier, 2004.

[12] M. Jaeger. Probabilistic decision graphs: Combining verification and AI techniques for probabilistic inference. In *Proceedings of the first European Workshop on Probabilistic Graphical Models*, 81–88, 2002.

[13] M. Jaeger, J.D. Nielsen and T. Silander. Learning probabilistic decision graphs. *International Journal of Approximate Reasoning*, 42:84–100, 2006.

[14] J. Kohonen and J. Corander. Computing exact clustering posteriors with subset convolution. *Communications in Statistics - Theory and Methods*, 45:3048–3058, 2016.

[15] D. Koller and N. Friedman. *Probabilistic Graphical Models: Principles and Techniques.* MIT Press, 2009.

[16] P. Parviainen, H.S. Farahani and J. Lagergren. Learning bounded tree-width Bayesian networks using integer linear programming. In *Proceedings of the 17th Conference on Artificial Intelligence and Statistics*, 751–759, 2014.

[17] J. Pensar, H. Nyman, T. Koski and J. Corander. Labeled directed acyclic graphs: a generalization of context-specific independence in directed graphical models. *Data Mining and Knowledge Discovery*, 29:503–533, 2015.

[18] J. Pensar, H. Nyman, J. Lintusaari and J. Corander. The role of local partial independence in learning of Bayesian networks. *International Journal of Approximate Reasoning*, 69:91–105, 2016.

[19] D. Poole. Probabilistic partial evaluation: Exploiting rule structure in probabilistic inference. In *Proceedings of the 15th International Joint Conference on Artificial Intelligence*, 1284–1291, 1997.

[20] D. Poole and N.L. Zhang. Exploiting contextual independence in probabilistic inference. *Journal of Artificial Intelligence Research*, 18:263–313, 2003.

[21] T. Silander and T.-Y. Leong A Dynamic Programming Algorithm for Learning Chain Event Graphs. In *Proceedings of Discovery Science: 16th International Conference*, 201–216, 2013.

[22] J.Q. Smith and P.E. Anderson. Conditional independence and chain event graphs. *Artificial Intelligence*, 172(1):42–68, 2008.

[23] J. Whittaker. *Graphical Models in Applied Multivariate Statistics.* Wiley, New York, 1990.

[24] N.L. Zhang and D. Poole. On the role of context-specific independence in probabilistic inference. In *Proceedings of the 16th International Joint Conference on Artificial Intelligence*, 1288–1293, 1999.

Geodetic inversion for spatial distribution of slow earthquakes under sparsity constraints

Takane Hori[1], Ryoko Nakata[1], Hideitsu Hino[2], Tatsu Kuwatani[3,4], Shoichi Yoshioka[5], Masato Okada[6]

1. Research and Development Center for Earthquake and Tsunami, Japan Agency for Marine-Earth Science and Technology, Japan
2. Department of Computer Science, University of Tsukuba, Japan
3. Department of Solid Earth Geochemistry, Japan Agency for Marine-Earth Science and Technology, Japan
4. PRESTO, Japan Science and Technology Agency, Japan
5. Research Center for Urban Safety and Security, Kobe University, Japan
6. Graduate School of Frontier Sciences, The University of Tokyo, Japan

E-mail: horit@jamstec.go.jp

Abstract. In geodetic data inversion, insufficient observational data and smoothness constraints for model parameters make it difficult to clearly resolve small-scale heterogeneous structures with discontinuous boundaries. We therefore applied sparse modelling to geodetic data inversion. In this paper, we reported two examples; one is developed a novel regularization scheme for the inversion problem that uses discontinuity, sparsity, and smoothness constraints. In order to assess its usefulness and applicability, the proposed method was applied to synthetic displacements calculated by a ring-shaped and sharply varying afterslip distribution on the plate interface beneath the Hyuga-nada region in southwest Japan. The discontinuous boundary was sharper than that obtained by using smoothness constraint only. The other is used the fused regularization, a type of sparse modelling suitable for detecting discontinuous changes in the model parameters. We estimated spatial distribution of the long-term slow slip events beneath the Bungo channel in southwest Japan. We found that the largest slip abruptly becomes zero at the down-dip limit of the seismogenic zone, and is immediately reduced to half at the up-dip limit of the deep low-frequency tremors, and becomes zero near its down-dip limit. Such correspondences imply that some thresholds exist in the generation processes for both tremors and SSEs. These results suggest that geodetic data inversion with sparse modelling can detect such abrupt changes and discontinuous boundaries in the slip distribution of slow earthquakes.

1. Introduction

Earthquake generation is related to the heterogeneous distribution of fault slips and locking conditions on the fault planes, including plate interfaces. The slip distribution of each earthquake varies in a piecewise smooth manner, but changes sharply across the boundaries between these heterogeneous areas and the surrounding area. Conversely, images obtained through geodetic inversion analysis of Global Navigation Satellite System (GNSS) data in most previous studies imposed smooth slip distributions with continuous boundaries. Commonly, when conducting geodetic data inversion, available observational data are insufficient, and the smoothness constraint is treated as a priori information based on methods such as Bayes' theorem [1] and Kalman filtering [2]. The smoothness constraint is introduced into the geodetic analysis because drastic and abrupt changes in the total slip distributions are the result of excessive concentration of stress, which does not seem to be reasonable considering the process of earthquake rupture. However, slip distributions obtained from geodetic data are significantly smoother than those from seismic data. It is suggested that heterogeneous distribution of slip and corresponding stress concentration may be removed from its original distribution by the prior constraint in geodetic data analyses. This is a critical problem to accomplish our purpose that we estimate preparation process of earthquake generation and forecast the following slip process using crustal deformation data during inter seismic periods. Therefore, we would like to extract information about slip distribution on a plate interface with adequate extent of variation between smoothness and heterogeneity from observation data. This corresponds to the "Computational theory" level , in Figure 1 [3]. For this purpose, we should create numerical model and represent algorithm for analysis. In data-driven science, we clearly distinguish between these two levels, "modeling" and "representation and algorithm" [3]. In this paper, we introduce two examples with different approaches to the different purpose of "computational theory" for seemingly two similar problems. The first example is directly represented above, and our goal is to extract adequately both smoothness and abrupt change in slip distribution to the extent of observation data quality. We analyze afterslip on a plate interface following a moment magnitude (Mw) 6.8 earthquake beneath the Hyuga-nada offshore region [4]. On the other hand, the goal of the second example is to determine the presence of abrupt changes and detect their locations without detail slip distribution. Here, we analyze long-term slow slip events (L-SSEs) beneath the Bungo Channel [5].

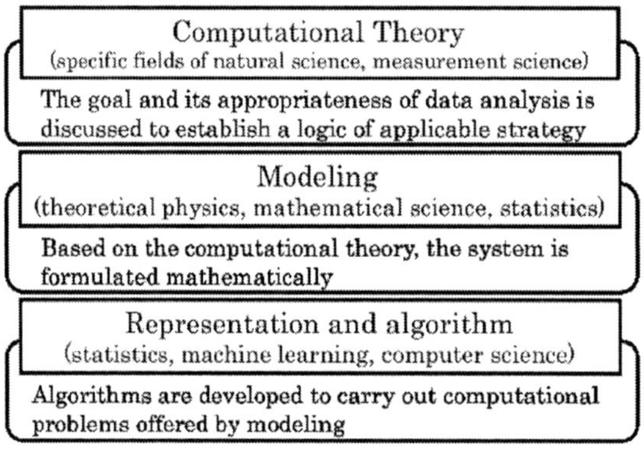

Figure 1. Three levels of data-driven science [3].

2. Geodetic data inversion

The relationship between the displacement on a free surface and the slip on a plate interface is expressed as:

$$d_k = \sum_{l=1}^{N} G_{kl} s_l + \varepsilon_k \qquad (k = 1, 2, \cdots, K), \qquad (1)$$

where K is the number of observed GNSS displacements. d_k is the observed displacements at the k-th station on the Earth's surface. s_l is the dip-slip of the l-th subfault on the plate boundary. N is the number of subfaults. G_{kl} is Green's function, which represents the displacement at station k due to a unit slip on subfault l. ε_k is the error (including observation noise) at the k-th station.

We divided the plate interface into small rectangular subfaults. Green's functions, which are the combined effect of three angular dislocations within a subfault in an elastic, homogeneous, and isotropic half space, were represented on a subfault. In southwest Japan, the Philippine Sea plate is subducting beneath the Amur plate. We used a realistic three-dimensional geometry of the Philippine Sea plate [6]. For simplicity, only dip-slips on the fault plane were considered. To shorter the calculation time, we did not consider strike-slip on the fault plane

3. Afterslip analyses under smoothness, discontinuity, and sparsity constraints

We assessed the accuracy of our proposed method using example data obtained by simulating an earthquake generation cycle based on the rate- and state-dependent friction law [7]. In the numerical simulation, source areas of large earthquakes were approximated to represent a circle of 10 km radius that was assumed to be frictionally heterogeneous. Frictional parameters at the boundary between the seismic source area and its surrounding area were discontinuous and sharp. This resulted in the area of the simulated afterslip enclosed the seismic source of the Mw 6.8 earthquake, and exhibited a ring-shaped afterslip distribution with a sharp boundary [7]. The afterslip depth-profile crossing the center of the seismic source area showed two peaks, and the local minimum corresponded to the large coseismic slip area (Figure 2). In addition, the boundaries were sharp between the locked area (zero and nearly-zero-slip within the afterslip area) and the afterslip area, and between the afterslip area and the outlying areas.

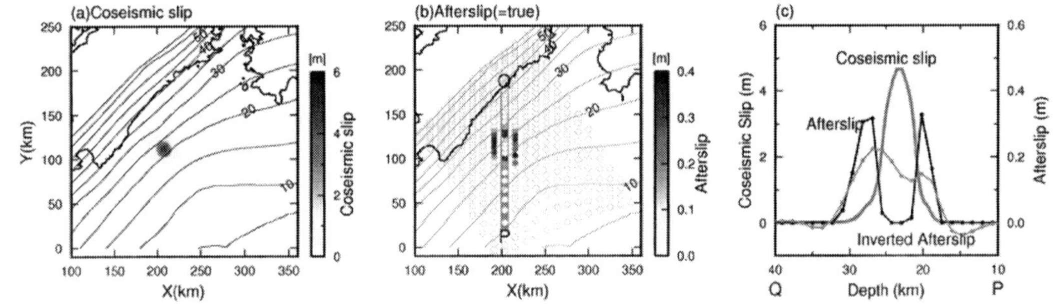

Figure 2. Slip distribution of the simulated coseismic slip (a) and afterslip (b) of a Mw6.8 earthquake. (c) Depth profile along X = 205 km (gray vertical shadow PQ in Figure 2b). Gray line and the left axis show coseismic slip (Figure 2a). Black line with dots and the right axis show afterslip (true slip, Figure 2b). Blue line shows afterslip estimated from previous inversion analysis [8].

3.1. Mehtods

We proposed a new evaluation function by incorporating three constraints: smoothness, discontinuity, and sparsity (SDS) constraints as prior information for inversion (Figure 3) [4]. The new evaluation function can be expressed as:

$$E(s;\alpha',\beta',\nu) = \frac{\beta'}{2}\sum_{k=1}^{K}\left(d_k - \sum_{l=1}^{N} G_{kl}s_l\right)^2 + \frac{\alpha'}{2}\sum_{i\sim j}(1-\delta'(s_i,s_j))(s_i-s_j)^2 + \nu\sum_{l=1}^{N}|s_l| \quad (2)$$

where β' is a precision hyperparameter, α' is a smoothness hyperparameter for the non-zero-slip area, ν is a sparsity hyperparameter. $\sum_{i\sim j}$ is the summation of all pairs of neighboring cells. $\delta'(s_i,s_j)$ behaves like a delta function, and is defined as follows:

$$\delta'(s_i,s_j) = \begin{cases} 1, & \text{if } s_i = 0 \text{ or } s_j = 0 \\ 0, & \text{if } s_i \neq 0 \text{ and } s_j \neq 0 \end{cases} \quad (3)$$

The proposed evaluation function in equation (2) consists of three terms: the first term on the right-hand side is the reproducibility between model parameters and observations, and the second term constrains the smoothness and discontinuity of model parameters ($\boldsymbol{s}$). For the neighboring pair, if there is zero slip in both cells, or if either of the neighboring cells is zero, then the function $\delta'(s_i, s_j)$ becomes equal to one; this results in the removal of the smoothness constraint term. The slip of both the neighboring cells must be non-zero for the smoothness constraint to be effective. The third term on the right-hand side of equation (2) represents the sparsity constraint.

In this case, K is equal to 531 as there are three components for 177 stations (note that this is not an underdetermined problem, but there are some low quality data from stations far from the afterslip area), and N is equal to 436, as dip-slip only occurs for 436 subfaults.

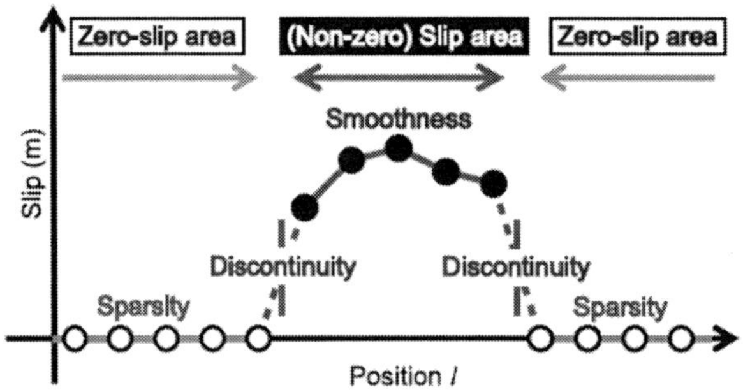

Figure 3. Schematic diagram for constraints of slip distribution [4].

3.2. *Results and Discussions*

Using the smoothness regularization, the estimated spatial distribution showed a wider slip area with a smoother boundary compared to the true slip distribution. By using the smoothness constraints, it was not possible to resolve the two peaks seen in the depth profile of the true slip distribution (Figure 4a). The estimated maximum slip was approximately 45% less than the true value. Local minimum slip within the slip area was significantly larger than the true values that are zero, or are approximately at the zero level. In addition, the boundary between the slip area and the outside area is too smooth, and small negative slips are falsely identified around the boundary. These discrepancies between the estimations and the true value result from excessive homogenization of the original heterogeneity of the slip area, which in turn results from the smooth regularization.

For the sparsity constraint only, the spatial distribution estimated by the sparsity regularization showed a jagged slip area compared to the true image (Figure 4b). However, the main slip area was roughly similar to the true slip distribution, when compared with the smoothness estimation. Several subfaults with non-zero-slip true values were mistakenly estimated as zero-slip subfaults and vice versa. In the depth profile, the sparsity estimation successfully resolved the two peaks present in the true slip distribution, and the coseismic slip area was successfully estimated to be almost zero-slip. However, the estimated slip amplitudes were less accurate. These characteristics are attributed to the sparsity constraint greatly amplifying the original heterogeneity.

The spatial distribution estimated using the SDS constraints exhibited a slip area of similar shape and extent to that of the true image (Figure 4c). The two peaks present in the depth profile were much more clearly resolved compared to those obtained via smoothness regularization. Slip in the coseismic slip area was estimated to be a non-zero value, which was smaller than that estimated through smoothness regularization. The maximum slip values of both peaks were approximately equal to the true values.

We also conducted the inversion test with a smoothly varying circular slip distribution. The SDS constraints accurately reproduced the slip values and the smooth distribution of the slip area.

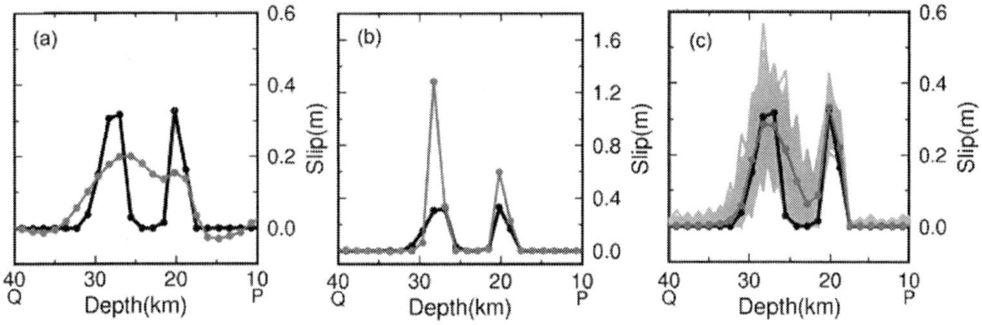

Figure 4. Slip distributions on the plate interface estimated from synthetic displacement data of ring-shaped slip distribution. (a) Depth profile along X = 205 km obtained using the smoothness constraint: red and black lines with dots represent results using smoothness constraint and true slip, respectively. (b) As for (a), but using sparsity constraint only. (c) As for (a), but using SDS constraints. Red line shows PM solution of the MCMC calculations; gray line shows the top 5% of solutions that minimize the evaluation function (equation 2).

4. Slow slip events using the Fused regularization

We estimated spatial distribution of the L-SSEs beneath the Bungo channel in southwest Japan. L-SSEs with durations of 0.5–5 years and recurrence interval of > 5 years occur within this gap of the Nankai subduction zone. L-SSEs in this region occurred repeatedly around 1997, 2003, and 2010. The total slip distributions of these three L-SSEs estimated by geodetic data have been smoothed to some extent due to prior constraints on inversion analyses [9]. The slip areas of these three events are estimated to nearly coincide, filling a spatial gap between deep ETS and seismogenic zones.

The upper (up-dip) boundary of the L-SSEs area was located close to the sources of two M > 7 interplate earthquakes (the 1946 M 8.0 Nankai earthquake and the 1968 M 7.5 Hyuga-nada earthquake). This implies partitioning between seismogenic and aseismic slip areas as highlighted from slow slip events [10]. The lower (down-dip) boundary of the L-SSEs area partly overlaps with the source areas of deep tremors. Temporal correspondence has been reported between the occurrence of L-SSEs and the acceleration of deep tremors beneath the Bungo Channel [11].

4.1. Method

Using generalised fused regularization [12], which is a kind of sparse modelling suitable for detecting discontinuous changes in the model parameters, the evaluation function can be expressed as:

$$\hat{s}^* = \arg\min\left\{\|\boldsymbol{d} - G\boldsymbol{s}\|_2^2 + \lambda \sum_{i\sim j}|s_i - s_j| + \gamma \cdot \lambda \|\boldsymbol{s}\|_1\right\} \tag{4}$$

where $\lambda \geq 0$ is a regularisation parameter for fused smoothness and $\gamma \geq 0$ is a hyperparameter to control the ratio between the fusion and the sparsity penalty terms. A large value of $\gamma \cdot \lambda$ decreases the non-zero components, while $\gamma \cdot \lambda = 0$ introduces no sparsity. The second term on the right-hand side of equation (4) represents the fused regularisation term, which indicates the smoothness and flatness of model parameters (s). The third term represents the sparsity constraint. Optimal solutions were obtained using a set of hyperparameters with minimum residuals via leave-one-out (LOO) cross validation [13].

We used digital data for the observed vertical and horizontal displacements of the 1997, 2003, and 2010 L-SSEs, which were in the Supporting Information of Yoshioka et al. [9]. Crustal displacements were observed at the positions of GNSS stations. In all, we used 33, 65, and 106 continuous GNSS stations as the number of stations increased over time. We used only the total displacement of each component during periods from 1996.7 to 1998.5, 2001.9 to 2004.5, and 2009.5 to 2011.2. In this study, we focused only on the spatial distribution to estimate the detailed distribution of the total slips during each L-SSE event. N is equal to 781 and the dip-slip only occurs for 781 subfaults.

4.2. Result and Discussions

The slip distribution on the plate interface was estimated using fused regularisation. For the 2010 L-SSEs, slips > 0.2 m were concentrated to a relatively narrow area, and slips < 0.2 m were mainly distributed on the deeper side (Figure 5a). There were no slips toward the shallower area (Figure 5a). The cross section of the estimated spatial distribution of the slips along line AB showed a clear step at the down-dip side and a steep boundary at the up-dip side (Figure 5b, red line). The 1997 and 2003 L-SSEs showed similar distributions to that of the 2010 L-SSE. However, due to fewer observation stations, the spatial resolutions were lower than that for 2010. Large (> 0.2 m) slips were estimated to have occurred at almost the same locations identified by Yoshioka et al. [9]. However, the slip distributions at the up-dip side significantly differed from the previous study [9].

The unique elements of our results are the up-dip boundary of the L-SSE area and the step-like distribution of the down-dip side. The L-SSE areas are clearly separated into two parts [the major (minor) slip area at the up-dip (down-dip) side] with two sharp boundaries (Figures 6 and 7). (i) The shallow boundary between the seismogenic zone and the major L-SSE area corresponds to the isotherm at 350 °C [14]. (ii) The deep boundary between the major L-SSE area and the minor L-SSE area corresponds to the up-dip limit of the tremor area. The sharp variation in the slip amount of L-SSEs and the correspondence of the boundaries with up-dip limit of the tremor area suggest that frictional properties, which control slip behaviour, may change sharply.

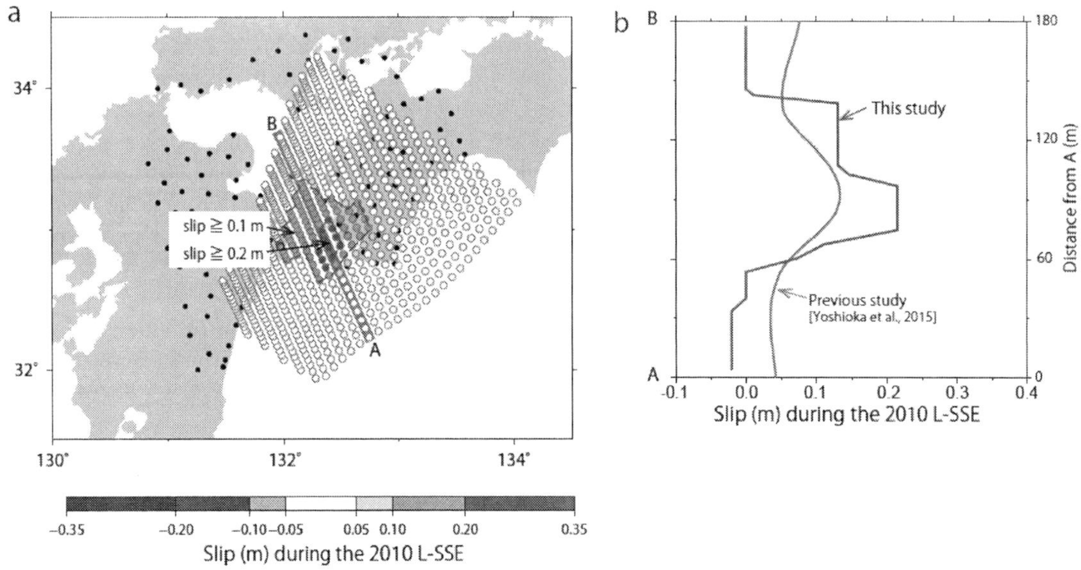

Figure 5. (a) Estimated slip distribution in the dip direction for the 2010 event. Black dots indicate GEONET stations. (b) Slip distribution along line AB. Red represent the 2010 event. Gray line represents the 2010 event estimated by smoothed inversion [9].

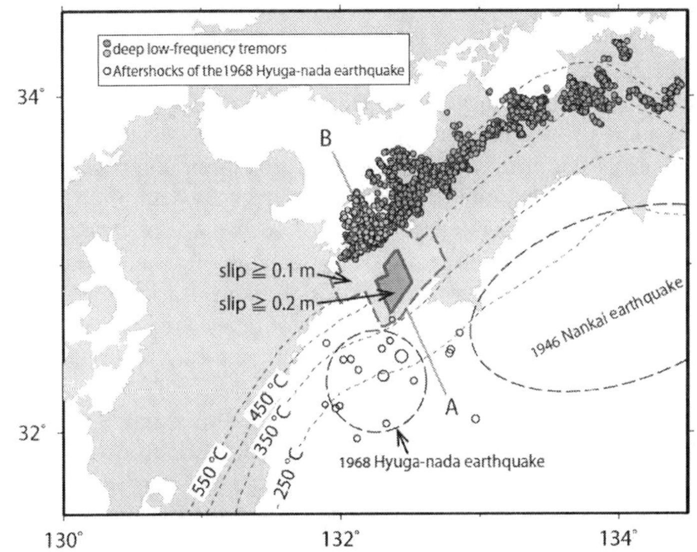

Figure 6. Comparison of L-SSE slip distribution in 2010 (red polygon with the solid and dashed lines) estimated in this study and spatial distribution based on findings from other seismological studies. Black dashed contours show the temperature (250–550 ° C) on the Philippine Sea plate [14].

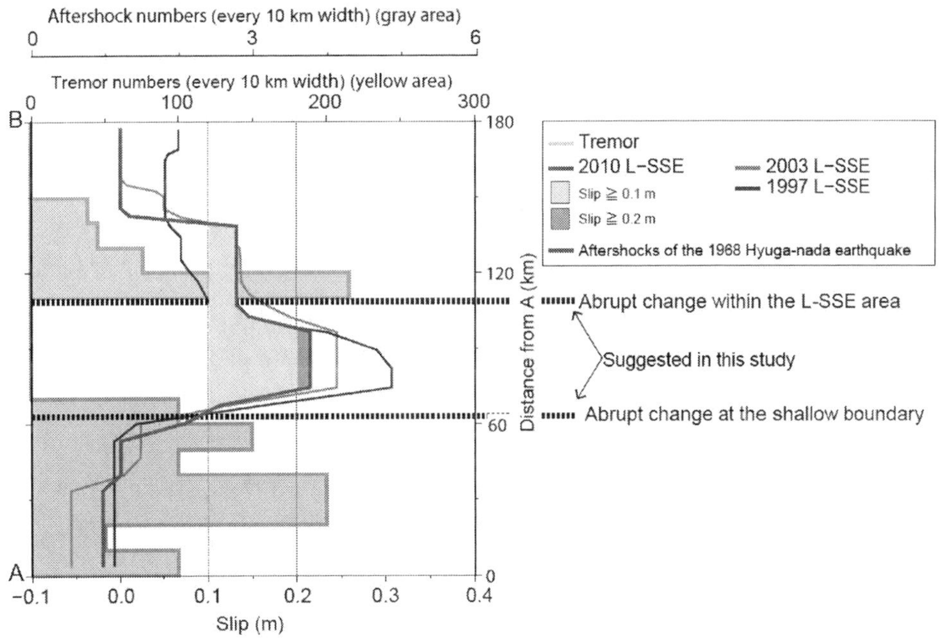

Figure 7. Blue, green, and red lines represent slip distributions of the 1997, 2003, and 2010 L-SSEs along line AB (as in Figure 5b). Yellow line shows the total tremor numbers in 10-km increments along line AB. Gray line shows the total number of aftershocks for the 1968 Hyuga-nada earthquake in 10-km increments along line AB.

5. Conclusions and future work

We demonstrated that the spatial distribution of both ring-shaped and smoothly varying circular slip distributions can be reproduced by the one evaluation function as shown in (2). That function can constrain simultaneously and properly smoothness, discontinuity, and sparsity depending on given data. The slip distribution that minimized the evaluation function better reproduced images in terms of both area and amplitude. Using the SDS constraints proposed here, we were able to avoid the difficulty caused by underdetermined problems, and clearly resolve the small heterogeneity with a discontinuous boundary, even if the grid size was not sufficiently small.

By using SDS constraints, the discontinuous boundary between the heterogeneous area and the surrounding area could be clearly visualized without any false slips at the plate edge. The heterogeneous distribution of the slip controls the processes of stress accumulation and release on a plate interface. In particular, a discontinuous boundary indicates areas of stress concentration, which lead to initiation of rupture. Further, it is critical to understand the time of occurrence of earthquakes, the rupture patterns, and nucleation processes, and to forecast slip transition in the future. The distribution of frictional properties on a plate interface were reflected in the slip distribution. Future work should aim to obtain precise slip distribution on the plate interface from measured observational geodetic data. More precise estimates of slip distribution may be helpful in monitoring plate coupling, and for making improvements in our forward simulation model.

On the other hand, using the fused regularization, we found that the largest slip abruptly becomes zero at the down-dip limit of the seismogenic zone, and is immediately reduced to half at the up-dip limit of the deep low-frequency tremors, and becomes zero near its down-dip limit. Such correspondences imply that some thresholds exist in the generation processes for both tremors and SSEs. It suggests that geodetic data inversion with sparse modelling can detect such abrupt changes in the slip distribution. We can use this method to monitor the position of the abrupt changes, which represents the stress concentration level at the seismogenic zone.

Furthermore, it should be noted that sparse modelling can examine the extent that sharp edges, steps and small slip zones can fit the geodetic data, which only showed a smooth distribution in previous studies. Our results indicate that the resolution of the position of sharp edges estimated from geodetic data analysed with sparse modelling is comparable to the spatial resolution of the hypocentre distribution determined from seismic data. Such independent datasets of the spatial distribution are useful to hypothesise about the physical processes of both megathrust and slow earthquakes. Furthermore, a method based on the generalised fused regularisation should be available for any slip and slip-deficit distribution analyses on various types of faults, promoting the consideration of physical processes.

Both proposed inversion methods are applicable not only to geodetic inversions, but also to other geophysical and geochemical inversions such as seismic tomography, electromagnetic tomography, and pressure–temperature inversion of metamorphic rocks. Using our methods, it is possible to obtain images that are clearer, sharper, and show appropriate smoothness. We will develop more efficient and high-speed methodology in order to expand the applicability of the proposed methods.

Acknowledgments

This work was partly supported by the "Initiative for High-Dimensional Data-Driven Science through Deepening of Sparse Modeling" by JSPS KAKENHI (Grant Numbers JP16H01564, 25120005, 25120009, and 25120011) and JSPS KAKENHI (Grant Numbers JP16K17798, 16K16108, 25280090, and 15K20864) and JST PRESTO (Grant Number JPMJPR1676). The Figures and graphs are created by Generic Mapping Tools software (GMT v4.5.12; http://gmt.soest.hawaii.edu/) [15].

References

[1] Yabuki, T. and M. Matsu'ura (1992), Geodetic data inversion using a Bayesian information criterion for spatial distribution of fault slip. Geophys J. Int. 109, 363-375.

[2] Segall, P. and M. Matthews (1997), Time dependent inversion of geodetic data. J. Geophys. Res. 102(B10), 22391-22409.

[3] Igarashi, Y. et al. (2016), Three levels of data-driven science. Journal of Physics Conference Series 699, 12001-1-012001-13.

[4] Nakata R., T. Kuwatani, M. Okada, and T. Hori (2016), Geodetic inversion for spatial

distribution of slip under smoothness, discontinuity, and sparsity constraints, Earth Planets Space, 68:20, doi: 10.1186/s40623-016-0396-2.

[5] Nakata, R., H. Hino, T. Kuwatani, S. Yoshioka, M. Okada, and T. Hori (2017), Discontinuous boundaries of slow slip events beneath the Bungo Channel, southwest Japan, Sci. Rep. 7, doi: 10.1038/s41598-017-06185-0.

[6] Baba, T, Y. Tanioka, P. R. Cummins, and K. Uhira (2002), The slip distribution of the 1946 Nankai earthquake estimated from tsunami inversion using a new plate model. Phys Earth Planet Inter 132, 59-73.

[7] Nakata R., M. Hyodo, and T. Hori (2012), Numerical simulation of afterslips and slow slip events that occurred in the same area in Hyuga-nada of southwest Japan, Geophys. J. Int., 109, 1213-1220.

[8] Nakata R., S. Miyazaki, M. Hyodo, and T. Hori (2014), Reproducibility of spatial and temporal distribution of aseismic slips in Hyuga-nada of southwest Japan, Mar. Geophys. Res., 35, 311-317.

[9] Yoshioka, S., Y. Matsuoka, and S. Ide (2015), Spatiotemporal slip distributions of three long-term slow slip events beneath the Bungo Channel, southwest Japan, inferred from inversion analyses of GPS data, Geophys. J. Int., 201, 1437-1455.

[10] Yagi, Y. and M. Kikuchi (2003), Partitioning between seismogenic and aseismic slip as highlighted from slow slip events in Hyuga-nada, Japan. Geophys. Res. Lett. 30(2), 1087, doi:10.1029/2002GL015664.

[11] Hirose, H. et al. (2010), Slow Earthquakes Linked Along Dip in the Nankai Subduction Zone. Science 330, doi:10.1126/science.1197102.

[12] Tibshirani, R., M. Saunders, S. Rosset, J. Zhu and K. Knight (2005), Sparsity and smoothness via the fused lasso, J. Royal. Statist. Soc. B., 67, 91-108.

[13] Bishop, C. M. (2006), Pattern recognition and machine learning. Springer, New York.

[14] Ji, Y., S. Yoshioka, and T. Matsumoto (2016), Three-dimensional numerical modeling of temperature and mantle flow fields associated with subduction of the Philippine Sea plate, southwest Japan. J. Geophys. Res. Solid Earth 121, doi:10.1002/2016JB012912.

[15] Wessel, P. and W. H. F. Smith (1998), New, improved version of Generic Mapping Tools released. Eos, Trans Am Geophys Union 79, 579-579, doi:10.1029/98EO00426.

IOP Conf. Series: Journal of Physics: Conf. Series **1036** (2018) 012019 doi:10.1088/1742-6596/1036/1/012019

Covariance Modeling in Applications of Data Assimilation to High-dimensional Earth and Geospace Systems

Tomoko Matsuo

Ann and H.J. Smead Department of Aerospace Engineering Sciences, University of Colorado Boulder, UCB 429, Boulder, CO 80309-0429, USA

E-mail: tomoko.matsuo@colorado.edu

Abstract. Data assimilation and inverse methods play a key role in integrating remote-sensing and in-situ Earth and Geospace observations into models of the Earth and Geospace system and subsystems, enabling weather prediction and climate projection of high societal relevance. The problem presents considerable methodological and computational challenges because of the high-dimensionality and non-linearity of the underling dynamics of the Earth and Geospace system and the large volumes of heterogeneous observations from various observing platforms, including Low-Earth-Orbit and geosynchronous-Earth-orbit satellites as well as ground-based and airborne platforms. To make the problem tractable, most data assimilation methods resort to scalable methods derived from well-established Gaussian process-based formulation of the sequential Bayesian estimation problem. One such approach is to use approximate Kalman filters that rely on low-rank approximations of the covariance matrices. In ensemble data assimilation, the covariance is expressed in the subspace spanned by the model ensemble that evolves dynamically. Though the low-rank ensemble-based approximation allows the covariance to track non-linear dynamics with a moderate computational cost, the impact of the observations that lie outside of the subspace represented by the ensemble cannot be effectively incorporated into the assimilation analysis. One heuristic remedy widely used in numerical weather prediction is to adopt a hybrid method that uses the full-rank stationary covariance and the ensemble-based covariance. As a scalable approach to modeling the full-rank nonstationary covariance, the use of a multiresolution-based covariance is discussed.

1. Introduction

From a statistical point of view, the process of data assimilation can be viewed as sequential Bayesian estimation of geophysical random processes in which the most likely states of the system are continuously updated as observations of time-evolving states become available sequentially over time. The states of our interest, denoted by a multivariate random vector $\mathbf{x} \in \mathbb{R}^n$, are usually observed incompletely in space and time, and the relationship of observations $\mathbf{y} \in \mathbb{R}^p$ to $\mathbf{x}$ at a given time t is characterized as

$$\mathbf{y}_t = \mathcal{H}(\mathbf{x}_t) + \epsilon_t \ , \tag{1}$$

where $\mathcal{H} : \mathbb{R}^n \rightarrow \mathbb{R}^p$ represents a non-linear forward operator and $\epsilon_t \in \mathbb{R}^p$ is a vector of the observational errors. The process of inferring $\mathbf{x}$ from $\mathbf{y}$ requires inverting the operation represented by $\mathcal{H}$ as well as incorporating the prior information provided by the model of the Earth and Geospace system or subsystems into the inference of $\mathbf{x}$. The relationship of $\mathbf{x}$ to its prediction by a forecast model is given as

$$\mathbf{x}_t = \mathbf{x}_t^f + \boldsymbol{\eta}_t \,, \tag{2}$$

where $\boldsymbol{\eta}_t \in \mathbb{R}^n$ is a vector of the forecast errors. The evolution of the state $\mathbf{x}$ from time $t-1$ to t can be expressed with a non-linear operator $\mathcal{M} : \mathbb{R}^n \to \mathbb{R}^n$ as

$$\mathbf{x}_t^f = \mathcal{M}(\mathbf{x}_{t-1}^a) + \gamma_t \,, \tag{3}$$

where $\gamma_t \in \mathbb{R}^n$ is a vector of the model errors.

In the application under consideration, in which the size of $\mathbf{x}$ can be as large as 10^8 and the size of $\mathbf{y}$ as large as 10^7, most data assimilation methods end up resorting to scalable methods based on well-established Gaussian processed-based formulation of the sequential Bayesian estimation problem to make the problem tractable [1]. The prior distribution of $\mathbf{x}$ is therefore given as $\mathbf{x} \sim \mathcal{N}(\mathbf{x}^f, \mathbf{P}^f)$ where $\mathbf{P}^f \in \mathbb{R}^{n\times n}$ is the covariance matrix of $\boldsymbol{\eta}$. The likelihood of $\mathbf{y}$ conditioned on $\mathbf{x}$ is $\mathbf{y}|\mathbf{x} \sim \mathcal{N}(\mathcal{H}(\mathbf{x}), \mathbf{R})$ where $\mathbf{R} \in \mathbb{R}^{p\times p}$ is the covariance of $\boldsymbol{\epsilon}$. Under the assumption that $\mathcal{H}$ can be linear or linearized as $\mathbf{H} = \left.\frac{\partial \mathcal{H}}{\partial \mathbf{x}}\right|_{\mathbf{x}=\mathbf{x}^f}$, the posterior distribution is given as $\mathbf{x}^a \sim \mathcal{N}(\mathbf{x}^a, \mathbf{P}^a)$, where

$$\begin{aligned} \mathbf{x}^a &= \mathbf{x}^f + \frac{\mathbf{P}^f\mathbf{H}^T}{\mathbf{H}\mathbf{P}^f\mathbf{H}^T + \mathbf{R}}(\mathbf{y} - \mathbf{H}\mathbf{x}^f) \\ &= \mathbf{x}^f + \mathbf{K}(\mathbf{y} - \mathbf{H}\mathbf{x}^f) \,, \qquad (4) \\ \mathbf{P}^a &= (\mathbf{I} - \mathbf{K}\mathbf{H})\mathbf{P}^f \,. \qquad (5) \end{aligned}$$

These constitute well-known formula for the Kalman filter state and covariance update. Under the assumption of linear forward and forecast models, the Kalman filter produces optimal analysis. The Kalman filter is however impractical for high-dimensional systems of the order of $O(n) = 10^6 - 10^8$, because it requires handling matrices of dimension $n \times n$. A variety of approximate Kalman filters have been developed for use in high-dimensional systems, and all of these approximate Kalman filters rely on *low-rank* approximations of the covariance matrices and yields sub-optimal analysis.

2. Reduce-rank ensemble-based approach

The reduced-rank methods approximate the Kalman filter by restricting the representation of covariances to a smaller subspace spanned by q eigenvectors. For instance, the eigenvalue decomposition of a $n \times n$ covariance matrix of rank q can be represented as, $\mathbf{P} = \mathbf{V}\boldsymbol{\Gamma}\mathbf{V}^T \approx \mathbf{V}_q\boldsymbol{\Gamma}_q\mathbf{V}_q^T$, where $\mathbf{V}_q$ is a $n \times q$ matrix that store q eigenvectors. This section shows that the ensemble Kalman filter is a type of the low-rank Kalman filter.

Suppose that $\mathbf{P}$ has rank $q << n$, and that the covariance can be expressed as

$$\mathbf{P} = \mathcal{D}\mathcal{D}^T \,, \tag{6}$$

where $\mathcal{D} \in \mathbb{R}^{n\times q}$. The Kalman Gain $\mathbf{K}$ can then be re-written as

$$\begin{aligned} \mathbf{K} &= \mathbf{P}^f\mathbf{H}^T(\mathbf{R} + \mathbf{H}\mathbf{P}^f\mathbf{H}^T)^{-1} \\ &= \mathcal{D}\mathcal{D}^T\mathbf{H}^T(\mathbf{R} + \mathbf{H}\mathcal{D}\mathcal{D}^T\mathbf{H}^T)^{-1} \\ &= \mathcal{D}(\mathbf{H}\mathcal{D})^T(\mathbf{R} + \mathbf{H}\mathcal{D}(\mathbf{H}\mathcal{D})^T)^{-1} \,. \qquad (7) \end{aligned}$$

The state update is given as

$$\begin{aligned} \mathbf{x}^a &= \mathbf{x}^f + \mathbf{K}(\mathbf{y} - \mathbf{H}\mathbf{x}^f) \\ &= \mathbf{x}^f + \mathbf{P}^f\mathbf{H}^T(\mathbf{R} + \mathbf{H}\mathbf{P}^f\mathbf{H}^T)^{-1}(\mathbf{y} - \mathbf{H}\mathbf{x}^f) \\ &= \mathbf{x}^f + \mathcal{D}(\mathbf{H}\mathcal{D})^T(\mathbf{R} + \mathbf{H}\mathcal{D}(\mathbf{H}\mathcal{D})^T)^{-1}(\mathbf{y} - \mathbf{H}\mathbf{x}^f) \,. \end{aligned} \tag{8}$$

The analysis increment is a linear combination of the columns of $\mathcal{D}$, and is thus confined to the subspace of dimension q.

The ensemble Kalman filters are reduced-rank Kalman filters, with their covariance approximated as the sample covariance C.

$$\mathbf{P} \approx C = \frac{1}{m-1}\mathcal{G}\mathcal{G}^T \,, \tag{9}$$

where m is the size of the ensemble and $\mathcal{G} \in \mathbb{R}^{n \times m}$ is a matrix with each column filled with a member of the ensemble of state vector $\mathbf{x}$ as

$$\mathcal{G} = \left[\ (\mathbf{x}^{(1)} - \boldsymbol{\mu}) \ \ (\mathbf{x}^{(2)} - \boldsymbol{\mu}) \ \ \cdots \ \ (\mathbf{x}^{(m)} - \boldsymbol{\mu}) \ \right] \,, \tag{10}$$

where $\mathbf{x}^{(m)}$ denotes the mth-member of $\mathbf{x}$ and $\boldsymbol{\mu} \in \mathbb{R}^n$ is the ensemble (sample) mean state. Note that the rank of C is $m - 1$ at most. Usually $m \ll n$, so $\mathbf{P}$ used in EnKF is rank deficient. For $m \ll n$ case, C is not uniquely determined. In actual implementation of EnKFs, the full matrix C is never explicitly computed and Equation (8) is not used to update the forecast ensemble members [2, 3]. The update step can be applied to each ensemble member in a stochastic or deterministic fashion to make the sample statistics consistent with the analysis update of the state and covariance stated in Equations (4) and (5). Deterministic methods have the advantage of avoiding sampling issues associated with the use of "perturbed observations" in stochastic analysis update methods. Though the low-rank ensemble-based approximation allows the covariance to track non-linear dynamics with a moderate computational cost, the impact of the observations that lie outside of the subspace represented by the ensemble cannot be effectively incorporated into the assimilation analysis. One heuristic remedy widely used in numerical weather prediction is to adopt a hybrid method that uses the full-rank stationary covariance and the ensemble-based covariance.

3. Full-rank wavelet-based approach

Modeling and estimation of a full-rank nonstationary covariance for high-dimensional systems pose computational and methodological challenges. For instance, representation of covariance with multiresolution compactly-supported basis functions such as wavelets offers a practical approach scalable to high-dimensional problems [4], and compliments the reduce-rank ensemble-based approach described in the previous section.

Main advantages of the wavelet-based approach are its flexibility in representing departures from stationarity, and sparsity in the covariance matrix among wavelet basis coefficients. Suppose that $\mathbf{x}$ is composed of one type of the geophysical state on n regular model grid locations and $\mathcal{W} \in \mathbb{R}^{n \times n}$ is a matrix where n wavelet basis functions are evaluated at n locations. If $\mathcal{W}$ has full rank it is always possible that the covariance matrix $\mathbf{P}$ can be decomposed as

$$\mathbf{P} = \mathcal{W}GG^T\mathcal{W}^T \,, \tag{11}$$

where $G \in \mathbb{R}^{n \times n}$ is a real nonsingular symmetric matrix, and GG^T is the covariance of wavelet coefficients. Because of the compact support of wavelet basis functions, most of the scaling and coarse-scale wavelet functions are uncorrelated, justifying a sparse approximation of the wavelet coefficient covariance, as well as G. As shown in [4] for wavelet coefficients of stationary processes, the covariance among coefficients decays geometrically with the scale of the coefficients where the rate of decay depends on the smoothness of the underlying random processes, and therefore the insignificant off-diagonal elements of G can be set to equal zero such that G has $O(n)$ rather than $O(n^2)$ nonzero elements, represented by $\boldsymbol{\theta}$, yet has full rank.

The estimation of covariance parameters $\boldsymbol{\theta}$ may not be trivial especially for high-dimensional problems. One way is to further parameterize $\boldsymbol{\theta}$ with a smaller number of hyper-parameters and estimate them using the maximum-likelihood method. Another strategy is a non-parametric approach that takes an advantage of the scalability of algorithms facilitated by the availability of fast transforms for wavelet bases on regular grids. As demonstrated in [4], the covariance parameters $\boldsymbol{\theta}$ can be estimated using an Monte-Carlo approach that draws on the expectation maximization algorithm. Let's suppose that $\mathbf{x}$ is partly observed, or composed of the observed state $\mathbf{x}_1$ and unobserved state $\mathbf{x}_2$, normally distributed as

$$\mathbf{x} = \begin{pmatrix} \mathbf{x}_1 \\ \mathbf{x}_2 \end{pmatrix} \sim \mathcal{N}(0, \mathbf{P}_\theta) \qquad \text{where } \mathbf{P}_\theta = \mathcal{W}\tilde{G}^2(\boldsymbol{\theta})\mathcal{W}^T. \tag{12}$$

Then $\boldsymbol{\theta}$ can be estimated iteratively by maximizing the following expression [5]:

$$\begin{aligned} Q(\boldsymbol{\theta}, \boldsymbol{\theta}^*) &= E[\mathcal{L}(\mathbf{x}, \boldsymbol{\theta})|\mathbf{x}_1, \boldsymbol{\theta}^*] \\ &\approx \frac{1}{M}\sum_{m=1}^{M} \mathcal{L}\left(\begin{pmatrix} \mathbf{x}_1 \\ \mathbf{x}_2^{(m)} \end{pmatrix}, \boldsymbol{\theta}\right), \end{aligned} \tag{13}$$

where $\boldsymbol{\theta}^*$ denotes an initial guess for the covariance parameters and $\mathcal{L}$ represents a (complete data) log-likelihood function. $\mathbf{x}_2^{(m)}$ is here sampled from $\mathbf{x}_2^{(m)} \sim [\,\mathbf{x}_2|\mathbf{x}_1, \mathcal{W}\tilde{G}^2(\boldsymbol{\theta}^*)\mathcal{W}^T]$. Specific application examples can be found for surface ozone data in [4] and for upper atmosphere plasma data in [6].

4. Summary

Applications of data assimilation and inverse methods to Earth and Geospace system present considerable methodological and computational challenges because of the high-dimensionality and non-linearity of the underling dynamics of the Earth and Geospace system, and the large volumes of heterogeneous observations. To make the problem tractable most data assimilation methods resort to scalable methods derived from well-established Gaussian process-based formulation of the sequential Bayesian estimation problem. A key component of a Gaussian process model is the covariance. The paper presents two scalable approaches to model the nonstationary covariance matrices: low-rank ensemble-based covariance and full-rank wavelet-based covariance.

Acknowledgments
The present study is supported by NASA Award NNX14AI17G.

References

[1] Nichols, N. K. 2010, *Mathematical Concepts of Data Assimilation, Data Assimilation*, eds. W. Lahoz et al., Springer-Verlag Berlin Heidelberg, 13-39.

[2] Evensen, G. 2009, *Data Assimilation: The Ensemble Kalman Filter*, Springer Berlin.

[3] Tippett, M. K., J. L. Anderson, C. H. Bishop, T. M. Hamill, and J. S. Whitraker 2003, *Ensemble Square Root Filters, Mon. Weather Rev.*, 131, 1485-1490.

[4] Matsuo, T., D. W. Nychka, and D. Paul 2011, *Nonstationary covariance modeling for incomplete data: Monte Carlo EM approach, Computational Statistics and Data Analysis*, 55, 2059 2073.

[5] Wei, G.C., and M. A. Tanner 1990, *A Monte Carlo implementation of the EM algorithm and the poor man s data augmentation algorithms*, *Journal of the American Statistical Association*, 85, 699 704.
[6] Sun, Y.-Y., T. Matsuo, E. A. Araujo-Pradere, and J. Y. Liu 2013, *Ground-based GPS observation of SED associated irregularities over CONUS*, *J. Geophys. Res. Space Physics*, 118, 2478-2489.

IOP Conf. Series: Journal of Physics: Conf. Series **1036** (2018) 012020 doi:10.1088/1742-6596/1036/1/012020

Sparse modeling for Quantum Monte-Carlo simulation

Masayuki Ohzeki

Graduate School of Information Sciences, Tohoku University, Sendai, 980-8579, Japan

E-mail: mohzeki@tohoku.ac.jp

Abstract. We show a new kind of applications of the sparse modeling to a traditional problem in the condensed-matter physics. In the quantum Monte-Carlo simulation, we observe a huge amount of data for investigation of the details of the low-energy behavior for interacting many-body systems. Although the real-time behavior is actually under investigation, the quantum Monte-Carlo simulation is performed on the imaginary time for restriction of the method. Thus we need a technique for analytical continuation connecting between the real and imaginary-time functions. However the analytical continuation can be problematic because the problem consists of solving the ill-conditioned equation. In the present study, by employing an adequate regularization, we solve efficiently the ill-conditioned equation in the analytical continuation. As a result, we have a novel way to perform the analytical continuation and find an intermediate representation between imaginary-time and real-frequency domains.

1. Introduction

In the present manuscript, we provide an elementary guide for our recent papers from many-body physics to fascinating applications of the sparse modeling [1, 2]. In the analysis on the condensed matter, it is harmful to directly deal with the theoretical model for investigating some nontrivial aspects. We usually employ various approximate method and numerical calculations. In the quantum many-body systems, most of the analyses are performed with recourse to the imaginary-time framework for computing the real-time behavior. Then one needs to utilize the analytical continuation to transform the estimated quantities in imaginary-time regions to real-frequency regions, which can be compared to the experimental results. The representative method to deal with the quantum many-body systems is the elaborate diagrammatic approach widely used for investigating both of static and dynamics responses of the systems [3, 4]. Variants of the quantum Monte-Carlo simulations (QMC) also take full advantage of imaginary-time descriptions [5, 6]. We consider the case of the QMC in the following.

The analytical continuation can be formulated as the inverse problem. Given the relationship between imaginary-time description $\mathbf{G}$ and the real-frequency behavior $\boldsymbol{\rho}$ as

$$\mathbf{G} = K\boldsymbol{\rho}, \tag{1}$$

we find a reasonable solution of ρ from $\mathbf{G}$ obtained by the QMC. However the inverse problem can be ill-conditioned one since the singular values of K decay very fast. Then most of the independent components in $\boldsymbol{\rho}$ lead to no contributions to $\mathbf{G}$. Furthermore any noise in $\mathbf{G}$ significantly affect $\boldsymbol{\rho}$. In order to obtain a reasonable solution, one often employs the maximum

IOP Conf. Series: Journal of Physics: Conf. Series **1036** (2018) 012020 doi:10.1088/1742-6596/1036/1/012020

entropy method (MaxEnt or MEM) [7, 8, 9]. By regarding the non-negative quantity $\boldsymbol{\rho}$ as the probability distribution, one maximize the entropy of the distribution function, while minimizing the consistency with Equation (1). The inclusion of an entropy term to find a reasonable solution leads to a bias toward spectra with large entropy. In some sense, the obtained solution can be most probable one given the QMC data. However, one do not know a priori how to induce the effect of the entropy in finding the solution. Despite extensive efforts from various aspects, the method remains arbitrariness how to solve the inverse problems.

Recent development of solving the inverse problem is the stochastic way [10, 11, 12]. In the stochastic analytical continuation, the desired quantity $\boldsymbol{\rho}$ is sampled according to the squared error between $\mathbf{G}$ and $K\boldsymbol{\rho}$ and the "temperature" controlling the variance of the sampling as the simulated annealing [13].

On the other hand, to solve the inverse problem on the QMC, we propose the application of the sparse modeling. In particular, we employ the procedure of the least-absolute value shrinkage and selection operators (LASSO) by considering the least square estimation with regularization of the L_1 norm. The L_1 is a core technology in the sparse modeling. The L_1 norm promotes the sparse solution compared to the case of the simple least square estimation. For instance, the relevant interaction from many snapshots of a vast number of components can be chosen by employing the L_1 norm as well as the standard cost function to analyze the data [14, 15, 16].

Here we assume that the relevant information, which can be obtained by the QMC data, should be sparse in the analytical continuation. The straightforward approach by use of the LASSO implies that some sparsity on the desired quantity can be expected. We do not stick this idea for the QMC data. Instead, we aim at efficiently cut down the effects from the noisy QMC data to the desired quantity by considering the feature of the Kernel matrix K.

The remaining of the paper consists of the following five sections. The second section shows the formulation of the Matsubara Green function for investigating the quantum many body systems. In the third section, we formulate the inverse problem to be solved and review the MEM in short. The fourth section explains the application of the sparse modeling to the QMC data following the literature [1]. We show how to solve the inverse problem in the fifth section. In the last section, we summarize our study.

2. Matsubara Green function

We first review the Matsubara Green function, which is a central quantity to be investigated in quantum many-body physics. In quantum many-body physics, one of the target is to estimate the density of states $\rho(\omega)$. From the density of states, we can compute the interested quantities as the internal energy, specific heat, thermal conductivity etc. of the system. In order to obtain the density of states, we compute the Matsubara Green function defined by the following thermal expectation:

$$G(\tau) = -\left\langle \mathrm{T}\hat{a}(\tau)\hat{a}^\dagger(0)\right\rangle, \tag{2}$$

where $\hat{a}(t)$ and $\hat{a}^\dagger(t)$ are the annihilation and creation operators of the particle in the system and the bracket denotes the thermal expectation defined as

$$\langle \hat{A}\rangle = \mathrm{Tr}\left\{\hat{\rho}\hat{A}\right\}. \tag{3}$$

The density operator $\hat{\rho}$ is

$$\hat{\rho} = \frac{1}{Z}\exp\left(-\beta\hat{H}\right). \tag{4}$$

where $\beta = 1/T$ is the inverse temperature and the symbol T denotes the time-ordered product. The quantum Monte-Carlo simulation is one of the standard tools to compute the Matsubara

Green function. The problem is how to reach the density of states from the observations in the quantum Monte-Carlo simulation.

Let us consider the Fourier transformation of the Matsubara Green function $G(\tau)$. Then we impose the aperiodic condition for the fermionic case and the periodic one for the bosonic one, respectively.

$$G(\tau) = \frac{1}{\beta}\sum_l \tilde{G}(i\omega_l)\exp(-i\omega_l\tau), \tag{5}$$

where ω_l is the Matsubara frequency $i(2l+1)\pi/\beta$ for the fermion case and $i2l\pi/\beta$. The density of states can be given by the imaginary components of the Fourier transformed Matsubara Green function as

$$\rho(\omega) = -\frac{1}{\pi}\mathrm{Im}\tilde{G}(\omega) \tag{6}$$

The residue theorem can reduce the above equality to

$$G(\tau) = \frac{1}{2\pi i}\oint_C dz \frac{\exp(-\tau z)}{1 \pm \exp(-\beta z)}\tilde{G}(z). \tag{7}$$

Here we perform the analytical continuation for $\tilde{G}(i\omega_l) \to \tilde{G}(z)$ and $+/-$ denotes the fermionic/bosonic case . In addition, following the definition of the Matsubara Green function, we find the relationship with the density of states as, by use of the Cauchy principal value,

$$\tilde{G}(z) = \int_{-\infty}^{\infty} d\omega \frac{\rho(\omega)}{z-\omega}. \tag{8}$$

When we derived the above relationship between the (Fourier-transformed) Matsubara Green function and the density of states, we imposed the causality of the dynamics of the particles created at 0 until annihilation at τ. Therefore

$$G(\tau) = \frac{1}{2\pi i}\int_{-\infty}^{\infty} d\omega \oint_C dz \frac{\exp(-\tau z)}{1 \pm \exp(-\beta z)}\frac{\rho(\omega)}{z-\omega}. \tag{9}$$

The integrand has the singular values at $z = \omega$ and $\exp(-\beta z) = \mp 1$, which leads to $z = i\pi(2l+1)/\beta$ for the fermionic case and $z = i2l\pi/\beta$ for the bosonic case, respectively. We take the singular values along the real axis and obtain

$$G(\tau) = \int_{-\infty}^{\infty} d\omega \frac{\exp(-\tau\omega)}{1 + \exp(-\beta\omega)}\rho(\omega). \tag{10}$$

for the fermionic case. On the other hand,

$$G(\tau) = \int_{-\infty}^{\infty} d\omega \frac{\exp(-\tau\omega)}{1 - \exp(-\beta\omega)}\rho(\omega). \tag{11}$$

for the bosonic case. As a consequence, we relate the density of states through the integral transformation to the Matsubara Green function.

3. Standard method: Maximum Entropy Method

The inverse problem to be solved in our problem is as follows:

$$G(\tau) = \int_{-\infty}^{\infty} d\omega K_{\pm}(\tau,\omega)\rho(\omega), \tag{12}$$

where $0 \leq \tau \leq \beta$ and $K(\omega)$ is the kernel, which is defined as

$$K_{\pm}(\tau, \omega) = \frac{\exp(-\tau\omega)}{1 \pm \exp(-\beta\omega)}. \tag{13}$$

Here, the $\pm$ stands for the fermionic case by $+$ and bosonic one by $-$. The analytical continuation problem is the inverse problem to obtain $\rho(\omega)$ by solving the above equation from the observations $G(\tau)$, which can be obtained by the quantum Monte-Carlo simulation. However the observation $G(\tau)$ suffers from the presence of the noise in the statistical error in the quantum Monte-Carlo simulation. In addition, the kernel decays exponentially in a region of the large absolute value of ω. Thus the quantity on the right-hand side of Equation (12) can be insenstive to variations of $\rho(\omega)$. As a consequnce, we have a plenty of number of plausible solutions satisfying Equation (12) within a given accuracy.

The maximum entropy method is a standard tool for solving the above ill-conditioned problem. In the following, let us deal with the discrete version of Equation (12) as in Equation (1), where $\mathbf{G} = \{G(\tau_l)\}$, $\boldsymbol{\rho} = \{\rho(\omega_l)\}$ and $(K)_{kl} = K(\tau_k, \omega_l)$. In addition, we define the squared error as

$$\chi^2(\mathbf{G}, \boldsymbol{\rho}) = \frac{1}{\sigma^2} \|\mathbf{G} - K\boldsymbol{\rho}\|_2^2 \tag{14}$$

where σ^2 is the variance of the noise in the observation, which is related to the statistical error in the quantum Monte-Carlo simulation. Hereafter we consider the case $\sigma^2 = 1$ for simplicity. The straightforward manipulation of the minimization of χ^2 yields a consistent results satisfying Equation (1). The resultant solution is not expected to be a real answer because we have enormous candidates of the solution if we do not employ any regularization term. To find a reasonable solution, we employ the Bayesian inference. In the Bayesian inference, we define a certain probabilistic model and select a prior distribution to estimate the desired quantity. We here construct the following probabilistic model. The likelihood function for the density of states given the QMC data is defined by the Gaussian distribution as

$$P(\mathbf{G}|\boldsymbol{\rho}) \propto \exp\left(-\frac{1}{2}\chi^2(\mathbf{G}, \boldsymbol{\rho})\right) \tag{15}$$

Second we set the prior distribution as

$$P(\boldsymbol{\rho}) \propto \exp\left(-\alpha S(\boldsymbol{\rho})\right) \tag{16}$$

where $S(\boldsymbol{\rho})$ is the "entropy" defined as

$$S(\boldsymbol{\rho}) = \sum_k \left(\rho(\omega_k) - m(\omega_k) - \rho(\omega_k)\log\left(\frac{\rho(\omega_k)}{m(\omega_k)}\right)\right) \tag{17}$$

Here $m(\omega)$ is called the "default" model, which represents some expected structure in $\rho(\omega)$ a priori. In addition, we often impose the following constraints on $\boldsymbol{\rho}$ as

$$1 = \sum_k \rho(\omega_k) \quad \rho(\omega_k) \geq 0. \tag{18}$$

We define the set satisfying the above constraints as $\mathcal{P}$. The parameter α controls the strength of the prior distribution in the posterior distribution. By using the Bayesian theorem, we reach the posterior distribution.

$$P(\boldsymbol{\rho}|\mathbf{G}) \propto P(\mathbf{G}|\boldsymbol{\rho})P(\boldsymbol{\rho}) \tag{19}$$

Therefore the maximum a posteriori estimation leads to the following optimization problem

$$\max_{\rho\in\mathcal{P}}\left\{\alpha S-\frac{1}{2}\chi^2(\mathbf{G},\boldsymbol{\rho})\right\}. \tag{20}$$

This is the MaxEnt or MEM [7, 8, 9].

The MEM is used in many fields rather than quantum many-body physics. For example, one of the applications of MEM is for deconvolution in radio astronomy [17].

The criticism of the MEM can be raised naturally. How can we determine the default model in the prior distribution? There is no definite answer.

4. Sparse modeling for analytical continuation

Instead of the MEM, we propose an alternative solver stemming from sparse modeling. We invent a new algorithm to solve the inverse problem (1) inspired by the compressed sensing (CS) in the framework of the sparse modeling. In the CS, we consider the L_1-norm minimization of the N-dimensional unknown vector $\mathbf{x}$ under an equality constraint $\mathbf{y}=A\mathbf{x}$, where $\mathbf{y}$ is the M-dimensional observed vector and A is the sensing matrix. The penalty method to solve the constrained minimization problem leads to an well-known L_1 and L_2-norm minimization problems, LASSO, defined as

$$\min_{\mathbf{x}}\left\{\frac{1}{2}\|\mathbf{y}-A\mathbf{x}\|_2^2+\lambda\|\mathbf{x}\|_1\right\} \tag{21}$$

In the CS, the unknown vector is assumed to be sparse. The minimization of the L_1 norm selects a relatively sparse solution from a plenty of the solutions satisfying $\mathbf{y}=A\mathbf{x}$ when $M<N$.

The situation in the inverse problem (1) is very similar to the case of the CS. The exponentially decaying components in the kernel matrix decrease the essentially relevant elements in the observation $\mathbf{G}$. To avoid the irrelevant elements of the observation $\mathbf{G}$, we perform the singular value decomposition (SVD) on the kernel matrix as $K=U\Lambda V^{\mathrm{T}}$, where U and V is the unitary matrix and Λ is the diagonal matrix containing the singular values. In addition, we define the transformed vectors as $\boldsymbol{\rho}'=V^{\mathrm{T}}\boldsymbol{\rho}$ and $\mathbf{G}'=U^{\mathrm{T}}\mathbf{G}$. Then the squared error is unchanged under the SVD as

$$\chi^2(\mathbf{G},\boldsymbol{\rho})=\left\|\mathbf{G}'-\Lambda\boldsymbol{\rho}'\right\|_2^2 \tag{22}$$

In this expression, Λ plays a roll of some filters. The components of $\boldsymbol{\rho}$ multiplied by small values in Λ do not affect significantly the observed vectors $\mathbf{G}'$. We then avoid several components of $\boldsymbol{\rho}$ automatically selected by using the L_1 norm. For this purpose, let us consider the following optimization problem

$$\min_{\rho\in\mathcal{P}}\left\{\frac{1}{2}\chi^2(\mathbf{G},\boldsymbol{\rho})+\lambda\left\|\boldsymbol{\rho}'\right\|_1\right\}. \tag{23}$$

By L_1 norm, decreasing the number of nonzero elements encourages the data fit in the significant contributions with the large values of Λ and avoids irrelevant contributions with the small values of Λ. Unless the case of the CS, we do not necessarily the sparsity of the vector representing the density of states. In order to avoid the irrelevant contributions for reconstruction of $\mathbf{G}$, we employ the L_1 norm. Although the part of the squared error takes a form satisfying the separability, the norm can be written as the simple summation as $\sum_l(G_l'-\Lambda_l\rho_l')^2$, the constraints on the density of states makes the optimization problem nontrivial. To solve the optimization problem, the alternating direction of the multiplier method (ADMM) is available and efficient.

Furthermore, we can generalize the case of the noise intensity dependent on the observed frequency τ_l and correlated with each other as

$$\chi^2(\mathbf{G},\boldsymbol{\rho})=(\mathbf{G}-K\boldsymbol{\rho})^{\mathrm{T}}\Sigma^{-1}(\mathbf{G}-K\boldsymbol{\rho}) \tag{24}$$

IOP Conf. Series: Journal of Physics: Conf. Series **1036** (2018) 012020 doi:10.1088/1742-6596/1036/1/012020

where Σ^{-1} is the covariance matrix of the noise. We similarly perform the SVD and transformation of $\mathbf{G}'$ and $\boldsymbol{\rho}'$ as

$$\chi^2(\mathbf{G},\boldsymbol{\rho}) = \left(\mathbf{G}' - \Lambda\boldsymbol{\rho}'\right)^{\mathrm{T}} U^{\mathrm{T}}\Sigma^{-1}U\left(\mathbf{G}' - \Lambda\boldsymbol{\rho}'\right) \tag{25}$$

The property of the kernel matrix is reflected on the covariance matrix as $U^{\mathrm{T}}\Sigma^{-1}U$.

5. ADMM procedure

We provide the ADMM procedure for readers below. The ADMM is to efficiently solve the optimization problem with the non-differential cost function and several constraints. In our problem, we utilize the augmented Lagrangian method for the constraints $\boldsymbol{\rho}'$ equals artificial variables $\mathbf{z}$ and $\mathbf{z}'$ for simply solving the original optimization problem and the standard Lagrangian multiplier method for the constraint $\sum_l \rho(\omega_l) = 1$, that is $\mathbf{1}^{\mathrm{T}}V\boldsymbol{\rho}' = 1$ as

$$\min_{\boldsymbol{\rho}}\left\{\frac{1}{2}\chi^2(\mathbf{G},\boldsymbol{\rho}) + \lambda\|\mathbf{z}\|_1 + \frac{\mu}{2}\|\boldsymbol{\rho}'-\mathbf{z}+\mathbf{u}\|_2^2 - \nu(\mathbf{1}^{\mathrm{T}}V\boldsymbol{\rho}'-1) + \lim_{\gamma_i\to\infty}\gamma_i\Theta(z_i') + \frac{\mu'}{2}\|V\boldsymbol{\rho}'-\mathbf{z}'+\mathbf{u}'\|_2^2\right\}. \tag{26}$$

Taking derivative with respect to $\boldsymbol{\rho}'$, we find the optimal solution on $\boldsymbol{\rho}'$ for the fixed values μ, ν and $\mathbf{z}$.

$$\begin{aligned}\boldsymbol{\rho} &= \left(\Lambda U^{\mathrm{T}}\Sigma^{-1}U\Lambda + (\mu+\mu')I\right)^{-1}\left(\Lambda U^{\mathrm{T}}\Sigma^{-1}U\mathbf{G}' + \mu(\mathbf{z}-\mathbf{u}) + \mu V^{\mathrm{T}}(\mathbf{z}'-\mathbf{u}') + \nu V^{\mathrm{T}}\mathbf{1}\right) &(27)\\ &\equiv \boldsymbol{\xi} + \nu\boldsymbol{\eta}. &(28)\end{aligned}$$

The Lagrange multiplier ν changes as satisfying the constraint $\sum_l \rho(\omega_l) = 1$. Thus

$$\nu = \frac{1-\sum_l \xi_l}{\sum_l \eta_l}. \tag{29}$$

Two artificial variables are solved simply as

$$\begin{aligned}\mathbf{z} &= S_{1/\mu}(\boldsymbol{\rho}'+\mathbf{u}) &(30)\\ \mathbf{z}' &= \mathcal{P}_+(V\boldsymbol{\rho}'+\mathbf{u}'), &(31)\end{aligned}$$

where $S_\lambda(x)$ is the Soft-Thresholding function defined as

$$S_\lambda(x) = \begin{cases} x-\lambda & (x>\lambda)\\ 0 & (-\lambda\le x\le\lambda)\\ x+\lambda & (x<-\lambda)\end{cases} \tag{32}$$

and $\mathcal{P}_+$ is a projection operator onto non-negative quadrant as $\mathcal{P}_+(z) = \max\{z, 0\}$. Following the auxiliary Lagrangian method, the Lagrangian multipliers are updated as

$$\begin{aligned}\mathbf{u} &= \mathbf{u} + (\boldsymbol{\rho}'-\mathbf{z}) &(33)\\ \mathbf{u}' &= \mathbf{u}' + (V\boldsymbol{\rho}'-\mathbf{z}'). &(34)\end{aligned}$$

When Σ^{-1} is diagonal, the computation of the inverse matrix can be simplified into the inverse of the diagonal elements.

6. Summary

In the present manuscript, we deal with the analytical continuation problem as an application of the sparse modeling. The analytical continuation includes a typical inverse problem and demands some regularization to be solved efficiently. The standard method was the MEM, which utilizes the prior distribution to infer the density of states via the MAP estimation. However it is difficult to choose an appropriate prior distribution in advance without any knowledge on the problem. Instead of the MEM, we employ the CS-like algorithm from the sparse modeling to elucidate the relevant contributions from observations. Because the sensing matrix has elements exponentially decaying, we perform the SVD in advance to filter out several irrelevant components. The components multiplied by small values of the singular values are irrelevant for reconstruction of the observations. We employ the L_1 norm to reduce the number of components and solve efficiently the inverse problem. As in the literature [1], the method can efficiently find reasonably the density of states without any prior knowledge. In addition, the proposed method elucidate the sparse representation for the QMC data [2, 18]. The sparse representation is useful for compression of the QMC data and finding the physical relevant feature from it. The resultant representation seems to consist of the highly nontrivial polynomial satisfying the integral equation stemming from the kernel. The future study aims at solving the nontrivial structure of the sparse representation of the kernel appearing in the analytical continuation in the quantum many-body physics. In addition, various techniques find inherent function from data as neural networks are available [19]. As other directions, in the stochastic analytical continuation, we may utilize the accelerated stochastic method [20, 21, 22, 23, 24]. In addition, the QMC data may be divided into several batch in order to efficiently optimize the cost function and avoid the overfit the noisy data. Then the Bayesian learning can be one of the useful techniques because it can predict the certainty of the estimation [25].

Acknowledgments

The present study was financially supported by MEXT KAKENHI No. 25120008, CREST, Japan Science and Technology Agency, JSPS KAKENHI No. 16H04382 and ImPACT Program of Council for Science, Technology and Innovation (Cabinet Office, Government of Japan).

References

[1] Otsuki J, Ohzeki M, Shinaoka H and Yoshimi K 2017 *Phys. Rev. E* **95**(6) 061302
[2] Shinaoka H, Otsuki J, Ohzeki M and Yoshimi K 2017 *Phys. Rev. B* **96**(3) 035147
[3] Abrikosov A, Gorkov L, Dzyaloshinski I and Dziyaloshinski I 1975 *Methods of Quantum Field Theory in Statistical Physics* Dover Books on Physics Series (Dover Publications) ISBN 9780486632285
[4] Baym G and Kadanoff L P 1961 *Phys. Rev.* **124**(2) 287–299
[5] Gubernatis J, Kawashima N and Werner P 2016 *Quantum Monte Carlo Methods: Algorithms for Lattice Models* (Cambridge University Press) ISBN 9781316483121
[6] Gull E, Millis A J, Lichtenstein A I, Rubtsov A N, Troyer M and Werner P 2011 *Rev. Mod. Phys.* **83**(2) 349–404
[7] Jarrell M and Gubernatis J 1996 *Physics Reports* **269** 133 – 195 ISSN 0370-1573
[8] Gunnarsson O, Haverkort M W and Sangiovanni G 2010 *Phys. Rev. B* **81**(15) 155107
[9] Bergeron D and Tremblay A M S 2016 *Phys. Rev. E* **94**(2) 023303
[10] Sandvik A W 1998 *Phys. Rev. B* **57**(17) 10287–10290
[11] Fuchs S, Pruschke T and Jarrell M 2010 *Phys. Rev. E* **81**(5) 056701
[12] Sandvik A W 2016 *Phys. Rev. E* **94**(6) 063308
[13] Kirkpatrick S, Gelatt C D and Vecchi M P 1983 *Science* **220** 671–680
[14] Yamanaka S, Ohzeki M and Decelle A 2015 *Journal of the Physical Society of Japan* **84** 024801
[15] Ohzeki M 2015 *Journal of the Physical Society of Japan* **84** 054801
[16] Ezaki T, Watanabe T, Ohzeki M and Masuda N 2017 *Philosophical Transactions of the Royal Society of London A: Mathematical, Physical and Engineering Sciences* **375**
[17] Starck J L, Pantin E and Murtagh F 2002 *Publications of the Astronomical Society of the Pacific* **114** 1051

[18] Shinaoka H, Otsuki J, Haule K, Wallerberger M, Gull E, Yoshimi K and Ohzeki M 2018 *Phys. Rev. B* **97**(20) 205111 URL `https://link.aps.org/doi/10.1103/PhysRevB.97.205111`
[19] Arai S, Ohzeki M and Tanaka K 2018 *Journal of the Physical Society of Japan* **87** 033001
[20] Ichiki A and Ohzeki M 2013 *Phys. Rev. E* **88**(2) 020101
[21] Ichiki A and Ohzeki M 2015 *Phys. Rev. E* **91**(6) 062105
[22] Ohzeki M and Ichiki A 2015 *Phys. Rev. E* **92**(1) 012105
[23] Takahashi K and Ohzeki M 2016 *Phys. Rev. E* **93**(1) 012129
[24] Ohzeki M and Ichiki A 2015 *Journal of Physics: Conference Series* **638** 012003
[25] Ohzeki M 2016 *Journal of Physics: Conference Series* **699** 012019

IOP Conf. Series: Journal of Physics: Conf. Series **1036** (2018) 012021 doi:10.1088/1742-6596/1036/1/012021

Combining Machine Learning and Physics to Understand Glassy Systems

Samuel S. Schoenholz
Google Brain
E-mail: schsam@google.com

Abstract. Our understanding of supercooled liquids and glasses has lagged significantly behind that of simple liquids and crystalline solids. This is in part due to the many possibly relevant degrees of freedom that are present due to the disorder inherent to these systems and in part to non-equilibrium effects which are difficult to treat in the standard context of statistical physics. Together these issues have resulted in a field whose theories are under-constrained by experiment and where fundamental questions are still unresolved. Mean field results have been successful in infinite dimensions but it is unclear to what extent they apply to realistic systems and assume uniform local structure. At odds with this are theories premised on the existence of structural defects. However, until recently it has been impossible to find structural signatures that are predictive of dynamics. Here we summarize and recast the results from several recent papers offering a data driven approach to building a phenomenological theory of disordered materials by combining machine learning with physical intuition.

1. Introduction

Over the last century, enormous progress has been made towards understanding physical systems in their condensed phases. For simple liquids, theoretical developments have resulted in liquid state theory [1] that assumes the system to have uniform local structure. For crystalline solids, theories based on the existence of topological defects that couple elastically through the bulk have been remarkably successful [2, 3]. Across both fields rapidly advancing numerical simulations of materials have typically been used to corroborate theories rather than inform novel theoretical work.

By contrast, a comprehensive theory of glasses and supercooled liquids has eluded researchers. In part, the struggle to understand these systems stems from two factors. First, amorphous systems have a large number of degrees of freedom and there are few heuristics that one can use to determine which (if any) may be neglected. Second, the range of temperatures where glassy liquids are in equilibrium is small since viscosity increases so rapidly with decreasing temperature. It is therefore difficult to test theories without grappling with non-equilibrium effects. Together these two effects have resulted in a field with too many theories that are not sufficiently constrained by experiments.

Mode coupling theory [4] and random first order theory [5, 6, 7, 8] extend liquid state theory to formally describe super-cooled liquids and glasses at the level of mean field theory. However, it is unclear to what extent these mean field results survive down to finite-dimensional realistic glass formers. Indeed, the dynamical glass transition predicted by mode-coupling theory is conspicuously absent in three-dimensions [9]. By contrast, theoretical frameworks such as shear transformation zones [10] and dynamical facilitation [11] are premised on the existence of defects analogous to those in a crystal. Placing these theories on firm experimental footing has been difficult because historically, only weak correlations were found between measures of local structure - such as bond orientational order [12], free volume [13], or local potential energy [14] - and dynamics.

In this article we will review a series of papers [15, 16, 17, 18] that combine traditional physical intuition with machine learning to analyze numerical simulations of molecular glasses and supercooled liquids and gain novel physical insight. To do this we will first introduce a quantity that we term the "softness". The softness leverages large amounts of data from molecular dynamics simulations or experiment to design a function of local-structure that most strongly correlates with dynamics. We proceed to investigate the relationship between softness and dynamics. Finally we leverage this connection to approach several open problems in glassy physics.

2. Background

We begin by presenting a brief recapitulation of the phenomenology of the glass transition. For a more detailed account see i.e. Debeneditti and Stillinger [19]. The freezing of a liquid into a crystalline solid proceeds via a first-order phase transition. At a critical temperature, T_m, the free energy of the disordered, isotropic, and homogeneous arrangement of particles in the fluid becomes equal to the free energy of the ordered anisotropy of the crystal. The ordered phase proceeds to nucleate in the fluid until the entire system freezes. In the view of Landau this process is relatively well understood and is described by the breaking of the continuous rotational and translational symmetries of the fluid down to the discrete set of crystallographic symmetries.

However, the process of nucleation is relatively slow. If the liquid is cooled quickly enough, crystallization can be avoided and the system enters a long-lived metastable state known as the supercooled liquid. As the temperature of the supercooled liquid is lowered further it becomes increasingly sluggish. At some temperature, known as the glass transition temperature T_g, the time required for the system to equilibrate exceeds any reasonable experimental timescale and it is said that the liquid has solidified to a glass. However, unlike in the case of the liquid to crystal transition there are no diverging length scales, critical exponents, or structural changes that have been associated with this change of state. Moreover, since the system has fallen out of equilibrium, the usual tools of statistical mechanics may not be applied once the temperature is lower than T_g. Finally, it is well known that the specifics of the glass transition depend on the cooling protocol: the slower the cooling rate the longer the system is able to remain equilibrated and the lower the glass transition temperature.

One of the biggest mysteries associated with these systems is the extremely rapid increase of the relaxation time (or viscosity) as temperature is lowered. Many systems feature a viscosity that increases by nearly ten orders of magnitude as T_g/T varies by only about 10% and which is accompanied by no obvious structural change. Glass formers are generically categorized into two groups based on the manner in which their viscosity varies with T_g/T. In strong glass-formers, such as SiO_2, viscosities scale exponentially with T_g/T as $\eta \sim \exp(A/T)$ and A can be interpreted as an energy scale. However, the many so-called "fragile" glass formers slow down significantly more quickly. The form of the viscosity in fragile systems is unknown and moreover it is not known whether or not the viscosity diverges at a genuine phase transition. As discussed above, this leads to a situation where the available data is insufficient to constrain the space of possible models.

A second phenomenon that supercooled liquids and glasses exhibit is non-exponential relaxation. To quantify the manner in which glasses relax one can use either the self-intermediate scattering function, $F_s(q,t)$, or the overlap function defined by the equation,

$$q(t) = \frac{1}{N}\sum_i \Theta(|r_i(t) - r_i(0)| - a). \tag{1}$$

The overlap function measures the fraction of the system that has moved greater than a characteristic distance, a after a time t. Both the overlap function and the self-intermediate scattering function behave similarly. In simple liquids, particles exhibit Brownian motion and $q(t)$ decays exponentially with a characteristic timescale known as the α-relaxation time. However, as supercooled liquids get closer to their glassy state the decay of the overlap function becomes stretched as $q(t) \sim e^{-(t/\tau_\alpha)^\beta}$. The origin of this stretched exponential form still lacks solid explanation in three-dimensions.

A final open question concerns the increasingly heterogeneous dynamics that emerge as supercooled liquids are cooled towards their glassy state (for a more thorough discussion see i.e. Keys *et al.* [20] or Widmer-Cooper *et al.* [21]). In a simple liquid the activity of different regions of the system are statistically identical and dynamics are homogeneous. By contrast in the supercooled liquid and glassy phase there are large heterogeneities in the mobility of particles. The origin of these large differences in particle motion is unknown and, indeed, it has yet to be determined whether or not they are a result of structural defects or whether they are purely dynamical in nature.

A number of other approaches have attempted to identify a structural signature of dynamics in supercooled liquids. These methods traditionally are divided into two approaches. The first approach correlates quasilocalized vibrational modes in the dynamical matrix with rearrangements [21, 13, 22, 23]. However, this method has a number of problems. First, diagonalizing the dynamical matrix is costly and is only applicable to systems where the Hamiltonian is available. Second, coupling between localized regions and the elastic background obfuscates the identification of defects. A second approach attempts to define locally preferred structures by hand and identify them in the glassy liquid [12, 24, 25, 14, 26]. However, the identification of locally preferred structures is far from trivial and must be done by hand for each new system. Both methods typically feature correlations between structure and dynamics that are too low to use in the construction of a theory. Several more recent papers have attempted to approach this question using the notion of local thermal energy [27].

3. Softness

We will summarize the results from four papers [15, 16, 17, 18] that discuss the three questions posed in the previous section by exploring the relationship between local-structure and dynamics in supercooled liquids and glasses. At the same time, we place some emphasis on being pedagogical. Our primary tool in this investigation will be a quantity that we introduced in two papers [15, 16] and term the "softness". We will describe the construction of the softness presently.

Suppose we have access to the time-dependent positions of particles in a glassy liquid, $\{x_i(t)\}$, either from computer simulation or experiment. Qualitatively, it seems that at sufficiently low temperatures, the position of particles in a supercooled liquid spend their time fluctuating in the vicinity of a local-minimum in the energy landscape. Thermal fluctuations intermittently drive the system over saddle-points to neighboring minima. It has been observed [28] that these transition events correspond to localized rearrangements involving a small fraction of the system. These intermittent dynamics can be seen at the single particle level in fig. 1 (a). If there is a connection between local-structure and dynamics then the local-neighborhood of particles just before a rearrangement should be distinguishable the that of particles that have not rearranged in a significant amount of time.

Instead of trying to intuit the relationship between structure and dynamics, we will take a machine learning approach using the large amounts of data from either molecular dynamics or experimental data [29]. We will take the following steps:

- Identify a population of local regions that are about to experience rearrangements and another population of regions that are unusually stable.
- Parametrize the degrees of freedom of each local region in a manner that is amenable to analysis.
- Learn the function of the parametrization that optimally separates the rearranging population from the stable population.

Each step outlined here can be performed in a number of ways. We find that our results are remarkably insensitive to the specific choices made. In light of this, we will choose to present a construction of the softness field that is easy to implement and physically motivated. This incurs a slight loss in our ability to differentiate between the two groups ($\sim 2\%$ difference in classification accuracy). For more details see the supplementary material in Schoenholz and Cubuk *et al.* [16].

Throughout this paper we will focus on results from simulations of a 10,000-particle, 80:20 binary Lennard-Jones (LJ) Kob-Andersen mixture [30]. We measure time in units of $\tau = \sqrt{\epsilon_{AA}/\sigma_{AA}^2}$ and take Boltzmann's constant to be $k_B = 1$. Finally, we cut off the LJ potential at $2.5\sigma_{AA}$ and smooth it so that the force varies continuously. Simulations were performed in LAMMPS [31] using and NVT ensemble with a Nosé-Hoover chain thermostat and a timestep of 0.0025τ. Every τ-timesteps we quench the system to its nearest inherent structure using a combination of conjugate gradient and FIRE minimization. We use inherent structure positions instead of instantaneous positions. However, as with the details of our approach our results are largely insensitive to these choices and indeed time-averaged positions or even instantaneous positions may be used in its place [15]. While we will not discuss it in detail here our method has also proven successful when applied to experimental systems and polymeric glasses [32].

3.1. Identifying Rearrangements

To identify rearrangements we have used both D^2_{min} as introduced by Falk and Langer [10] as well as p_{hop} discussed in Candelier *et al.* [33, 34]. While D^2_{min} is better suited for systems that are under external load, p_{hop} gives slightly better results for quiescent systems and so we will focus on it here.

We define p_{hop} by first picking a timescale $t_R = 10\tau$ that is approximately the time it takes for the system to complete a rearrangement. At a time t, this choice then gives two intervals $A = [t - t_R/2, t]$ and $B = [t, t + t_R/2]$. With reference to these two intervals we define for a single particle,

$$p_{\text{hop}}(i,t) = \sqrt{\langle (x_i - \langle x_i \rangle_B)^2 \rangle_A \langle (x_i - \langle x_i \rangle_A)^2 \rangle_B}. \tag{2}$$

where $\langle\rangle_A$ and $\langle\rangle_B$ are averages over the intervals A and B respectively. In practice these averages are computed as $\langle f(x_i)\rangle_A = \frac{2}{t_R}\sum_{t'=t-t_R/2}^{t} f(x_i(t'))$ with an analogous expression holding for $\langle f(x_i)\rangle_B$.

If no rearrangement takes place between intervals A and B then eq. (2) reduces to computing the variance of the particle position over time, which will be intimately related to the volume of caged motion that the particle experiences. In the event that the inherent structure is used instead of the instantaneous position this amounts to computing the variance of the inherent structure position (which will be nonzero both because the the inherent structure position might not be exactly constant and because the minimization has some inherent uncertainty associated with it). If a rearrangement does occur then p_{hop} will be proportional to the square of the

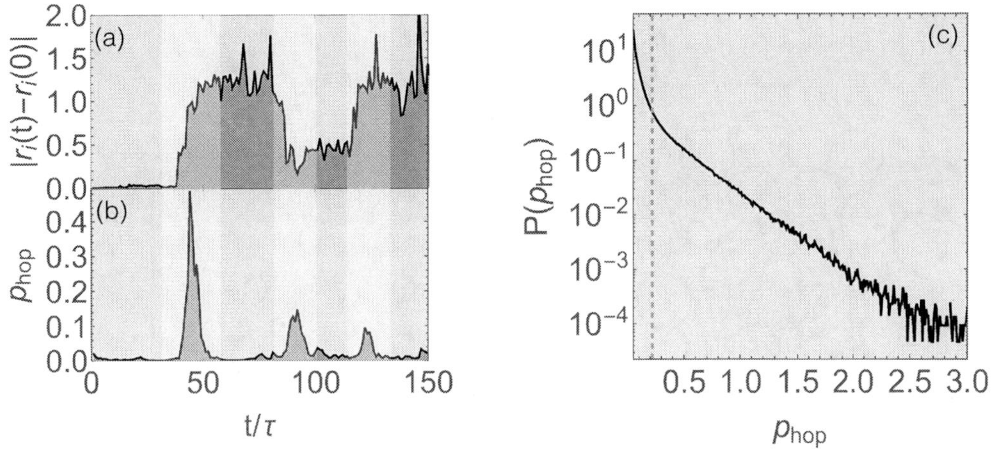

Figure 1. Quantifying rearrangements. (a) The distance in the inherent structure position of a particle as a function of time. Several well-defined hopping events can be seen. (b) The p_{hop} indicator function of this same trajectory. We see that p_{hop} is very sensitive to rearrangements of the inherent structure position. (c) The probability distribution of p_{hop}. We see a clear crossover to an exponential distribution at a well-defined value of p_{hop}.

distance the particle moves as the state transitions between the two minima. The behavior of p_{hop} over the course of a rearrangement can be seen in fig. 1 (a) and (b).

Together these two limits result in a separation of scales that emerge between rearrangements and caged motion. This can be seen by referring to the distribution of p_{hop} in fig. 1 (c) which exhibits a clear crossover between exponential and non-exponential behavior. This allows us to cleanly identify rearrangements in disordered systems.

To do this we select two thresholds $p_L \approx 0.05$ which defines a lower bound on what we consider to be a rearrangement. We then identify, from one or a series of particle trajectories, all rearrangement events which we define to be instances where p_{hop} exceeds p_L. For each event we define p^* to be the maximum value of p_{hop} achieved during the event. We also define t_{start} and t_{end} to be the time when p_{hop} first is greater and less than p_L respectively. Empirically this gives a very precise definition of rearrangements. One can then study quantities such as the duration of rearrangements by considering $t_{\text{end}} - t_{\text{start}}$ or the displacement of particles during rearrangements, $\Delta x_i = x_i(t_{\text{end}}) - x_i(t_{\text{start}})$.

With rearrangements well-defined we construct two populations of particles. The first group are particles that are about to experience a large rearrangement. We define large rearrangements to be those with $p^* > p_c \approx 0.6$. For each event we consider the particle configuration just before the rearrangement at $t_{\text{start}} - 2\tau$. The second group contains particles that have not rearranged in a long time. Thus we take particles that experience no rearrangement events for approximately $\tau_c \approx \tau_\alpha$. Here we consider particle configurations in the middle of the interval. In general the larger p_c and τ_c the better the results so long as each population of particles contain at least 3000 particles.

3.2. Parametrizing Local Structure

We now present a simple and physically motivated parametrization of the local-structure around a central atom i. There are, however, a wide-variety of methods that can be used instead [35, 36, 37, 38, 39]. Qualitatively, our results are insensitive to the choice of parametrization so long as it is sufficiently expressive. We typically find that using more sophisticated methods for describing local-structure results in an improvement in our ability to differentiate the two populations by about 2-4%.

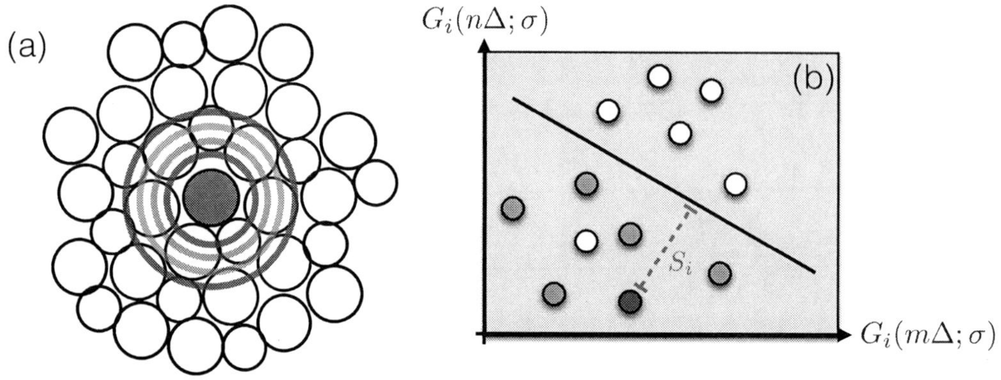

Figure 2. Parametrizing local structure. (a) We show schematically the functions $G_i^X(r;\sigma)$ for several different values of r. (b) A schematic showing how the local-structure around a central atom is embedded into a high dimensional space. We then find a hyperplane that separates particles that are about to rearrange from those that have not rearranged in a long time. Once the hyperplane has been established we call the softness the distance of that local configuration to the hyperplane.

We begin by defining the function,

$$G_i^X(r;\sigma) = \frac{1}{\sqrt{2\pi}} \sum_{j\in X} e^{-\frac{1}{2\sigma^2}(R_{ij}-r)^2} \tag{3}$$

where $R_{ij} = |x_i - x_j|$ is the distance between particles i and j and $X \in \{A, B\}$ refers to the species of atom. $G_i^X(r;\sigma)$ essentially counts the number of particles of a given species whose distance from a central particle is within about σ of r. This can be seen schematically in fig. 2 (a). In principle if we could access the entire function $G_i(r;\sigma)$ we would be able to perform inferences on the local structure that depended on the entire density profile around an atom.

In practice we discretize the set of radii into $r_n = n\Delta$ and consider the vector $G_i = (G_i^A(\Delta;\sigma), G_i^A(2\Delta;\sigma), \cdots, G_i^B(N\Delta;\sigma)) \in \mathbb{R}^N$. This provides an embedding of the density profile into $\mathbb{R}^N$ where many more techniques have been developed to help perform inference. A good selection of Δ that we found is $\Delta \approx \sigma$ and $\sigma \approx 0.1\sigma_{AA}$. However, note that with this selection $G_i^X(n\Delta;\sigma)$ and $G_i^X((n+1)\Delta;\sigma)$ will be highly correlated. We will discuss the implications of this correlation more in the next section. By choosing $\Delta \approx 3\sigma$ we get a set of $G_i^X(n\Delta;\sigma)$ that are not as correlated with a minimal drop in accuracy (usually $\lesssim 1\%$).

3.3. Constructing the Softness

With a population of rearranging and stable particles defined and a parametrization of the local-structure in the neighborhood of each particle through an embedding into $\mathbb{R}^N$, we can proceed to discuss the construction of the softness field itself. The question of finding functions of features (in this case G_i) that best separates points in different classes (in this case about to rearrange and stable) is well studied in statistics and machine learning. We will use a method from machine learning called support vector machines [40] with a linear kernel, however as with most other pieces of the method a number of different techniques can be used with similar effect (e.g. logistic regression, nonlinear kernels, neural networks).

Using linear support vector machines gives us a hyperplane defined by a normal $\hat{w}$ and a bias b so that $\hat{w} \cdot G_i + b$ is positive when a particle is about to rearrange and negative when a particle is stable. For a schematic representation see fig. 2 (b). When we apply this technique to systems at a temperature $T = 0.47$ we find that our hyperplane is able to correct separate about 88% of the points into their correct classes.

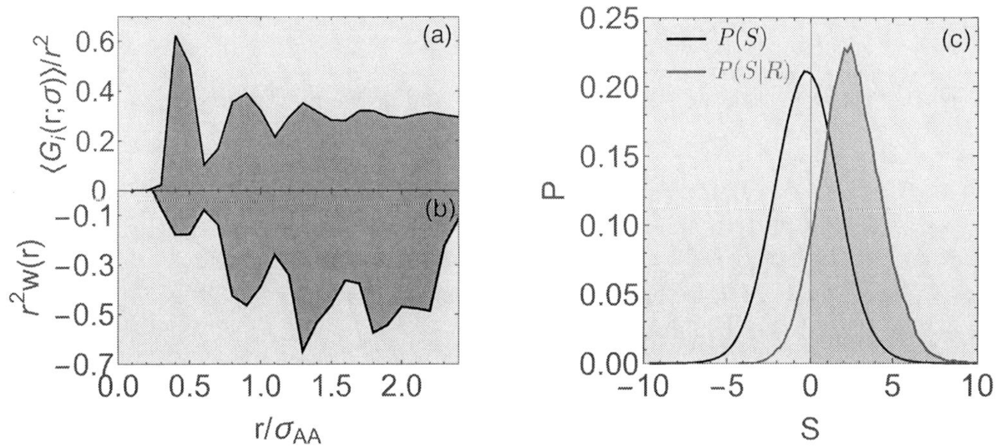

Figure 3. The softness. (a) The function $r^{1-d}\langle G_i(r;\sigma)\rangle$ averaged over a whole system. We see that it is qualitatively similar to $g(r)$. (b) The function $\tilde{w}(r)$ learned by the support vector machine. We see that it is sensitive to $g(r)$ up to about $2.5\sigma_{AA}$. (c) The softness distribution and the softness distribution for particles that were about to rearrange. We see that the two distributions are well-separated and 90% of the rearranging particles get assigned $S > 0$.

Once this has been done new local structures can be classified using this hyperplane and we can make predictions about whether or not the particle is about to rearrange from its structure. However, we found it more useful to consider the signed-distance to the hyperplane, $S_i = \hat{w} \cdot G_i + b$ which we call the softness. The rest of this paper will be concerned with understanding properties of the softness.

One motivation for our simplified choice of parametrization as well as our choice of linear support vector machine as opposed to some more complicated nonlinear method is interpretability. We will now provide a physical picture for softness that is related, on average, to the radial distribution function. For the purposes of this discussion we will let $\Delta = \sigma$ and we will consider a hypothetical system composed of only one species of atom however the result generalizes trivially to multiple species. Let us also assume we fix the maximum radius at which we probe local structure so that the number of features is $N = R_c/\sigma$ with R_c the radial cutoff. It follows that, neglecting the bias, the softness can be written as,

$$S_i = \sum_{n=1}^{N} G_i(n\sigma;\sigma)w(n\sigma) \tag{4}$$

where $w(i)$ are the weights from the SVM. In the limit of zero width we find that,

$$S_i = \lim_{\sigma\to 0}\sum_{n=1}^{R_c/\sigma} G(n\sigma; i, \sigma)w(n\sigma) \tag{5}$$

$$= \lim_{\sigma\to 0}\sum_{n=1}^{R_c/\sigma} \sigma\left[\frac{1}{\sqrt{2\pi\sigma^2}}\sum_j e^{-(R_{ij}-n\sigma)^2/2\sigma^2}\right] w(n\sigma). \tag{6}$$

This is exactly the definition of a Riemann integral and so we may write,

$$S_i = \int dr\left[\sum_j \delta(r - R_{ij})\right] w(r) \tag{7}$$

where we have made the assignments $\lim_{\sigma\to 0}\sum_{n=1}^{R_c/\sigma}\sigma \to \int_0^{R_c} dr$, $n\sigma \to r$, and $\lim_{\sigma\to 0}\frac{1}{\sqrt{2\pi\sigma^2}}e^{-x^2/2\sigma^2} \to \sqrt{2\pi}\delta(x)$.

We can use this formula to understand the softness in two different but equivalent ways. We define the "local density" operator ,

$$g_i(r) = \frac{1}{r^{d-1}}\sum_j \delta(r - R_{ij}) \tag{8}$$

that measures the density of particles a distance r away from a tagged particle i. In this case, the softness can be written as,

$$S_i = \int dr g_i(r)\tilde{w}(r) \tag{9}$$

where we have defined $\tilde{w}(r) = r^{d-1}w(r)$. It follows that we can think of S_i as the functional that integrates the local density weighted by some weight function $\tilde{w}(r)$. The purpose of the SVM is to then find an appropriate weight function over the local density. We note further that - when the system is isotropic - our symmetry functions are related to the radial distribution function in the sense that $g(r) \sim \lim_{\sigma\to 0} r^{1-d}\langle G_i(r;\sigma)\rangle$.

These two relationships can be seen in fig. 3 (a) and (b). In (a) we see that when σ is sufficiently small $r^{1-d}\langle G_i(r;\sigma)\rangle$ does indeed give the radial distribution function. In (b) we see the weight function, $r^{d-1}w(r)$, that is learned by our network. We see that the weight function features many of the characteristic features of $g(r)$. Moreover, we see that $r^{d-1}w(r)$ has support only within a region where $r \lesssim 2.5\sigma_{AA}$, this gives a characteristic volume that is important for predicting rearrangements.

To investigate the effectiveness of softness at predicting rearrangements it is useful to consider the distribution of softnesses over an entire molecular dynamics simulation, $P(S)$, and to compare it with the distribution of softness for those particles that are currently involved in a rearrangement, $P(S|R)$. This is shown in fig. 3 (c). We see, in particular, that both $P(S)$ and $P(S|R)$ are approximately Gaussian distributed. Moreover we notice that the classification accuracy of our model is given by $P(S > 0|R) \approx 0.88$ which means that our model is able to correctly predict approximately 90% of rearrangements correctly.

4. Relating Softness with Dynamics

4.1. The Probability of Rearrangement

The success of support vector machines at differentiating particles that are about to rearrange from those that are stable offers definitive proof that there is a strong structural component to dynamics in glassy systems. However, the definition of softness unto itself is not terribly informative. As with previous predictors of dynamics from local structure it is not obvious how to relate classification accuracy with a useful theory of dynamics of glassy systems. This issue is further complicated by the fact that softness itself is not a physical quantity. Indeed, both the scale and the shift of the hyperplane identified by the support vector machine are arbitrary and we could easily define a transformation of softness $\tilde{S}_i = \alpha S_i + \beta$ that would have essentially the same physical significance.

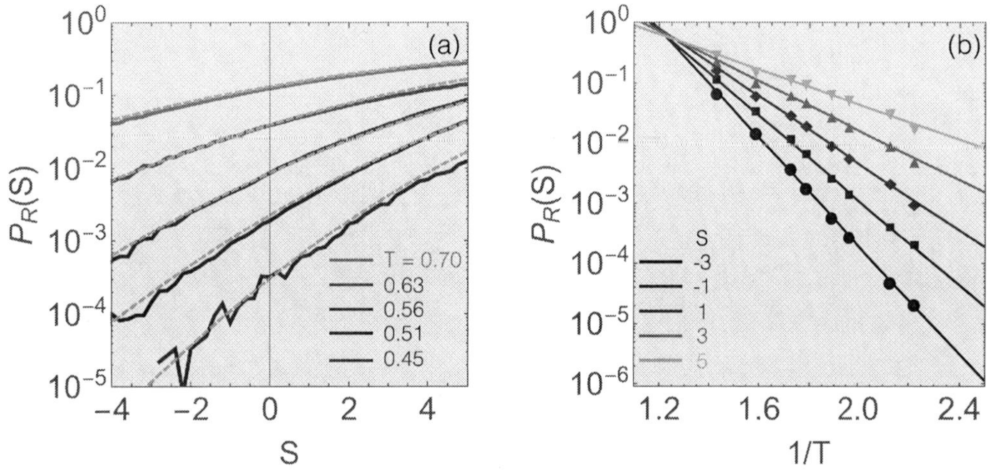

Figure 4. The probability of rearrangement. (a) We plot the probability of rearrangement at different temperatures. Overlaid is the prediction from the Arrhenius form of $P_R(S)$. (b) Plotting $P_R(S)$ for different softness values as a function of $1/T$. We see that the probability of rearrangement viewed in this way is approximately Arrhenius.

There is a far more useful quantity that we can construct from an understanding of $P(S)$ and $P(S|R)$. Indeed using Bayes theorem we can compute the probability that particles of a given softness will rearrange,

$$P_R(S) \equiv P(R|S) = \frac{P(S|R)P(S)}{P(R)}. \tag{10}$$

This allows us to predict, from local-structure, the probability that a particle will experience a rearrangement. In practice it is usually easier to compute $P_R(S)$ by evaluating using p_{hop} at each instant of the simulation what fraction of particles are rearranging binned by softness. To define rearrangements here we can be less restrictive than when constructing the training set and take $p_{\text{hop}} \gtrsim 0.2$. We will briefly discuss the choice of cutoff later (and it is discussed more thoroughly in the supplementary material of Schoenholz *et al.* [16]) but the results are qualitatively insensitive to the selection. Note that $P_R(S_i)$ will be invariant to the transformation discussed above.

We plot the probability of rearrangement at a number of different temperatures in fig. 4 (a). We see that there is a strong softness dependence on the probability of rearrangement with high-softness particles having a probability of rearrangement nearly two orders of magnitude larger than their low-softness counterparts.

The probability of rearrangement can be interpreted in a slightly different way by plotting $P_R(S)$ against $1/T$ for different values of softness. The result of this transformation can be seen in fig. 4 (b) and we see, in particular, that for a given value of softness $P_R(S)$ depends exponentially on $1/T$. Surprisingly, this implies that, separated by softness, the probability of rearrangement is Arrhenius. With reference to transition state theory we can therefore write down what we believe to be a fundamental equation of glassy dynamics,

$$P_R(S) = Z(S)e^{-E(S)/kT}. \tag{11}$$

Here $E(S)$ and $Z(S)$ is a structure-dependent energy scale and multiplicity scale respectively that together govern local rearrangements. We will frequently refer to a structure dependent entropy, $\Sigma(S) = \log Z(S)$.

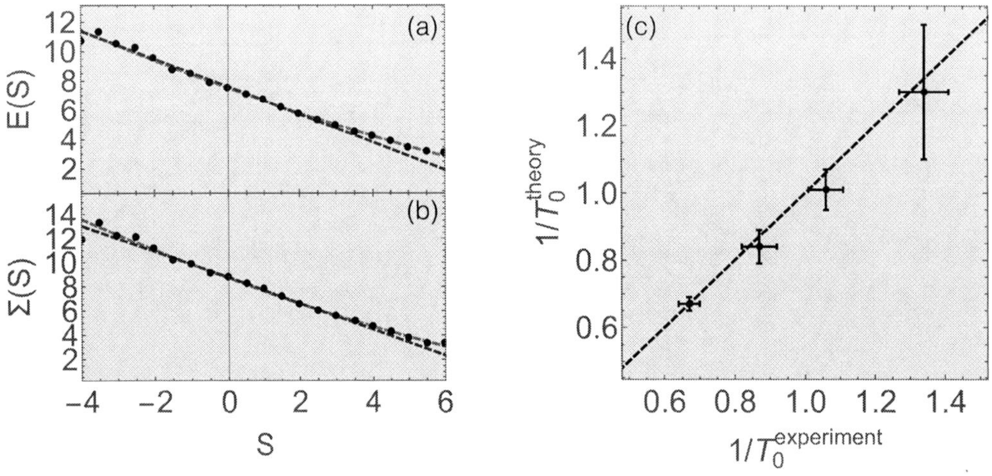

Figure 5. Energy scales. (a) The energy scale governing local rearrangements as a function of softness with a linear fit (blue) and a quadratic fit (red). (b) The entropy scale Σ also with linear and quadratic fits. (c) A comparison between the theoretical onset temperature and the onset temperature identified by Keys *et al.* [28].

Using eq. (11) we can extract energy and entropy scales that control local rearrangements as a function of softness by fitting to the Arrhenius form in fig. 4 (b). The result of these fits can be seen in fig. 5 (a) along with linear (blue) and quadratic (red) approximations to $E(S)$ and $\Sigma(S)$. We see that both the energy scale and entropy scale are approximately linear near $S = 0$. This is corroborated by considering the fits themselves; we find, $E(S) \approx E_0 - E_1 S + E_2 S^2$ with $E_0 \approx 7.50$, $E_1 \approx 0.90$, $E_2 \approx 0.02$ giving $E_1/E_2 \sim 50$. This implies that nonlinear effects are relatively small in the regime with most of the probability mass of S. We will therefore typically take a linear approximation to the energy and entropy scales. We also find that $\Sigma(S) \approx \Sigma_1 - \Sigma_1 S + \Sigma_2 S^2$ with $\Sigma_0 \approx 8.70$, $\Sigma_1 \approx 1.1$, $\Sigma_2 \approx 0.029$ which gives a similar ratio. In fig. 4 (a) we show $P_R(S)$ predicted by the quadratic approximation to the energy scale and entropy scale combined with the Arrhenius form in eq. (11).

For Lennard-Jones glasses we therefore have that,

$$P_R(S) \approx \exp\left[E_0\left(\frac{1}{T_0} - \frac{1}{T}\right) - E_1\left(\frac{1}{T_1} - \frac{1}{T}\right)S\right] \tag{12}$$

where $T_0 = E_0/\Sigma_0$ and $T_1 = E_1/\Sigma_1$. Note that $T_0 \approx T_1 \approx 0.86$ and so we can further approximate eq. (11) by,

$$P_R(S) \approx \exp\left[\left(\frac{1}{T_0} - \frac{1}{T}\right)(E_0 - E_1 S)\right]. \tag{13}$$

When $T = T_0$ we find that $P_R(S) = 1$ independent of structure and the structural component of dynamics vanishes. We can compare this temperature with the onset of glassy dynamics found independently [28]. We see the result of this comparison in fig. 5 (b) for the Lennard-Jones mixture at densities, $\rho = 1.15, 1.20, 1.25$, and 1.30. The excellent agreement between these two quantities lead us to the conclusion that we can actively define the onset of glassy dynamics by the onset of correlations between local-structure and dynamics. In addition to providing an explanation for the onset temperature, eq. (13) also explains the heterogeneous dynamics of glassy liquids as the heterogeneous distribution of softness leads a corresponding spatial distribution of local energy barriers to rearrangement.

4.2. Relaxation

We will now discuss the relationship between softness and relaxation in glassy systems both in-and-out- of equilibrium. To do this we first establish a connection between the probability of rearrangement and the relaxation time of a glassy liquid. We begin by defining a per-softness overlap function in analogy to eq. (1) to be the overlap after a time t for particles that had softness S at time $t = 0$,

$$q(S,t) = \frac{1}{N_S}\sum_i \Theta(|r_i(t) - r_i(0)| - a)\delta(S_i(0) - S) \tag{14}$$

where N_S is the number of particles with softness S at $t = 0$. The full overlap function is related to the softness dependent overlap function by $q(t) = \int dSq(S,t)P(S)$.

To develop a connection between eq. (11) and $q(S,t)$ we must take into account facilitation which will manifest itself as a particle whose softness changes without a rearrangement (i.e. because some nearby rearrangement changed its local structure). For a more detailed discussion of this derivation see the supplementary material in Schoenholz *et al.* [16]. We define the "softness propagator", $G(S, S_0, t)$ to be the fraction of particles that have softness S at a time t if they had softness S_0 at time $t = 0$ and did not rearrange in the interval $[0,t]$. Note that $\int dSG(S, S_0, t) = 1$.

Using the softness propagator we can write down the fraction of particles that had a softness S_0 at $t = 0$ that rearrange at a time t,

$$f_t(S) = \int dS'G(S', S, t)P_R(S'). \tag{15}$$

We then consider a simple model of the glass that proceeds in discreet steps of time τ_R, the average duration of a rearrangement. In terms of $f_t(S)$ we can write down the softness dependent overlap function,

$$q(S,t) = 1 - \sum_{t'=0}^{t}\prod_{t''=0}^{t'-1}(1 - f_{t''}(S))f_{t'}(S). \tag{16}$$

Thus, if we understand $P_R(S)$ and $G(S', S, t)$ then we can compute the relaxation of the entire glassy system.

To study the relaxation time we introduce a simple mean-field model where we replace our system by a system with uniform local structure with softness equal to $\langle S\rangle$. This is equivalent to setting $P(S) = \delta(S - \langle S\rangle)$ and $G(S, S', t) = \delta(S - \langle S\rangle)$. In this approximation it follows that $q(t) = q(\langle S\rangle, t)$ and $f_t(S) = P_R(\langle S\rangle)$. It follows that,

$$q(t) = 1 - \sum_{t'=0}^{t}(1 - P_R(\langle S\rangle))^{t'}P_R(\langle S\rangle) = (1 - P_R(\langle S\rangle))^t \tag{17}$$

using the properties of geometric series. This implies that

$$\tau_\alpha \sim \frac{-1}{\log(1 - P_R(\langle S \rangle))} \approx \frac{1}{P_R(\langle S \rangle)} \tag{18}$$

when $P_R(\langle S \rangle)$ is small. For our Lennard-Jones system this implies that at the level of our mean-field theory,

$$\tau_\alpha \sim \exp\left[-\left(\frac{1}{T_0} - \frac{1}{T}\right)(E_0 - E_1\langle S \rangle)\right]. \tag{19}$$

If correct, this implies that the bulk of the super Arrhenius scaling of τ_α comes from changes in $\langle S \rangle$.

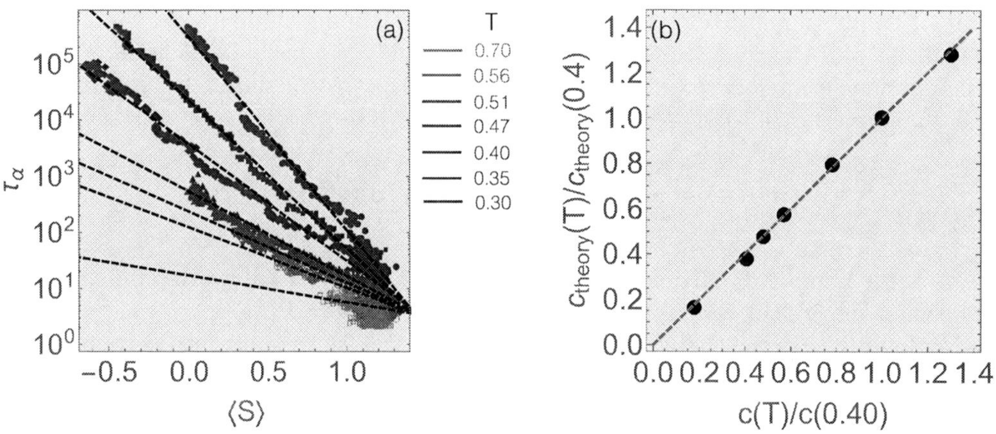

Figure 6. The relaxation time. (a) The relaxation time, τ_α, for an aging system as a function of the average softness of the system. We see a definitive exponential relationship. (b) A comparison of the exponential pre-factor multiplying softness with the theoretical prediction. We see strong agreement between the two.

We can check the validity of this equation by considering non-equilibrium glassy systems as they age. For a more detailed discussion see Schoenholz *et al.* [18]. We let a Kob-Andersen mixture come to equilibrium at a high temperature, $T_I \approx 1.0$. and then instantaneously quench to some final temperature T_F. We then measure how the average softness and relaxation time evolve with time. Plotting this relationship in fig. 6 (a) we see that indeed the relaxation time appears to scale exponentially with the average softness, $\tau_\alpha \sim e^{-c(T)\langle S \rangle}$. We then plot the measured ratio of $c(T_F)/c(0.4)$ against the theoretical prediction from eq. (19) for a number of different final temperatures. The result of this plot is in fig. 6 where we see excellent agreement.

Together these results show that our mean field model successfully captures the fundamental features of the relaxation time and moreover that eq. (19) seems to be accurate. This gives hope to mean field models of the glass transition since it seems as though the bulk relaxation time does not depend on variations in local structure. Indeed we see that, $\langle S \rangle = \int dr g(r)\tilde{w}(r)$ and so the relaxation time will be a function of $g(r)$ alone. Moreover, the success of eq. (19) places severe constraints on the possible forms for the functional relaxation time [18]. It would be interesting to attempt to formally relate this formalism to mode coupling theory.

Unlike the case of the relaxation time, it appears that local structure and facilitation are necessary to properly capture the full relaxation profile of the overlap function, $q(t)$. To see this in fig. 7 (a) we compare the overlap function found empirically to the mean field overlap function. We see that as the temperature is lowered the form of the true overlap function

IOP Conf. Series: Journal of Physics: Conf. Series **1036** (2018) 012021 doi:10.1088/1742-6596/1036/1/012021

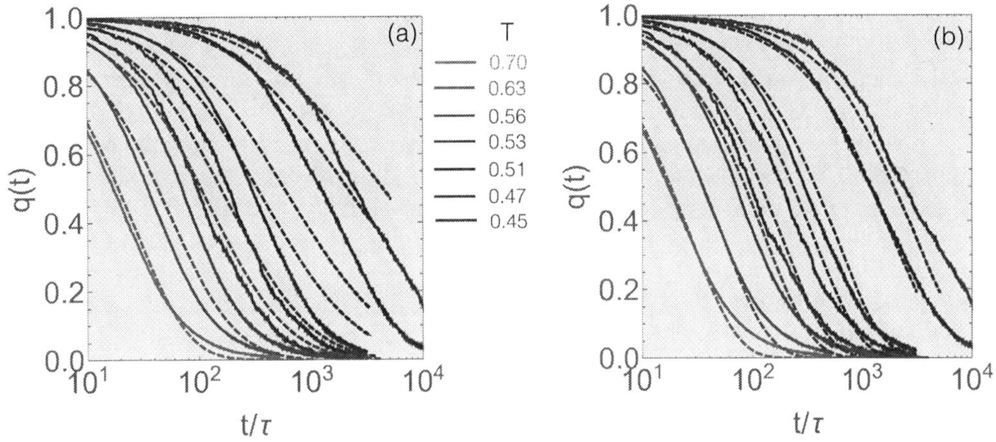

Figure 7. Non-exponential behavior of the overlap function. (a) The overlap function as well as mean-field estimates from our model. The mean field relaxation clearly fails to accurately describe the behavior of $q(t)$. (b) The same plot but compared with the prediction of our full model including facilitation. There is clear agreement once local effects are taken into account.

diverges from the mean-field prediction. In fig. 7 (b) we plot the same overlap function against theoretical prediction found by evaluating eq. (16) directly using a measured function for the softness propagator and numerically integrating it against the rearrangement probability. We see that the agreement improves drastically; indeed it seems as though the full model does successfully predict the full form of the overlap function.

5. Conclusions and Future Directions

We have offered a data-driven approach to understanding the behavior of supercooled liquids and glasses from their structure through the construction of the softness. We are able to use softness to connect structure to dynamics and in the process offer explanations for a number of long-standing problems in glass physics with strong experimental evidence.

The distribution of softness, the probability of rearrangement, and the softness propagator are, at this point, fundamentally empirical constructions without solid theoretical underpinning. Therefore it seems that future work should focus on making connection between these quantities and *ab initio* theory. Most pressing is probably the development of a model for the dynamics of softness from more fundamental measurements. Furthermore, it would be interesting to explore the connections between the formalism presented here and the mean-field theories of random first order theory and mode-coupling theory. This is especially true of the relaxation time which we have shown explicitly does not depend on fluctuations in the local structure to an excellent approximation. Ultimately, the results presented here should act as a strong constraint on any theoretical work on the glass transition.

It is possible that at lower temperatures the Arrhenius form for $P_R(S)$ will break down. We have never been able to accurately enough estimate $P_R(S)$ at such low temperatures. An interesting direction for future work might be to probe the very low temperature behavior of softness. To do this very long simulations would have to be performed on glasses. Recent work [41] has shown that it might be possible to stitch together many trajectories to do very long time simulations. It would be interesting to combine this approach with softness to perform very long simulations of glasses and in turn to measure the Arrhenius form of $P_R(S)$ at very low temperatures.

Another interesting avenue for future work is to use the softness to investigate different properties of systems by measuring the distribution of energy scales and entropy scales. Already we have pursued this approach to successfully investigate the role of structure in enhancing the mobility of particles near a free surface in a polymeric glass [32]. There we found minimal structural involvement which shows that these methods can be used to prove both positive and negative results about the role of structure in dynamics. We have also begun to use this method to investigate grain boundaries where it has been remarkably successful.

References

[1] D. Henderson, J. A. Barker, and S. Kim. Theory of the liquid state. *International Journal of Quantum Chemistry*, 3(S3A):265–292, 1969.

[2] N.W. Ashcroft and N.D. Mermin. *Solid State Physics*. Saunders College, Philadelphia, 1976.

[3] L. D.avidovitch Landau and E. M. Lifshitz. *Course of theoretical physics: theory of elasticity*. Butterworth-Heinemann, 1986.

[4] David R Reichman and Patrick Charbonneau. Mode-coupling theory. *Journal of Statistical Mechanics: Theory and Experiment*, 2005(05):P05013, 2005.

[5] T. R. Kirkpatrick and D. Thirumalai. Dynamics of the structural glass transition and the p-spin interaction spin-glass model. *Phys. Rev. Lett.*, 58:2091–2094, May 1987.

[6] Giorgio Parisi and Francesco Zamponi. Mean-field theory of hard sphere glasses and jamming. *Reviews of Modern Physics*, 82(789):789, 2010.

[7] Ludovic Berthier, Giulio Biroli, Jean-Philippe Bouchaud, Luca Cipelletti, and Wim van Saarloos. *Dynamical Heterogeneities in Glasses, Colloids, and Granular Media*. Oxford Scholarships Online, 2011.

[8] Patrick Charbonneau, Jorge Kurchan, Giorgio Parisi, Pierfrancesco Urbani, and Francesco Zamponi. Exact theory of dense amorphous hard spheres in high dimensions. iii. the full rsb solution. *J. Stat. Mech.*, page 10009, 2014.

[9] Wolfgang Götze. Recent tests of the mode-coupling theory for glassy dynamics. *Journal of Physics: Condensed Matter*, 11(10A):A1, 1999.

[10] M. L. Falk and J. S. Langer. Dynamics of viscoplastic deformation in amorphous solids. *Phys. Rev. E.*, 57:7192, 1998.

[11] Glenn H. Fredrickson and Hans C. Andersen. Kinetic ising model of the glass transition. *Phys. Rev. Lett.*, 53:1244–1247, Sep 1984.

[12] Takeshi Kawasaki and Hajime Tanaka. Structural evolution in the aging process of supercooled colloidal liquids. *Phys. Rev. E*, 89:062315, Jun 2014.

[13] M. L. Manning and A. J. Liu. Vibrational modes identify soft spots in a sheared disordered packing. *Phys. Rev. Lett.*, 107:108302, 2011.

[14] C Patrick Royall, Stephen R Williams, Takehiro Ohtsuka, and Hajime Tanaka. Direct observation of a local structural mechanism for dynamic arrest. *Nature materials*, 7(7):556–561, 2008.

[15] Ekin Dogus Cubuk, Samuel Stern Schoenholz, Jennifer M. Rieser, Brad Dean Malone, Joerg Rottler, Douglas J. Durian, Efthimios Kaxiras, and Andrea J. Liu. Identifying structural flow defects in disordered solids using machine learning methods. *Phys. Rev. Lett.*, 114(10):108001, 2015.

[16] S. S. Schoenholz, E. D. Cubuk, D. M. Sussman, E. Kaxiras, and A. J. Liu. A structural approach to relaxation in glassy liquids. *Nat Phys*, 12(5):469–471, 05 2016.

[17] Ekin D. Cubuk, Samuel S. Schoenholz, Efthimios Kaxiras, and Andrea J. Liu. Structural properties of defects in glassy liquids. *The Journal of Physical Chemistry B*, 120(26):6139–6146, 2016. PMID: 27092716.

[18] Samuel S. Schoenholz, Ekin D. Cubuk, Efthimios Kaxiras, and Andrea J. Liu. Relationship between local structure and relaxation in out-of-equilibrium glassy systems. *Proceedings of the National Academy of Sciences*, 114(2):263–267, 2017.

[19] Pablo G Debenedetti and Frank H Stillinger. Supercooled liquids and the glass transition. *Nature (London)*, 410(6825):259–267, 2001.

[20] A. S. Keys, J. P. Garrahan, and D. Chandler. Calorimetric glass transition explained by hierarchical dynamic facilitation. *PNAS*, 110(12), 2013.

[21] A. Widmer-Cooper, H. Perry, P. Harrowell, and D. R. Reichman. Localized soft modes and the supercooled liquid's irreversible passage through its configuration space. *J. Chem. Phys.*, 131(194508), 2009.

[22] S. S. Schoenholz, A. J. Liu, R. A. Riggleman, and J. Rottler. Understanding plastic deformation in thermal glasses from single-soft-spot dynamics. *Phys. Rev. X*, 4:031014, Jul 2014.

[23] J. Rottler, S. S. Schoenholz, and A. J. Liu. Predicting plasticity with soft vibrational modes: From dislocations to glasses. *Phys. Rev. E.*, 89:042304, 2014.

[24] J. Gilman. Metallic glasses. *Phys. Today*, 28(5):46, 1975.

[25] Ludovic Berthier and Robert L Jack. Structure and dynamics of glass formers: Predictability at large length scales. *Phys. Rev. E*, 76(4):041509, 2007.
[26] R. L. Jack, A. J. Dunleavy, and C. P. Royall. *Phys. Rev. Lett.*, 113:095703, 2014.
[27] Jacques Zylberg, Edan Lerner, Yohai Bar-Sinai, and Eran Bouchbinder. Local thermal energy as a structural indicator in glasses. *Proceedings of the National Academy of Sciences*, 114(28):7289–7294, 2017.
[28] A. S. Keys, L. O. Hedges, J. P. Garrahan, S. C. Glotzer, and D. Chandler. Excitations are localized and relaxation is hierarchical in glass-forming liquids. *Phys. Rev. X.*, 1:021013, 2011.
[29] K. Chen, T. Still, S. Schoenholz, K. B. Aptowicz, M. Schindler, A. C. Maggs, A. J. Liu, and A. G. Yodh. Phonons in two-dimensional soft colloidal crystals. *Phys. Rev. E.*, 88:022315, 2013.
[30] Walter Kob and Hans C Andersen. Scaling behavior in the β-relaxation regime of a supercooled lennard-jones mixture. *Phys. Rev. Lett.*, 73(10):1376, 1994.
[31] S. Plimpton. *J. Comp. Phys.*, 117:1–19, 1995.
[32] D. M. Sussman, S. S. Schoenholz, E. D. Cubuk, and A. J. Liu. Disconnecting structure and dynamics in glassy thin films. *ArXiv e-prints*, October 2016.
[33] R. Candelier, A. Widmer-Cooper, J. K. Kummerfeld, O. Dauchot, G. Biroli, P. Harrowell, and D. R. Reichman. Spatiotemporal hierarchy of relaxation events, dynamical heterogeneities, and structural reorganization in a supercooled liquid. *Phys. Rev. Lett.*, 105:135702, 2010.
[34] Anton Smessaert and J org Rottler. Distribution of local relaxation events in an aging three-dimensional glass: Spatiotemporal correlation and dynamical heterogeneity. *Physical Review E*, 88(022314):022314, 2013.
[35] Jörg Behler and Michele Parrinello. Generalized neural-network representation of high-dimensional potential-energy surfaces. *Phys. Rev. Lett.*, 98(14):146401, 2007.
[36] Ekin D Cubuk and Efthimios Kaxiras. Theory of structural transformation in lithiated amorphous silicon. *Nano letters*, 14(7):4065–4070, 2014.
[37] Ekin D. Cubuk, Brad D. Malone, Berk Onat, Amos Waterland, and Efthimios Kaxiras. Representations in neural network based empirical potentials. *The Journal of Chemical Physics*, 147(2):024104, 2017.
[38] J. Gilmer, S. S. Schoenholz, P. F. Riley, O. Vinyals, and G. E. Dahl. Neural Message Passing for Quantum Chemistry. *ArXiv e-prints*, April 2017.
[39] F. A. Faber, L. Hutchison, B. Huang, J. Gilmer, S. S. Schoenholz, G. E. Dahl, O. Vinyals, S. Kearnes, P. F. Riley, and O. Anatole von Lilienfeld. Machine learning prediction errors better than DFT accuracy. *ArXiv e-prints*, February 2017.
[40] Corinna Cortes and Vladimir Vapnik. Support-vector networks. *Mach. Learn.*, 20(3):273–297, 1995.
[41] Danny Perez, Ekin D. Cubuk, Amos Waterland, Efthimios Kaxiras, and Arthur F. Voter. Long-time dynamics through parallel trajectory splicing. *Journal of Chemical Theory and Computation*, 12(1):18–28, 2016. PMID: 26605853.

IOP Conf. Series: Journal of Physics: Conf. Series **1036** (2018) 012022 doi:10.1088/1742-6596/1036/1/012022

Bayesian spectroscopy in solid-state photo-physics

Ichiro Akai[1], Kazunori Iwamitsu[2], Masato Okada[3,4]

[1]Institute of Pulsed Power Science, Kumamoto University, 2-39-1 Kurokami Chuo-ku, Kumamoto 860-8555, Japan.

[2]Faculty of Science, Kumamoto University, 2-39-1 Kurokami Chuo-ku, Kumamoto 860-8555, Japan.

[3]Department of Complexity Science and Engineering, The University of Tokyo, 5-1-5 Kashiwanoha Kashiwa, Chiba 277-8561, Japan.

[4]Research and Service Division of Materials Data and Integrated System, National Institute for Material Science, 1-2-1 Sengen, Tsukuba, Ibaraki, 305-0047, Japan.

E-mail: `iakai@kumamoto-u.ac.jp`

Abstract. We propose Bayesian spectroscopy (BS) both for the spectral analyses of optical spectra and for normal mode analyses of coherent phonon (CP) signals in solid-state materials. In optical spectra due to elementary electronic excitations, multiple excitonic absorption peaks appear and are overlapped with other background optical transitions, which makes their spectral decomposition difficult. We demonstrate that BS can be used to estimate the band gap and the excitonic binding energies with high accuracies by the introduction of physical laws related with the excitonic transitions. For normal mode analyses of CP signals, we show that BS can estimate the frequency and initial vibrating phase of the normal mode with high accuracies. In contrast to the Fourier transform spectrum which has a broad spectral width on account of the damping behaviors of the CP signals, BS with a physical model of a damped oscillation improves the estimation accuracy of the normal mode frequency by two orders of magnitude compared to the Fourier transform method. These results indicate the significant advantages of BS over conventional methodologies such as the least-squares and Fourier transform methods.

1. Introduction

Optical spectroscopy is a powerful method to investigate various physical properties of materials because electrons and related elementary excitations interact with the incident light. Any substances absorb and/or emit light of specific wavelengths; therefore, optical spectra can be measured that includes various information related to their physical properties. Recent advances in spectroscopic and ultrashort pulsed laser technologies have led to improvements in spectroscopic energy resolution and time resolution of temporal measurements, and have contributed significantly to frontier research in photo-physics.

The least-squares method [1] has typically been used for spectral analyses. The method is based on causal determinism, and it considers that the measured data will be scattered due to random noise. However, the spectral parameters that represent the physical properties are nonlinearly included in the spectral functions that represent the causality of the optical spectra; therefore, analyses by the least-squares method frequently encounter difficulties. The difficulties increase, particularly in solid-state materials, because the optical spectra become complicated due to overlapping of several types of discrete and continuous optical transitions. In such

IOP Conf. Series: Journal of Physics: Conf. Series **1036** (2018) 012022 doi:10.1088/1742-6596/1036/1/012022

complicated spectra, it cannot be guaranteed that the analysis results are correct, and the estimation accuracies of the spectral parameters cannot be discussed with statistical confidence.

Recently, a Bayesian methodology to decompose such complicated spectra has been developed. Nagata *et al.* have succeeded in the deconvolution of X-ray photoelectron and optical absorption spectra that include multi-spectral components, and have statistically evaluated the probability distributions of their spectral parameters [2]. More recently, a virtual measurement analysis (VMA) methodology based on the Bayesian inference has also been proposed [3], which has enabled evaluation of the detection limits and optimization of the experimental setups.

In this paper, we propose Bayesian spectroscopy for solid-state photo-physics, which has significant advantages over the conventional least-squares method. In Sec. 2, we survey the fundamental methodology of the Bayesian inference. In Sec. 3, Bayesian spectroscopy of an optical spectrum of elementary electronic excitations is described. From the absorption spectrum of a Cu_2O thin crystal, we demonstrate that the band gap energy E_g, can be estimated with extremely high accuracy by the introduction of physical laws of excitonic transitions [4]. In addition, a VMA is performed to evaluate the variation of the estimation accuracy for E_g with respect to the spectral widths of exciton transitions. Bayesian spectroscopy has significant advantages, not only for such spectral decomposition, but also in normal mode analyses for coherent phonon (CP) signals. In solid-state substances, we can coherently excite phonons by femtosecond (fs) laser pulses. To resolve normal modes from such coherently-excited phonon signals, Fourier and wavelet transformations have been employed to date. In Sec. 4, we demonstrate that Bayesian spectroscopy can be used to estimate the normal mode frequency and initial vibrating phase with high accuracy beyond such conventional methods. In the final sub-sections of Secs. 3 and 4, we summarize the future prospects of such Bayesian spectroscopy.

2. Bayesian inference

In Bayesian spectroscopy, we attempt to fit a measured data set $D(y_i; x_i)$ $(i = 1, \cdots, N)$ by an appropriate model function $G(x_i; \Theta)$ specified by a parameter set Θ. However, noise n_i is randomly superposed during measurements. Therefore, y_i is expressed as $y_i = G(x_i; \Theta) + n_i$ and the mean square error function $E(\Theta)$ of the model function is defined as.

$$E(\Theta) = \frac{1}{2N} \sum_{i=1}^{N} [y_i - G(x_i; \Theta)]^2 . \tag{1}$$

A joint probability $P(D, \Theta)$ for D and Θ is equivalent to $P(D|\Theta)P(\Theta)$, where the $P(D|\Theta)$ and $P(\Theta)$ are a conditional probability of D under Θ given and a prior probability of Θ, respectively. By Bayes' theorem [5], the $P(D, \Theta)$ is also written as $P(\Theta|D)P(D)$, where $P(\Theta|D)$ is a conditional probability of Θ under the measured data D given, which is what must be evaluated. When noise n_i is dispersed by a normal distribution with a standard deviation σ_{noise}, and a mean value of zero, the noise distribution $P(\{n_i\})$ is the same as $P(D|\Theta)$ and it is proportional to $\exp\left[-NE(\Theta)/\sigma_{\text{noise}}^2\right]$. Consequently, the posterior probability distribution $P(\Theta|D)$ can be evaluated using Eq. (2) [2].

$$P(\Theta|D) = \frac{P(D|\Theta)P(\Theta)}{P(D)} \propto \exp\left[-\frac{N}{\sigma_{\text{noise}}^2} E(\Theta)\right] P(\Theta), \tag{2}$$

where $P(D)$ becomes constant after measurement of the data set D. When $P(\Theta)$ is a continuous uniform distribution, $P(\Theta|D)$ is proportional to $\exp\left[-NE(\Theta)/\sigma_{\text{noise}}^2\right]$. Consequently, the respective values in Θ will be distributed nearby the minimum point of $E(\Theta)$ in the parameter space, according to the σ_{noise} of the noise.

Markov chain Monte Carlo (MCMC) methods [2, 6] have been employed to efficiently obtain the distributions of $P(\Theta|D)$. In this paper, the parameters in Θ were updated and recorded using a Metropolis algorithm [7, 8] after sufficient burn-in [6] phases. To perform Bayesian spectroscopy appropriately, the Metropolis samplings were conducted repeatedly with changing $\sigma_{\rm noise}$ and the value of $\sigma_{\rm noise}$ was determined based on a criterion that the root mean squared error of $\{G(x_i;\Theta)\}$ obtained from the data set D is balanced with $\sigma_{\rm noise}$.

3. Bayesian spectroscopy for elementary electronic excitations

In solid-state materials, dipole-allowed excitonic transitions [9, 10] play important roles in properties such as optical nonlinearity [11, 12] and luminescence [13, 14]. In contrast, dipole-forbidden excitonic states have also attracted much attention from a recent experimental result [15]. Such excitonic states are important candidates for the realization of excitonic Bose-Einstein condensation [16] (X-BEC) because of their long lifetimes. Investigations of X-BEC have been conducted intensively on Cu_2O crystals due to the finding of unusual and ballistic exciton transport [17]. However, a valid manifestation of X-BEC has only recently been reported [15] at sub-Kelvin temperatures. In addition, the optical responses (e.g., absorption and luminescence) of such excitonic states are very weak because of the dipole-forbidden transitions. Therefore, we encounter difficulties in their spectral analyses, even though we can detect such signals.

The conventional approach to decompose such complicated and weak optical spectra is the least-squares method. After decomposition, the properties of such excitonic state are investigated for each spectral component. However, it is difficult to evaluate the accuracies of the respective spectral parameters because such parameters are included nonlinearly in each spectral function. To overcome such a limitation, we have recently employed Bayesian spectroscopy to decompose an excitonic absorption spectrum in a Cu_2O thin crystal sandwiched by paired MgO substrates [4].

3.1. Excitonic absorption spectrum in the Cu_2O thin crystal

The open circles in Fig. 1(a) shows an absorption spectrum measured for a Cu_2O thin crystal [4]. The absorption intensity is quite weak, as shown in the ordinate scale, due to the thinness (ca. 2.8 μm [18]) of the crystal; therefore, this spectrum is obtained through wavelength integration of a wavelength modulated absorption spectrum [18, 19]. To realize X-BEC in such thin crystals, spatial accumulation of excitons beyond the critical density and suppression of spatial diffusion of excitons are essential. To achieve this goal, we must therefore confirm the formation of a shallow potential to trap excitons. Therefore, it is important to quantitatively evaluate the differences in the band gap energy E_g, and the excitonic Rydberg constant R_y, between single and thin Cu_2O crystals.

The absorption spectrum in Fig. 1(a) includes spectral components of excitonic resonance transitions for the yellow excitonic series X_n $(n = 2, 3, \cdots, \infty)$, excitonic continuous and band-to-band absorption bands. The spectral functions of X_n are asymmetric Lorentzian functions [20] in Eq. (3) for the resonance energy E_n of the n–th exciton state, and they have three spectral parameters of absorption strength f_n, asymmetric degree A_n, and homogeneous broadening width, Γ_n.

$$X_n(E;E_n) = f_n \times \frac{1}{\pi}\frac{\Gamma_n + 2A_n(E-E_n)}{(E-E_n)^2+\Gamma_n^2}. \tag{3}$$

Spectral functions $X_{\rm C}(E;E_g)$ and $B(E;E_g)$ for the excitonic continuous and band-to-band absorption bands have been described in [4]. For analysis of the measured absorption spectrum, it is necessary to convolve inhomogeneous broadenings in the respective spectral components $g(E;E_0)$ as expressed in Eq. (4), because of the spectral resolution and sample inhomogeneity.

$$G\Big(g(E;E_0)\Big) = \int g(E;E_0+{\rm d}\varepsilon)\exp\left(-\ln 2\frac{\varepsilon^2}{\gamma_g^2}\right){\rm d}\varepsilon \Big/ \int \exp\left(-\ln 2\frac{\varepsilon^2}{\gamma_g^2}\right){\rm d}\varepsilon. \tag{4}$$

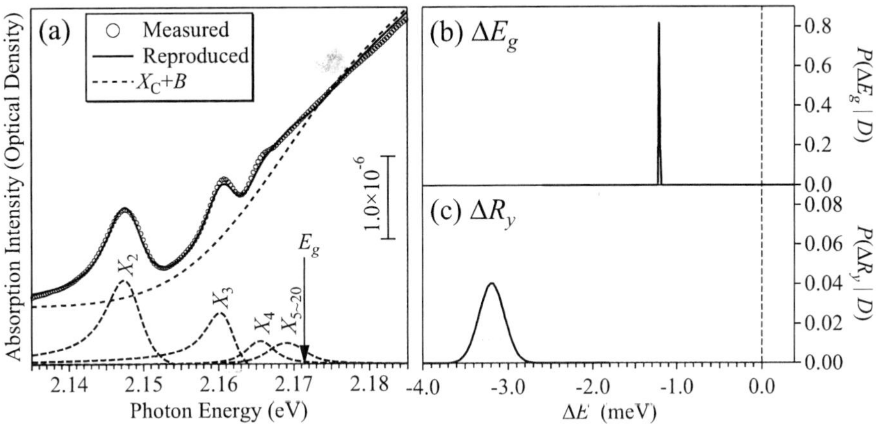

Figure 1. (a) Measured (○) and reproduced (—) absorption spectra for a Cu_2O thin crystal sandwiched by paired MgO substrates. The dashed line spectra show the spectral components included in the reproduced spectrum [4]. Posterior probability distributions of (b) ΔE_g and (c) ΔR_y.

3.2. Physical laws for excitonic transitions

In our recent work [4], we have claimed that even if each spectral component has broad spectral width, it is possible to estimate spectral parameters with high accuracy by the introduction of appropriate physical laws between these spectral parameters. For the absorption spectrum in Fig. 1(a), a physical law was introduced in Eq. (5) among the excitonic resonance energies E_n, the band gap energy E_g, and the excitonic Rydberg constant R_y [9].

$$E_n = E_g - \frac{R_y}{n^2}, \quad (n = 2, 3, \cdots, n_\infty). \tag{5}$$

In addition, a physical law was also introduced in Eq. (6) for the absorption intensities of the dipole-forbidden direct n–th excitonic resonances [21].

$$f_n = f_0 \left(\frac{n^2 - 1}{n^5} \right), \quad (n = 2, 5, \cdots, n_\infty), \tag{6}$$

where f_3 and f_4 do not obey this physical law, as reported in [18], so that they are dealt with as independent spectral parameters. n_∞ in Eqs. (5) and (6) is the number of excitonic excited states to be considered and this is set to 20 because f_{20}/f_2 becomes sufficiently small.

3.3. Bayesian spectroscopy of excitonic absorption spectrum

The parameter set Θ for our Bayesian spectroscopy is expressed in Eq. (7).

$$\Theta = \{E_g, R_y, A_2, A_3, f_0, f_3, f_4, \gamma_X, \Gamma_2, \Gamma_X\}, \tag{7}$$

where it was assumed that all excitonic resonances X_n $(n = 2, \cdots, 20)$, have the same inhomogeneous broadening factor γ_X. Γ_2 and Γ_X are homogeneous broadening factors for the excitonic resonance peaks, where Γ_2 for the broad absorption peak of X_2 was implemented separately from the other peaks X_n $(n = 3, \cdots, 20)$. Asymmetric factors of the excitonic resonances higher than X_4 are neglected because their absorption strengths become weak at large n. We have successfully obtained posterior probability distributions $P(\Theta|D)$, through 2.5 million samplings after a sufficient burn-in phase. Using the mean values $\hat{\Theta}$ of $P(\Theta|D)$, we can reproduce the measured absorption spectrum well, as depicted by the solid line in Fig. 1(a).

For the Cu_2O thin crystal sandwiched by paired MgO plates, a small lattice mismatch between Cu_2O and MgO induces two-dimensional stresses at the epitaxial interface between Cu_2O and MgO [22]. It is thus expected that the transition energies of the excitonic resonance are changed by such stresses. Therefore, we evaluate both energy shifts ΔE_g and ΔR_y in the band gap energy and the effective Rydberg constant from those in the single crystal [23]. Figures 1(b) and (c) show the posterior probability distributions $P(\Delta E_g|D)$ and $P(\Delta R_y|D)$, respectively. $P(\Delta E_g|D)$ has a very sharp profile in the negative abscissa region and it is obvious that E_g for the Cu_2O thin crystal is significantly reduced compared with that for the single crystal [4]. In Fig. 1(c), although $P(\Delta R_y|D)$ has a rather broad profile in the same abscissa scale as that with Fig. 1(b), it is also clear that the effective Rydberg constant, i.e. the binding energy of electron and hole in the exciton, becomes small [4].

3.4. Virtual measurement analysis of excitonic absorption spectrum

The dashed-line spectrum overlapped on the measured spectrum in Fig. 1(a) indicates the background spectrum of the excitonic resonance transitions. The background comes from X_C and B, and has only a gentle slope rising to the higher energy side. Therefore, the spectral profiles $X_C(E;\hat{E}_g)$ and $B(E;\hat{E}_g)$ [4] do not change significantly with a small change of $\hat{E}_g$ and do not have a significant contribution to the estimation of $\hat{E}_g$. In contrast, the excitonic resonance transitions X_2, X_3 and X_4 provide clear peaks and a shoulder structure on the measured spectrum in Fig. 1(a). Consequently, these transitions make it possible to estimate $\hat{E}_g$ and $\hat{R}_y$ with high precision because of the strong constraint among the transition energies E_n, in Eq. (5).

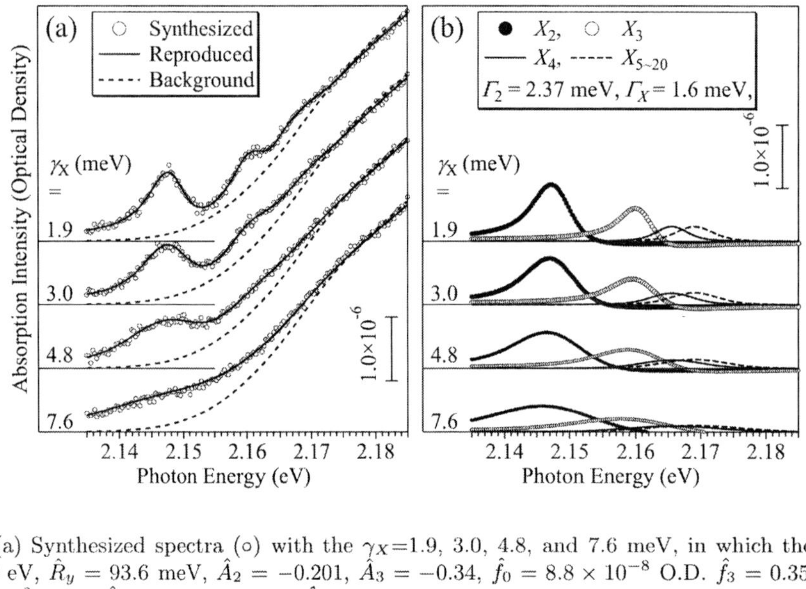

Figure 2. (a) Synthesized spectra (○) with the γ_X=1.9, 3.0, 4.8, and 7.6 meV, in which the parameters of $\hat{E}_g$ = 2.171392 eV, $\hat{R}_y$ = 93.6 meV, $\hat{A}_2$ = −0.201, $\hat{A}_3$ = −0.34, $\hat{f}_0 = 8.8 \times 10^{-8}$ O.D. $\hat{f}_3 = 0.355 \times 10^{-8}$ O.D., $\hat{f}_4 = 0.179 \times 10^{-8}$ O.D., $\hat{\Gamma}_2$ = 2.37 meV and $\hat{\Gamma}_X$ = 1.6 meV are used. The dashed and solid lines represent the background component ($X_C + B$) in the synthesized spectra and the reproduced spectra obtained from Bayesian spectroscopy. (b) Spectral profiles of the excitonic resonances with the respective γ_X.

To verify how the introduction of the physical law in Eq. (5) is effective for Bayesian spectroscopy, a VMA [3] was performed to evaluate the estimation accuracies of E_g and R_y from the synthesized data, which were made with different inhomogeneous broadening factors γ_X in Eq. (7). Spectra plotted by open circles in Fig. 2(a) are synthesized absorption spectra with the respective γ_X. The mean values $\hat{\Theta}$ of $P(\Theta|D)$ are used for other spectral parameters as they are.

Figure 2(b) shows the variation of the spectral profiles of X_2, X_3, X_4 and $X_{5\sim20}$. The standard deviation of noise in the synthesized spectra is the same as that in the measured spectrum of Fig. 1(a). The higher excitonic resonance peaks X_n ($n \geq 3$) fade away with increasing γ_X and the lowest X_2 peak only shows a shoulder structure at the largest γ_X (=7.6 meV).

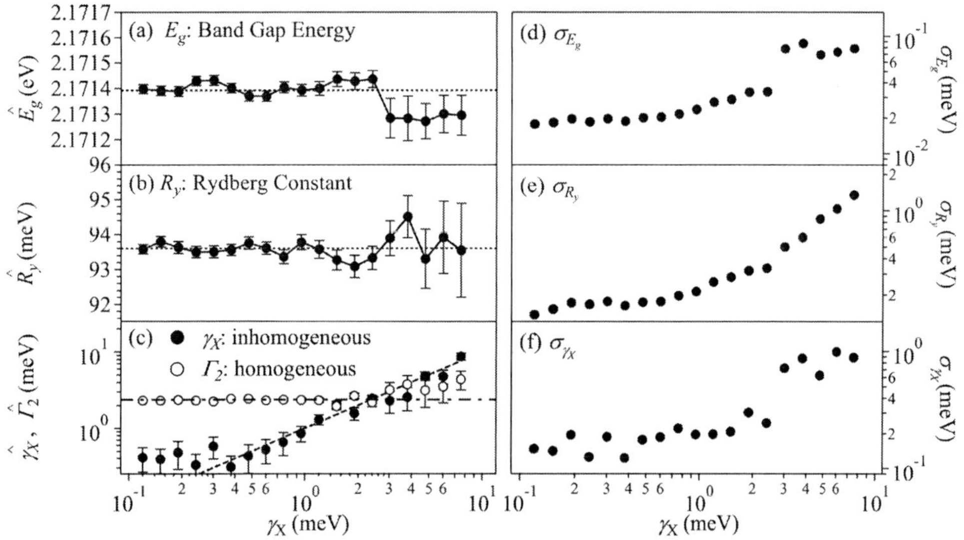

Figure 3. (a-c) Variations of the mean values $\hat{E}_g$, $\hat{R}_y$ and $\hat{\gamma}_X$ with the change in γ_X, which were obtained by Bayesian spectroscopy using the synthesized spectra in Fig. 2. γ_X are denoted in the logarithmic abscissa scale. The ordinate for $\hat{\gamma}_X$ is also a logarithmic scale. Error bars indicate the standard deviations σ_{E_g}, σ_{R_y} and σ_{γ_X} of $P(E_g|D)$, $P(R_y|D)$ and $P(\gamma_X|D)$. Dashed lines indicate the true values of the respective parameters. The horizontal dot-dashed line in (c) shows the true value of Γ_2. (d-f) Variations of σ_{E_g}, σ_{R_y} and σ_{γ_X} with the change in γ_X, where the abscissa and ordinate axes are logarithmic scales.

The variations of the mean values $\hat{E}_g$, $\hat{R}_y$ and $\hat{\gamma}_X$, and the distribution widths of the respective $P(\Theta|D)$ are plotted by closed circles and error bars in Figs. 3(a), (b) and (c), respectively, which were obtained in the same manner as the measured spectrum. It should be noted that $\hat{E}_g$ and $\hat{R}_y$ can be estimated with high accuracy for $\gamma_X < 3$ meV. Figures 3(a) and (b) show that $\hat{E}_g$ and $\hat{R}_y$ do not significantly deviate from their true values (horizontal dashed lines), and $\sigma_{E_g}(\gamma_X)$ and $\sigma_{R_y}(\gamma_X)$ do not noticeably increase, as shown in Figs. 3(d) and (e). In the measured spectrum of Fig. 1, $\hat{\gamma}_X$ was estimated to be 1.2 ± 0.1 meV [4]. Therefore, the accuracies of the estimated $\hat{E}_g$ and $\hat{R}_y$ depicted in Figs. 1(c) and (d) are sufficiently guaranteed.

In contrast to the cases with $\gamma_X < 3$ meV, $\hat{E}_g$ and $\hat{R}_y$ suddenly change with $\gamma_X \geq 3$ meV, and $\sigma_{E_g}(\gamma_X)$ and $\sigma_{R_y}(\gamma_X)$ begin to increase, as shown in Fig. 3. It is considered that such a change at the boundary of $\gamma_X \sim 3$ meV is due to the change in the spectral profiles of the synthesized spectra in Fig. 2. In the synthesized spectra with $\gamma_X \geq 3$ meV, the intensity of the second excitonic resonance peak X_3 becomes weak and its peak structure fades into the higher energy tail of the X_2 peak, and the absorption bands of $X_{5\sim20}$ and the background (X_C and B).

3.5. Future prospects in Bayesian spectroscopy of elementary electronic excitations

As noted in Sec. 3.1, measured optical spectra always include inhomogeneous broadenings in addition to homogeneous broadenings that represent physical properties such as lifetimes of the elementary electronic excitations. Although the inhomogeneous broadenings are convolved into the optical spectra by Eq. (4), decomposition of the inhomogeneous and homogeneous broadenings is generally a difficult problem. However, it is strongly considered that Bayesian spectroscopy will exhibit good performance for such decomposition. The closed and open circles in Fig. 3(c) show the variations of the inhomogeneous ($\hat{\gamma}_X$) and homogeneous ($\hat{\Gamma}_2$) broadening factors with the change in the inhomogeneous γ_X. Below the crossing point between these broadening factors, the decomposition is fully succeeded. The open circles for $\hat{\Gamma}_2$ lie on the horizontal dot-dashed line that represents the true value of Γ_2. On the other hand, $\hat{\gamma}_X$ increases linearly; therefore, under the conditions where the spectrum has sufficient data-point numbers, Bayesian spectroscopy can appropriately distinguish the difference between the Lorentzian and Gaussian spectral functions, and can thus decompose the spectral width into the homogeneous and inhomogeneous broadening factors.

Dipole-forbidden excitonic transitions were analyzed in the previous sections however, Bayesian spectroscopy is also effective for other spectral analyses. Spatially indirect excitons in semiconductor quantum wells and their related elementary electronic excitations in the high density regime have recently attracted much attention in the field of condensed matter physics [24] due to controversy regarding the existence of a stable electron-hole liquid phase [25]. In such quantum well systems, photoluminescence bands of the excitons, excitonic molecules, electron-hole droplets and their phonon-side bands are overlapped under high-density excitation conditions. These spectral components must be decomposed using several types of physical laws for the spectral parameters based on the established data; however, the potential of Bayesian spectroscopy for the spectral decomposition of such complicated spectra is high [26].

4. Bayesian spectroscopy for normal modes in coherent phonons

The spectroscopy of phonon normal modes is also an important issue in solid-state physics. Such normal mode spectroscopy has been performed with Raman [27] and infrared (IR) absorption [28] spectroscopy measurements. The Bayesian spectroscopy described in Sec. 3 is also useful for the spectral decomposition of such Raman and IR spectra.

4.1. Coherent phonon spectroscopy

Recent developments of fs-laser technologies have been remarkable [29], and measurements of CP signals [30, 31] using fs-laser pulses have opened a new frontier for normal mode spectroscopy. CP signals are detected as time-series intensity variations of weak probe-pulses after illumination with intense pumping pulses. At the time-origin, phonons in solid-state substances can be excited coherently by intense fs-pumping pulses and subsequent phonon vibrations periodically change the reflectivity and/or transmittance of substances, which can then be detected as CP signals using the delayed probe-pulses from the time origin.

To analyze normal modes and their dynamics from CP signals, Fourier [32] and wavelet [33] transformations have been universally employed. However, the oscillating amplitude of CP signals varies dynamically with time for such damped oscillations. When the decay time constant of the damped oscillation is τ, the normal mode peak in the Fourier spectrum has spectral width uncertainty of ca. $1/\tau$. In addition, it is difficult to extract important information such as the initial vibrating phase by such conventional methods. In contrast, Bayesian spectroscopy can solve these problems through the use of an appropriate physical model that represents CP signals, and can thus significantly improve the accuracy of physical parameter estimation [34].

4.2. Bayesian spectroscopy of coherent phonons in a bismuth thin film

The dashed line in Fig. 4(a) shows the CP signal for a bismuth thin film evaporated on a Si (100) substrate [34]. The data set $D \equiv \{(t_1, y_1), \cdots, (t_N, y_N)\}$ of this CP signal can be explained by a model function $g(t_i; \Theta)$ in Eq. (8) with a parameter set $\Theta \equiv \{A, f, \phi, \tau_\mathrm{r}, \tau_\mathrm{d}\}$, in which A, f, ϕ, τ_r and τ_d are the time-integrated intensity, vibration frequency, initial vibrating phase, the

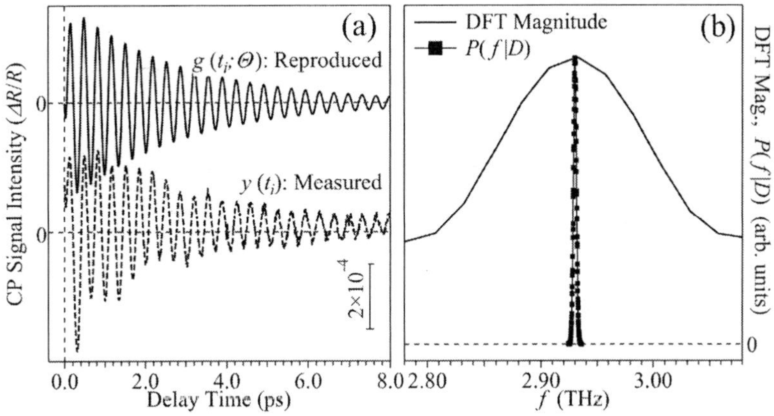

Figure 4. (a) CP signal in a bismuth thin film and reproduced data by Bayesian inference, which were reported in [34]. Dashed and solid lines represent the measured and reproduced data, respectively. (b) Magnitude spectrum of the CP signal in (a) obtained by the DFT method and the posterior probability distribution $P(f|D)$, of the normal mode frequency obtained by Bayesian spectroscopy. The solid line shows the DFT magnitude spectrum. Solid squares with lines represent $P(f|D)$.

rising time-constant and the damping constant, respectively.

$$g(t;\Theta) = \frac{A}{\tau_{\rm d}-\tau_{\rm r}}\left[\exp\left(-\frac{t}{\tau_{\rm d}}\right)-\exp\left(-\frac{t}{\tau_{\rm r}}\right)\right]\sin\left(2\pi ft+\phi\right), \qquad (t\geq 0). \tag{8}$$

Their posterior probability distributions $P(\Theta|D)$, have been obtained successfully through nine hundred thousand samplings after a sufficient burn-in phase. $\sigma_{\rm noise}$ of noise included in the measured data was set to 3.1×10^{-5} for the CP signal intensity and was confirmed to be balanced with the root mean squared error of $g(t;\hat{\Theta})$ obtained from the data set, D. The reproduced $g(t;\hat{\Theta})$ is depicted by the solid line in Fig. 4(a), which is well consistent with the measured data.

The most impressive results of Bayesian spectroscopy for CP signals are that the $P(f|D)$ of the normal mode frequency has an extremely narrow distribution and the initial vibrating phase ϕ, can also be estimated with high accuracy. The $P(f|D)$ is plotted by solid squares in Fig. 4(b) and has a distribution width of 0.93 GHz in the probable error criterion. The solid line in Fig. 4(b) shows a magnitude spectrum calculated by discrete Fourier transformation (DFT), which has a band width of ca. 160 GHz. The noise intensity included in the measured data corresponds to a signal-to-noise (S/N) ratio of 16 dB compared with the maximum amplitude of the $g(t;\hat{\Theta})$. As shown by the measured data in Fig. 4(a), even if the S/N ratio is not good (16 dB), the estimation accuracy of the normal mode frequency can be improved by more than two orders of magnitude compared with that for the Fourier transform. The reason for such improvement is to introduce the causality of the appropriate physical model in Eq. (8). In addition, this is the specialty of Bayesian spectroscopy for vibrational wave form analyses.

4.3. VMA of coherent phonon spectroscopy

A VMA [3] was performed to study the noise immunity of such high accuracy estimation ability of Bayesian spectroscopy for CP signals. Synthesized CP signals were prepared, as shown in Fig. 5(a), with variation of the noise intensity. In the synthesized data, the mean values $\hat{\Theta}$ obtained in Sec. 4.2 are used as true values. In the case of −0.95 dB, the oscillatory waveform is buried in noise; therefore, it is difficult to resolve a clear normal mode peak in the conventional Fourier and wavelet transformations.

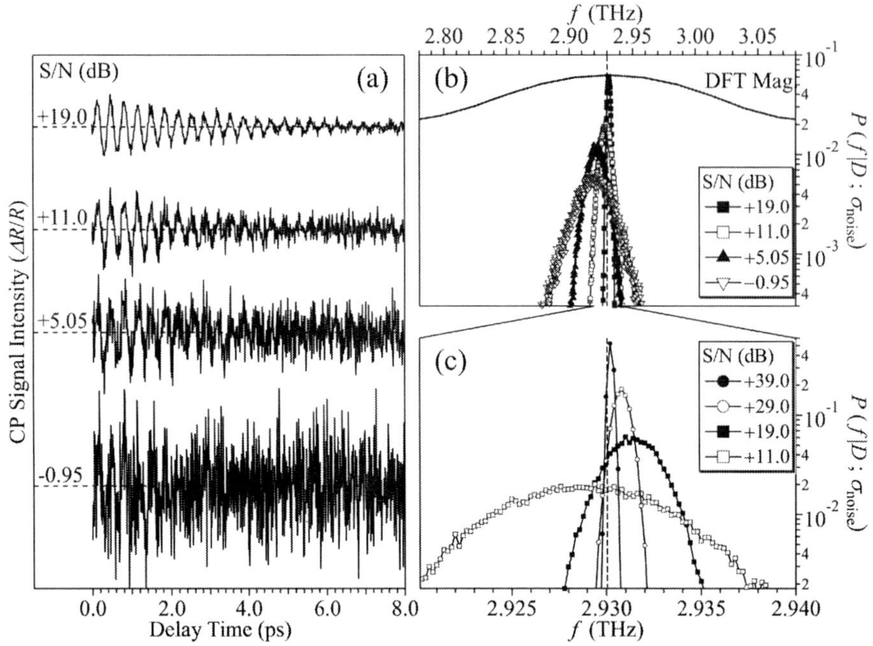

Figure 5. (a) Synthesized CP signals for the VMA, in which parameters in Θ are set to be the $\hat{\Theta}$ obtained by Bayesian spectroscopy in Sec. 4.2. The noise intensities in the respective synthesized data increase from top to bottom, as indicated by the S/N ratios. (b) and (c) Posterior probability distributions, $P(f|D;\sigma_{\text{noise}})$. The solid line in (b) represents the DFT magnitude spectrum of the synthesized data with +19.0 dB. The vertical dashed line shows the true value of f commonly adopted in the synthesized data. (c) is plotted with an expanded abscissa scale.

Figures 5(b) and (c) show the variations of $P(f|D;\sigma_{\text{noise}})$ with respect to σ_{noise}, where the noise intensities are denoted as S/N ratios in the legends. The ordinates are given as logarithmic scales. A DFT magnitude spectrum of the synthesized data at +19.0 dB is superimposed by the solid line in Fig. 5(b) with the same logarithmic scale. As reported in [34], $P(f|D;\sigma_{\text{noise}})$ at +19.0 dB was confirmed to have a predominantly narrow distribution width compared to the band width of the DFT spectrum, as shown in Fig. 5(b). With further decrease in the noise intensity, the estimation accuracy of f, i.e., the distribution width of $P(f|D;\sigma_{\text{noise}})$, was improved, as shown in Fig. 5(c). In addition, even when the S/N ratio becomes negative (S/N= −0.95 dB), the distribution width of $P(f|D;\sigma_{\text{noise}})$ is sufficiently smaller than the bandwidth of the DFT magnitude spectrum, as shown in Fig. 5(b). $P(f|D;\sigma_{\text{noise}})$ also includes the true value of f, which is indicated by the vertical dashed line in Figs. 5(b) and (c). In experimental measurements, the amplitude of the oscillatory waveform is frequently rather small and buried in noise. However, this result implies that Bayesian spectroscopy can estimate the normal mode frequencies correctly in such rigorous situations.

The noise immunity of Bayesian spectroscopy for the CP signals is summarized in Fig. 6. Closed circles and error bars in Figs. 6(a)-(e) represent the variations of the mean values $\hat{\Theta}$, and the standard deviations σ_Θ of the respective $P(\Theta|D)$ for the parameters in Eq. (8). Horizontal dashed lines show the true values of the corresponding parameters used in the synthesized CP signals. Here, attention should be paid to the ordinate ranges of the respective parameters. In these figures, the ranges indicated by double-arrowed vertical lines are given as percentages of the respective true values. It should be first noted that the estimations of the normal mode frequency f and the initial vibrating phase ϕ have strong noise immunity, as shown in Figs. 6(b)

IOP Conf. Series: Journal of Physics: Conf. Series **1036** (2018) 012022 doi:10.1088/1742-6596/1036/1/012022

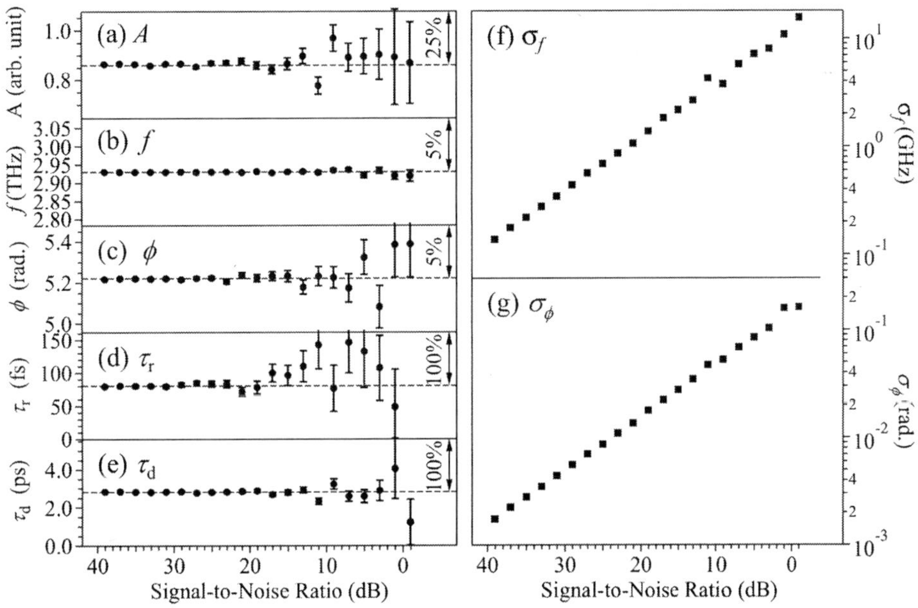

Figure 6. (a-e) Closed circles show the variations of $\hat{\Theta}$ according to the noise intensity included in the corresponding synthetic data. The noise intensity on the abscissa is expressed by the S/N ratio in decibels. Error bars denote the distribution widths of $P(\Theta|D)$ represented in standard deviations. Double-arrowed vertical lines and percentages indicate the ratios of the plotting ranges with respect to the corresponding true values. (f) and (g) Noise intensity dependence of the distribution widths, σ_f and σ_ϕ, of $P(f|D;\sigma_{\text{noise}})$ and $P(\phi|D;\sigma_{\text{noise}})$, respectively, shown with logarithmic ordinate scales.

and (c), respectively. Even at the right end of these figures, where the oscillatory amplitude is buried in noise, the estimated values of $\hat{f}$ and $\hat{\phi}$ are confined within the range of $\pm5\%$ from their true values, and it is understood that sufficient precision required in experiments is satisfied even with the worst case (S/N=-0.95 dB). In contrast, the amplitude A and the time constants of τ_{r} and τ_{d} deviate significantly from the corresponding true values at S/N ratios of less than 10 dB, as shown in Figs. 6(a), (d) and (e), respectively. Such differences in the noise immunity among these parameters are discussed in the next paragraph.

Although the optical spectrum discussed in Sec. 3 contains multiple spectral components, these components generally appear at different abscissa positions and their superposition describes the entire spectrum. In the case of the CP signals, the signal intensity change vibrationally in a successive time series. As a result, even small deviations of f and ϕ give large residuals on $g(t;\Theta)$ at all data points where the oscillatory amplitude remains. The large residuals significantly increase the fitting error $E(f,\phi;\hat{A},\hat{\tau}_{\text{r}},\hat{\tau}_{\text{d}})$ in Eq. (1). Therefore, $E(f,\phi;\hat{A},\hat{\tau}_{\text{r}},\hat{\tau}_{\text{d}})$ will have a deep and steep valley at the minimum point ($\Theta=\hat{\Theta}$) in the parameter space of Θ, so that f and ϕ can be estimated correctly with significantly narrow distributions of $P(f|D)$ and $P(\phi|D)$. In the parameters A, τ_{r} and τ_{d}, the behavior that these small deviations contribute to $E(f,\phi;\hat{A},\hat{\tau}_{\text{r}},\hat{\tau}_{\text{d}})$ are different from those of f and ϕ because these parameters only change the envelope (temporal) profile of the oscillatory waveform. Therefore, these parameters will be strongly disturbed by an increase in the noise intensity.

The effect of the noise immunity on f and ϕ is confirmed by the variations of the distribution widths of $P(f|D)$ and $P(\phi|D)$ with respect to the noise intensity. The results are plotted in Figs. 6(f) and (g), respectively, where their distribution widths are expressed by the standard deviations σ_f and σ_ϕ, of the distributions. As shown in these figures, σ_f and σ_ϕ are proportional to the noise intensity, as expected.

4.4. Future prospects in Bayesian spectroscopy for coherent phonons

As described in Secs. 4.2 and 4.3, Bayesian spectroscopy for CP signals has significant advantages for estimation of the normal mode frequency f and the initial vibrating phase ϕ. Regardless of the superiority of frequency estimation, the advantage of Bayesian spectroscopy is the high accuracy for the estimation of ϕ for the CP signal analyses. In the conventional methods, information on ϕ obtained by Fourier transformation has been mostly discarded. However, the initial vibrating phase ϕ provides insight with respect to the mechanism for the excitation of coherent phonons because the initial phases are different depending on the excitation mechanisms [35, 36, 37]. In the case of the impulsive stimulated Raman scattering (ISRS) mechanism, atoms located at the equilibrium points in the unit cells start to vibrate from the equilibrium points by impulsive light irradiations. Therefore, the CP signals excited by the ISRS mechanism have temporal profiles of the sine function [35, 36]. On the other hand, in the mechanism for the displacive excitation of coherent phonons (DECP), the CP signals change according to cosine functions with time [37].

On the other hand, from the perspective of experimental research, it is necessary to improve the measurement accuracy of the time-origin to make full use of this Bayesian spectroscopy performance. The initial vibrating phase can also be estimated with a high accuracy of 0.012 radian [34] from the measured CP signal in Fig. 4(a). This accuracy of ϕ corresponds to a time-origin accuracy of ca. 0.65 fs in this normal mode. It is generally difficult to determine the time-origin with such high accuracy in experimental setups of the pump-probe methods. Therefore, when the determination of an absolute value of ϕ with high accuracy is required, it is necessary to improve the experimental setup.

5. Conclusion

We have proposed Bayesian spectroscopy for both spectral analyses of optical spectra and for normal mode analyses of CP signals in solid-state materials.

In the optical absorption spectrum of excitonic series in a Cu_2O thin crystal sandwiched by paired MgO substrates, it is an important issue to elucidate small changes of the band gap energy E_g and the binding energy of the excitons R_y due to stress effects from the small mismatch of the lattice constants between Cu_2O and MgO. Although such highly accurate estimation is strongly desired, it has been impossible by conventional analysis that decomposes the entire excitonic absorption spectrum into individual excitonic absorption peaks of X_2, X_3 and X_4, $\cdots$ because of the broad spectral widths of the absorption peaks. For highly accurate estimation beyond this limitation, Bayesian spectroscopy was performed by introducing physical laws with respect to the transition energies and transition strengths of the excitonic series in Cu_2O. It has been confirmed that the changes of E_g and R_y by the stress effect have very sharp posterior probability distributions, and E_g and R_y are significantly reduced compared with that for stress-free single crystals. In VMA for spectral analysis of the excitonic absorption, the effectiveness of Bayesian spectroscopy has been examined with synthesized spectra by changing the inhomogeneous broadening factor γ_X. As a result, the clear appearance of the excitonic X_2 and X_3 absorption peaks is necessary for high accurate estimation of E_g and R_y. Furthermore, when the inhomogeneous broadening factor γ_X is smaller than the homogeneous broadening factor Γ_X, it has been clearly shown that the inhomogeneous and homogeneous broadening factors can be distinguished and both γ_X and Γ_X can be estimated with high accuracy. These results demonstrate the potential of Bayesian spectroscopy and it is strongly expected that Bayesian spectroscopy will be employed to decompose complicated optical spectra and to estimate material parameters with high accuracy.

For the analysis of time series data in CP signals, the Fourier transform method has been traditionally used to decompose the normal mode with CP signals and to estimate the vibrating frequency f. In the Fourier transformation, continuous plane wave functions (sine and cosine functions) are employed as expanding basis functions. In contrast, the amplitude of CP signals changes with time and the normal mode vibrations can be modeled by damped oscillating waveforms. As a result, the normal mode has broad peaks in the magnitude spectrum obtained by Fourier transformation. In addition, the damped oscillating waveforms are expanded by inappropriate basis functions (continuous plane wave functions); therefore, the initial vibrating phase ϕ cannot be estimated by the Fourier transform method. Such limitations of the Fourier transform method can be overcome by Bayesian spectroscopy, in which the temporal variation of the oscillating amplitude has been modeled appropriately by exponential functions with rising τ_r and decaying τ_d time constants. In the CP signal of a bismuth thin film with an S/N ratio of 16 dB, Bayesian spectroscopy can be used to estimate the normal mode frequency with a posterior probability distribution width of 0.93 GHz, which is two orders of magnitude smaller than the bandwidth of the normal mode in the Fourier magnitude spectrum. In VMA for CP signal analysis, the noise immunity of the highly accurate estimation has been examined by changing the noise intensity. Although the estimation accuracy for the intensity A, and for τ_r and τ_d degrades with an increase in the noise intensity, Bayesian spectroscopy can correctly estimate both the vibrating frequency f and the initial phase ϕ within 5% deviations from the respective true values, even when the S/N ratio is negative (-0.95 dB) and where the normal mode vibration is buried by the noise. Through such advantages of Bayesian spectroscopy, it will become possible to elucidate the excitation mechanism of CPs by the initial phase, and both detection and analysis of extremely weak CP signals will be possible.

The advantages of Bayesian spectroscopy clearly demonstrated here are expected to be applied to realize significant progress in solid-state physics.

Acknowledgments

The present study was financially supported by a KAKENHI Grant-in-Aid (Nos. 25280090, 16H01552, 16H04002 and 16K13824) from the Japan Society for the Promotion of Science (JSPS).

References

[1] Squires G L 2001 *Practical Physics 4th Ed.* (Cambridge University Press)
[2] Nagata K, Sugita S and Okada M 2012 *Neural Networks* **28** 82
[3] Igarashi Y, Takenaka H, Nagata K and Okada M 2016 *Ouyou Toukeigaku* **45** 75
[4] Iwamitsu K, Aihara S, Okada M and Akai I 2016 *J. Phys. Soc. Japan* **85** 094716
[5] Bayes T and Price R 1763 *Phil. Trans. Roy. Soc.* **53** 370
[6] Green P J 1995 *Biometrika* **82** 711
[7] Metropolis N, Rosenbluth A W, Rosenbluth M N, Teller A H and Teller E 1953 *J. Chem. Phys.* **21** 1087
[8] Andrieu C, Freitas N D, Doucet A and Jordan M I 2003 *Machine Learning* **50** 5
[9] Knox R S 1963 *Theory of Excitons* (Academic Press)
[10] Toyozawa Y 2003 *Optical Processes in Solids* (Cambridge University Press)
[11] Mysyrowicz A, Hulin D, Antonetti A, Migus A, Masselink W T and Morkoç H 1986 *Phys. Rev. Lett.* **56** 2748
[12] Akai I, Karasawa T, Komatsu T and Kaifu Y 1991 *Phys. Rev. B* **43** 4484
[13] Landsberg P T 2003 *Recombination in Semiconductors* (Cambridge University Press)
[14] Nakamura S, Pearton S and Fasol G 2000 *The Blue Laser Diode: The Complete Story* (Springer)
[15] Yoshioka Y, Chae E and Kuwata-Gonokami M 2011 *Nature Commun.* **2** 328
[16] Moskalenko S A and Snoke D W 2005 *Bose-Einstein Condensation of Excitons and Biexcitons: and Coherent Nonlinear Optics with Excitons* (Cambridge University Press)
[17] Fortin E, Fafard S and Mysyrowicz A 1993 *Phys. Rev. Lett.* **70** 3951
[18] Iwamitsu K, Aihara S, Shimamoto T, Fujii A and Akai I 2013 *Eur. Phys. J.* **86** 194
[19] Iwamitsu K, Aihara S, Shimamoto T, Fujii A and Akai I 2012 *Rev. Sci. Instrum.* **83** 073101
[20] Toyozawa Y 1964 *J. Phys. Chem. Solids* **25** 59

[21] Elliott R J 1957 *Phys. Rev.* **108** 1384
[22] Iwamitsu K, Aihara S, Ota A, Ichikawa F, Shimamoto T and Akai I 2014 *J. Phys. Soc. Japan* **83** 124714
[23] Gross E F 1956 *Nuovo Cimento Suppl.* **3** 672
[24] Jeffries C D, editor 1983 *Electron-Hole Droplets in Semiconductors* (Elsevier)
[25] Furukawa Y and Nakayama M 2016 *J. Phys. Soc. Japan* **85** 034701
[26] Iwamitsu K, Furukawa Y, Nakayama M, Okada M, Akai I 2018 *J. Luimn. in press*
[27] Tobin M C 1971 *Laser Raman Spectroscopy (Chemical Analysis: A Series of Monographs on Analytical Chemistry and Its Applications)* (Wiley)
[28] Nakanishi K and Solomon P H 1998 *Infrared Absorption Spectroscopy* (Emerson Adams Pr Inc.)
[29] Diels J-C and Rudolph W 2006 *Ultrashort Laser Pulse Phenomena, 2nd Ed. (Optics and Photonics Series)* (Academic Press)
[30] Merlin R 1997 *Solid State Commun.* **102** 207
[31] Kawamoto T, Yoshizaki Y, Yamashiro K, Iwamitsu K, Shimamoto T and Akai I 2014 *JPS Conf. Proc.* **1** 012071
[32] Hase M, Kitajima M, Nakashima S and Mizoguchi K 2002 *Phys. Rev. Lett.* **88** 067401
[33] Hase M, Kitajima M, Constantinescu A M and Petek H 2003 *Nature* **426** 51
[34] Aihara S, Hamamoto M, Iwamitsu K, Okada M and Akai I 2017 *AIP Adv.* **7** 045107
[35] Yan Y-X and Nelson K A 1987 *J. Chem. Phys.* **87** 6240
[36] Yan Y-X and Nelson K A 1987 *J. Chem. Phys.* **87** 6257
[37] Zeiger H J, Vidal J, Cheng T K, Ippen E P, Dresselhaus G and Dresselhaus M S 1992 *Phys. Rev. B* **45** 768